CHRISTIAN SARDET

PLANKTON

Wonders of the Drifting World

EDITED BY
RAFAEL D. ROSENGARTEN AND THEODORE ROSENGARTEN

TRANSLATED FROM THE FRENCH BY
CHRISTIAN SARDET AND DANA SARDET

PROLOGUE BY MARK OHMAN

THE UNIVERSITY OF CHICAGO PRESS · CHICAGO AND LONDON

CONTENTS

UNICELLULAR CREATURES: From the Origins of Life 31

CTENOPHORES AND CNIDARIANS: Ancestral Forms 93

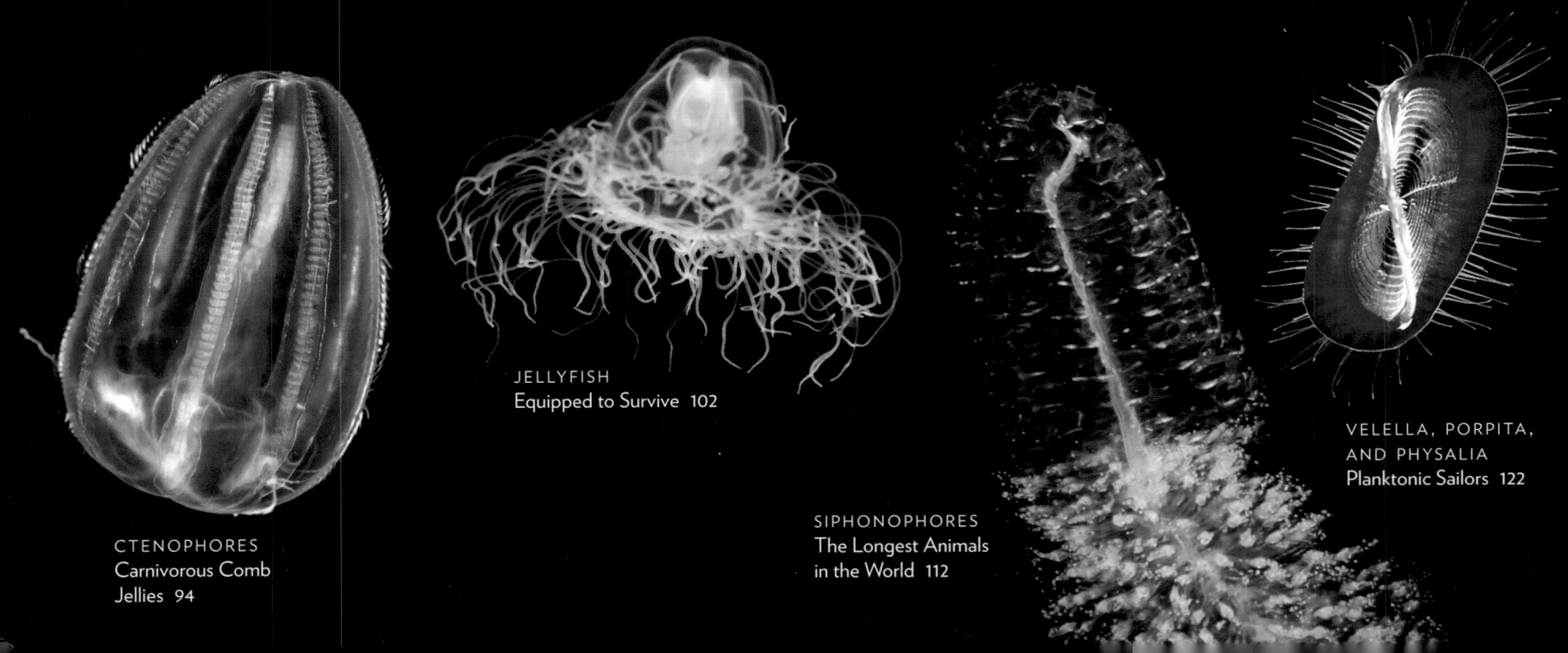

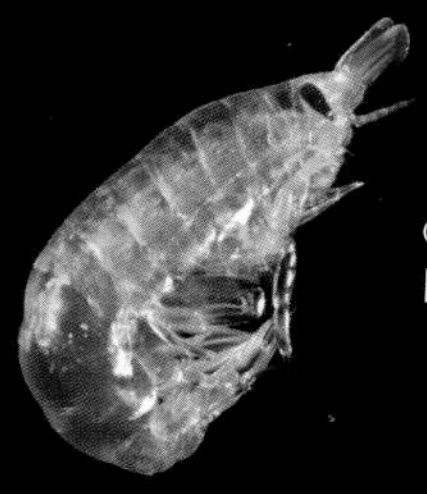

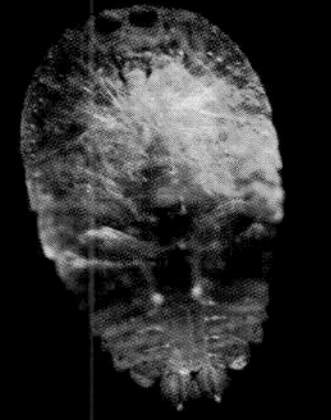

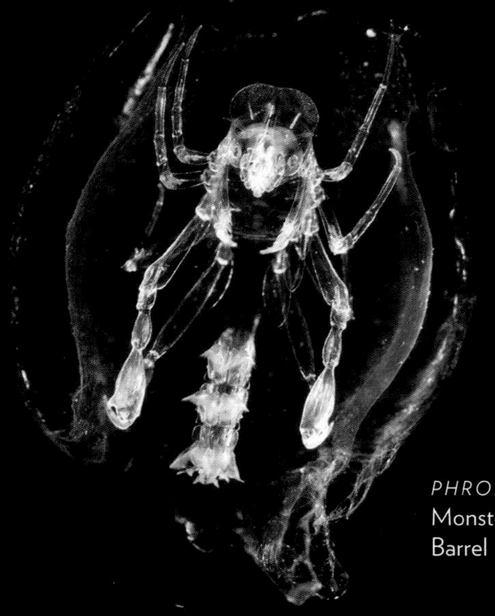

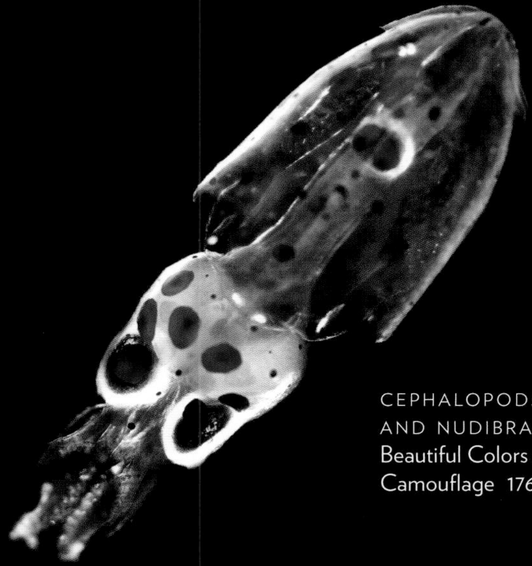

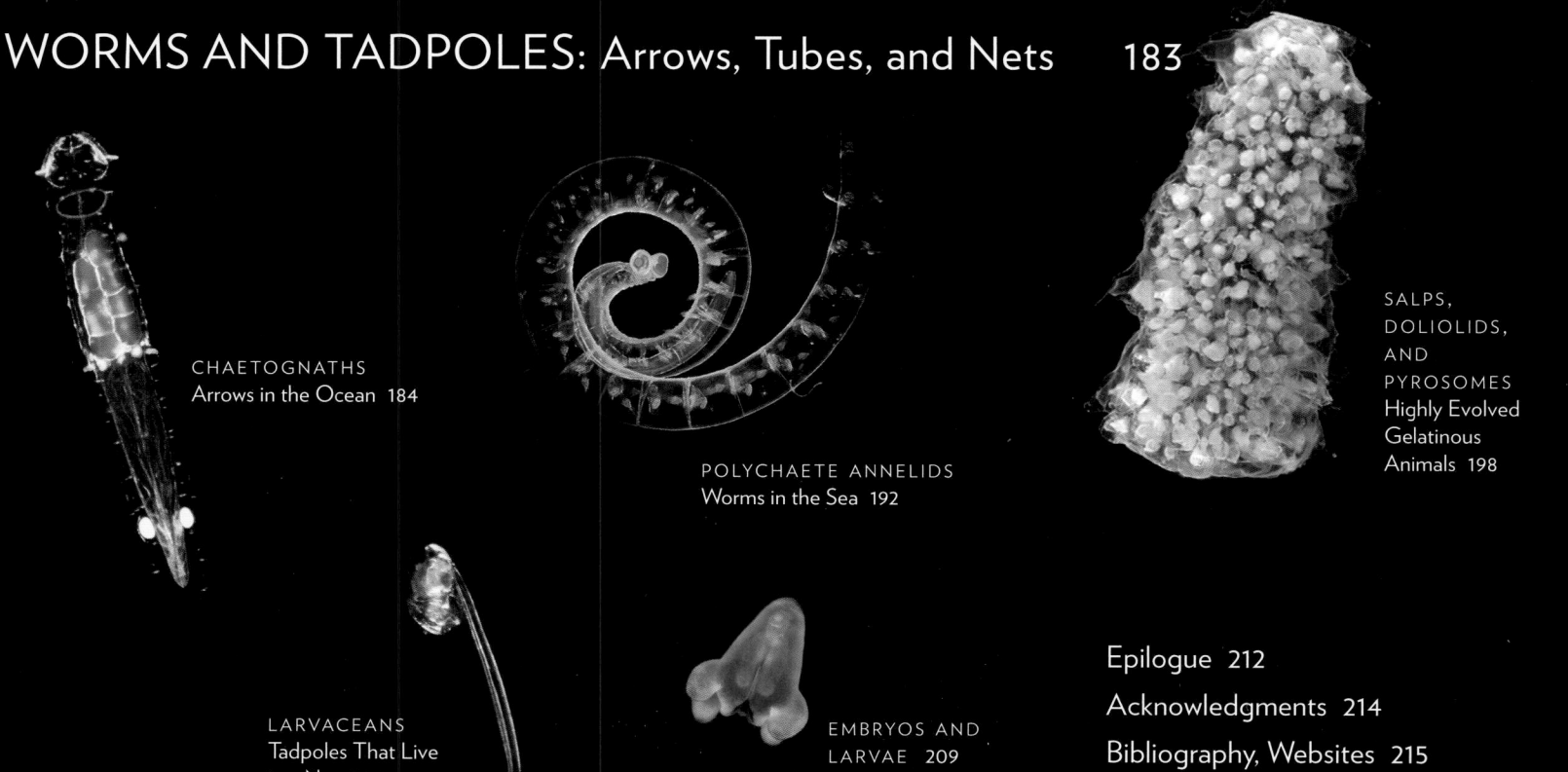

PROLOGUE

Take two breaths. For one of them, you can thank the plankton, in particular the single-celled photosynthetic drifters that comprise the phytoplankton of the world ocean. Remarkably, these elegant, microscopic cells perform nearly half of the photosynthesis and consequent oxygen production on Earth—equivalent to the total amount of photosynthetic activity of land plants combined. These tiny single cells have transformed the ocean, atmosphere, and terrestrial environment and helped make the planet habitable for a broad spectrum of other organisms, including ourselves. In many cases, blooms of phytoplankton reach such densities that they change the color of ocean surface waters and are even visible from satellites orbiting Earth. As you turn the pages of Christian Sardet's superb book, you will soon get a taste of the rich diversity of different types of phytoplankton, together with associated bacteria and drifting animals (zooplankton), their variations of form, and their exceptional beauty. Yet this is only the beginning.

Every schoolchild knows that baleen whales, the biggest animals in the sea, subsist on huge quantities of krill, which are small zooplankton. But ocean food webs (the linkages between predators and prey) are far more intricate than this familiar example. Many types of plankton eat other plankton. In this book you will meet some of the key members of these elaborate food webs, including zooplankton that graze on single-celled or colonial phytoplankton, carnivorous jellyfish, and specialized carnivores that eat only other types of predatory zooplankton. Some plankton have the ability to function as plants (carrying out photosynthesis) and animals at the same time. Others secrete elaborate mineral skeletons of calcium carbonate or silica. Still others live in complex symbiotic relationships with partner organisms. One type of gelatinous zooplankton—the appendicularians—has remarkably fine mesh feeding filters that trap the smallest bacteria in the ocean, leading to a size difference between the consumer and prey comparable to the size difference between whales and krill. Most fishes also eat some types of planktonic prey, especially in the crucial larval stages when availability of just the right kind of zooplankton at the right time and place can determine their survival.

How much of the carbon dioxide accumulating in the atmosphere can be buried deep in the ocean? The ocean's capacity to sequester carbon depends first and foremost on the freely drifting plankton. You will be introduced to some of these creatures in this book: the single-celled primary producers that synthesize new organic carbon, the bacteria that consume both dissolved matter and particles, single-celled grazers that can grow as fast as the organisms they consume, and the spectacular array of drifting multicellular zooplankton that both consume other organisms and accelerate the sedimentation of organic matter into deeper ocean waters. This "biological pump" of carbon from the upper ocean to the deep sea is the primary means by which carbon is buried in the ocean. The rate of future climate change depends, in part, on the waxing and waning of the planktonic communities. And past (and future) formation of oil and gas deposits on the ocean floor also depends largely on these planktonic organisms.

All of these processes—photosynthesis, fish and shellfish production, and carbon sequestration and climate regulation—depend on the right balance of biological diversity in the ocean. This biodiversity is the principal focus of Sardet's book. It does not attempt to conquer the vast bodies of scientific inquiry that address plankton ecology, oceanography, or climate science. Instead, through remarkable images and lucid prose, *Plankton: Wonders of the Drifting World* serves as a unique introduction to the captivating flora and fauna of this planktonic world.

The book can be appreciated from many different perspectives. The beautiful, diverse forms of plankton are illustrated here as never before, an aesthetically pleasing collection of fascinating creatures that most people have never encountered—at least, not knowingly. In many respects this book harkens back to the spirit of the early twentieth-century contribution by the naturalist Ernst Haeckel, who sought to illustrate the symmetry and beauty of the living world in his landmark work *Art Forms in Nature*. But Christian Sardet also situates the diversity of the planktonic world in an evolutionary context that illustrates how life on Earth has radiated over time, bearing the scars of the five great extinctions. He highlights some of the regional differences in plankton communities in different ocean provinces, drawing on the epic *Tara Oceans Expedition*—of which he was a cofounder—that circumnavigated the globe. And the book can be appreciated for its rich entrée to swimming, feeding, reproduction, bioluminescence, and sensory biology of organisms of the planktonic world.

The book is certain to excite its audience about plankton and energize them to want to learn more, and

A bloom of phytoplankton mostly consisting of coccolithophores and diatoms off the Atlantic coast of Patagonia.

Observed by the NASA Aqua satellite in December 2010.

even to do more. We confront many daunting challenges in the twenty-first century, many of which will require a better understanding of the significance of plankton in the ocean and in our lives. Changes in climate and ocean chemistry, and the indisputable decline of world fisheries, are linked to the fate of plankton.

Plankton photosynthesis and respiration affect the accumulation of greenhouse gases and aerosols in the atmosphere. As ocean warming displaces some plankton species toward the colder poles, it also alters the density stratification, or layering, of ocean waters. In many regions these changes have made it more difficult for deep nutrients to be transported toward the sea surface to stimulate phytoplankton blooms. The absorption of atmospheric carbon dioxide by the ocean has already resulted in ocean acidification, with potentially dire consequences for some shell-bearing organisms. Coastal eutrophication—the discharge of excess nutrients—has been linked to growing regions of depleted oxygen in the ocean, due to the blooms and subsequent decomposition of plankton populations. While most blooms of phytoplankton are beneficial, proliferation of certain species of harmful algae can be toxic to humans or to other members of the food web. In some regions, the species and sizes of harvestable fish have been markedly altered by overfishing. Global shipping has resulted in introductions of exotic species of plankton into many coastal environments, and some of these invaders can markedly disrupt natural food webs. And there is growing recognition that small, suspended plastic particles—the "synthetic plankton"—can concentrate not only in coastal environments, but in the great midocean gyres, with as yet unknown consequences for ocean ecosystems. Addressing these challenges will require enhanced understanding of the processes regulating plankton populations and their adaptions to environmental change.

For many years I have sought a book that communicates to the nonscientist the exceptional diversity of form and function in the planktonic world and that conveys the sense of fascination that many of us have when working with these creatures. *Plankton: Wonders of the Drifting World*—and the accompanying website *Plankton Chronicles* (www.planktonchronicles.org)—achieves that goal par excellence. It cannot help but whet your appetite for the magnificent organisms of this drifting world, upon which so much of life on Earth depends. Prepare yourself for the thrill of discovery.

Mark Ohman

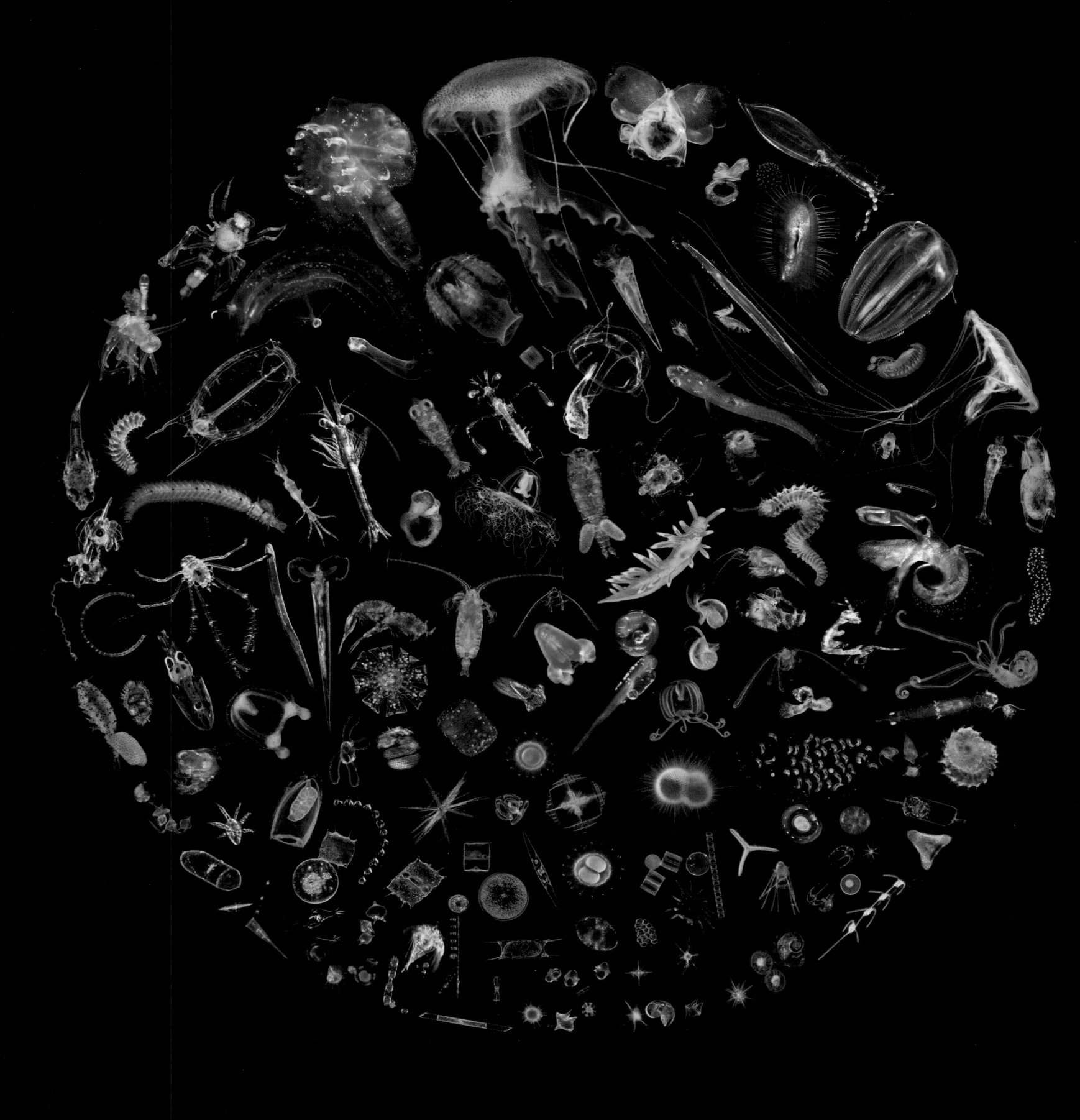

PLANKTON MANDALA

Plankton include a multitude of organisms from across all domains of life. This mandala depicts more than 200 different kinds of plankton, most of them described in this book. In the upper part of the mandala are the largest creatures of zooplankton: jellyfish, siphonophores, ctenophores, salps. In the center are a mix of chaetognaths, annelids, mollusks, and crustaceans. Also included are big larvae and juveniles, ranging in size from a few millimeters to several centimeters. The lower part of the mandala shows microscopic organisms (measuring less than 1 mm), mostly single-cell protists: radiolarians, foraminifera, diatoms, and dinoflagellates. This microscopic world also includes many multicellular organisms, including the embryos and larvae of the animals above. (Note: individual images are not to scale).

PLANKTON
A DRIFTING
WORLD

What Are Plankton?

Plankton are the multitudes of living creatures afloat in the world's bodies of water. The name comes from *planktos*, the Greek word for wandering or drifting. Vast communities of organisms comprise the planktonic ecosystem, an ode to the origins and diversity of life that has evolved for more than three billion years in the ocean. Microscopic bacteria and archaea, and all kinds of unicellular and multicellular organisms, cooperate and compete to survive.

The animals we can see, like fish, squid, octopus, and marine mammals, comprise only 2 percent of the ocean's living biomass. The remaining 98 percent consists of the mostly invisible multitudes of plankton. Tiny crustaceans and mollusks, and gelatinous animals such as jellyfish and salps, drift their entire lives in ocean currents. *Velella* and Portuguese man-of-war, the most famous of the siphonophores, are driven by winds at the surface. Complex interactions occur between very different planktonic organisms. Parasitism and symbiosis are the rule more often than the exception.

Diversity and abundance of plankton vary with the currents, geography of seas and ocean basins, and atmospheric conditions. Plankton composition also changes with the seasons, climate, and pollution. Organisms grow and proliferate when conditions of temperature, salinity, and nutrients are favorable for them. The most numerous organisms among the plankton are bacteria and archaea, unicellular organisms lacking a nucleus and other organelles, which are classified as prokaryotes. Organisms possessing a nucleus and organelles are known as eukaryotes, and include everything from complex animals to unicellular protists. Typical protists—diatoms, dinoflagellates, and coccolithophores—divide and multiply so fast that the populations explode in blooms. These microscopic creatures can be so numerous that they color the seas and

A whale feeding on a plankton bloom.
Photo by Wayne Davis, www.oceanaerials.com

show up in photographs taken by satellites. While typically beneficial, some types of plankton blooms can devastate fish or mollusks in that habitat. Certain species also secrete substances that can provoke the formation of clouds and influence climate.

The microorganisms that most affect our climate are those capable of photosynthesis. Called phytoplankton, these photosynthetic creatures capture the sun's energy just as terrestrial plants do. They use this energy to convert carbon dioxide (CO_2), water (H_2O), and minerals into organic matter and oxygen. Phytoplankton such as diatoms and dinoflagellates

Dynamics of Phytoplankton in the Ocean

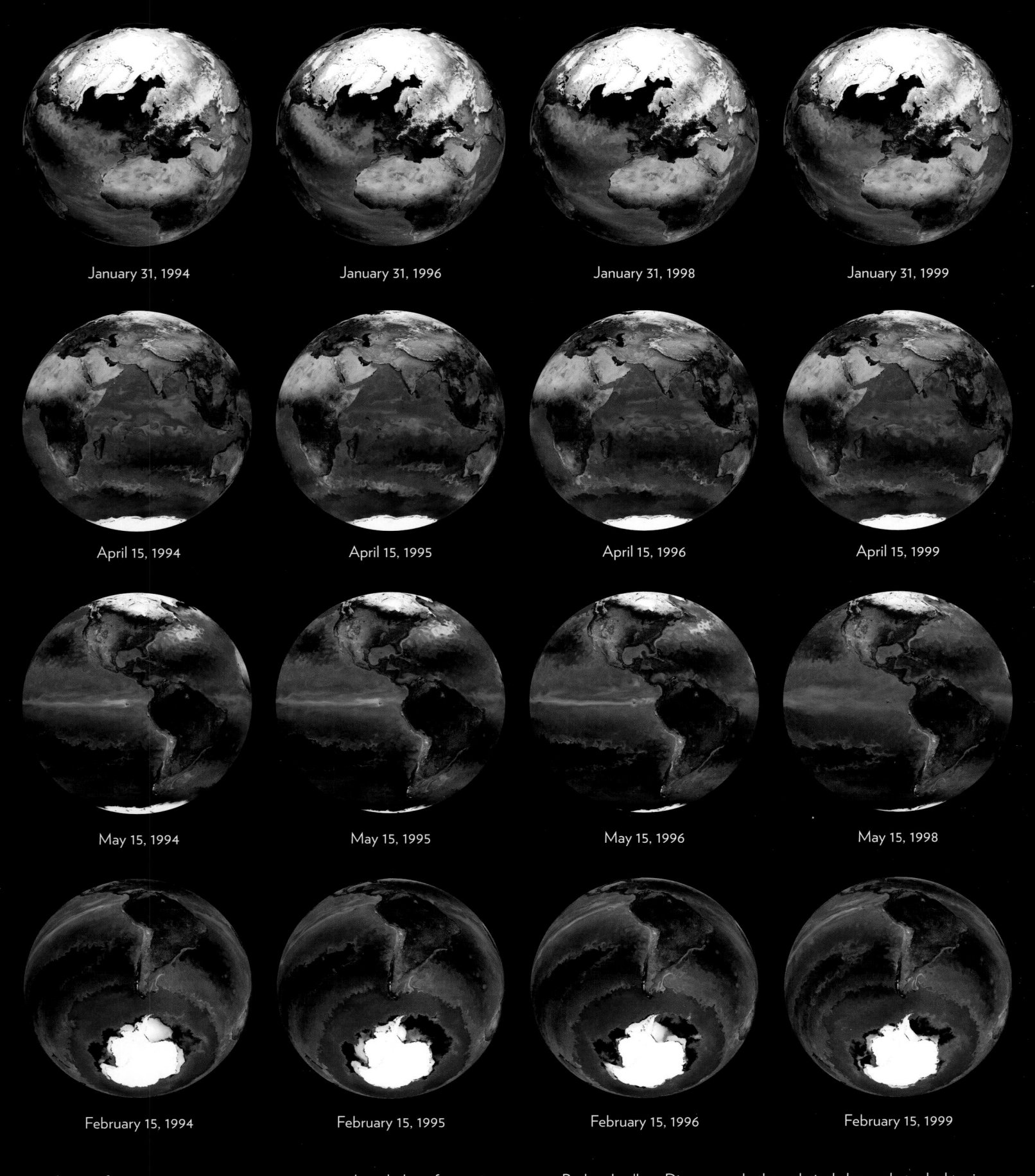

January 31, 1994   January 31, 1996   January 31, 1998   January 31, 1999

April 15, 1994   April 15, 1995   April 15, 1996   April 15, 1999

May 15, 1994   May 15, 1995   May 15, 1996   May 15, 1998

February 15, 1994   February 15, 1995   February 15, 1996   February 15, 1999

Images from a computer animation sum up our knowledge of oceanic currents and microscopic phytoplankton. Mick Follows and his colleagues at MIT (Cambridge, Massachusetts) simulated the abundance and distribution of different sizes of phytoplankton in the ocean, and their changes from 1994 to 1999.

Red and yellow: Diatoms and other relatively large phytoplanktonic organisms.
Green and blue: Cyanobacteria Prochlorococcus and Synechococcus and other very small phytoplanktonic organisms.

Mick Follows, Oliver Jahn, ECCO2 and Darwin Project, Massachusetts Institute of Technology

An aerial view of a plankton bloom, or "red tide," near Heron Island, the Great Barrier Reef, Australia. Photo by Gary Bell, OceanwideImages.com

form the base of the food chain, and are consumed by other unicellular organisms such as radiolarians and foraminifera. Protists are themselves food for the zooplankton, countless tiny animals and their young. Corals, echinoderms, mollusks, and crustaceans inhabit the coasts and sea bottoms as adults, but produce huge numbers of planktonic gametes, embryos, larvae, and juveniles.

For the majority of marine animals, reproduction is sexual, and in many species it happens without mating. Zooplankton shed large quantities of eggs and sperm or release massive quantities of embryos into the sea. Fertilization most often takes place in open water, followed by rapid development: the larvae hatch, feed, and drift with the currents, and some eventually attach themselves to rocks or seaweed. Larvae and juveniles that are not eaten become adults, often after multiple metamorphoses. Once mature, some planktonic species may drift all their lives while others—fish or cephalopods—follow their adult kin moving in and out of the currents.

Plankton and Man

Human lives are intimately entwined with plankton. Every breath we take is a gift of oxygen from the phytoplankton. In fact photosynthetic bacteria and protists produce as much oxygen as all the forests and terrestrial plants combined. And for the last three billion years, phytoplankton have absorbed huge amounts of carbon dioxide. Plankton regulate the productivity and acidity of the ocean through the carbon cycle, and exert a major influence on climate.

Plankton are also our largest supplier of fossil fuels. The bodies of planktonic organisms and their waste products fall to the bottom of the ocean in the form of bacteria-rich particles known as "marine snow." These particles have accumulated on the seabed for hundreds of millions of years. Buried, compressed, metabolized by microorganisms, they eventually produce a kind of viscous rock that ultimately turns into reserves of oil and gas. Man has drawn on this carbon resource to provide energy for electricity and transportation fuels and raw material for a great variety of manufacturing. Every year we consume the oil equivalent of a million years of plankton buried beneath the ocean floor.

Throughout the ages, the shells and skeletons of some protists—foraminifera, diatoms, radiolarians, and coccolithophores, for example—deposited thick layers of calcareous or siliceous material, forming sedimentary rocks. Heaved up by great displacements of the Earth's crust, these sedimentary rocks became mountains. Microscopic shells and skeletons are now found in cliffs and deserts and in the stone used for our buildings and monuments. Finally, plankton feed us. They form the foundation of the food chain where the larger creatures eat the smaller ones, from fish and shrimp to microorganisms. Without plankton there would be no fish!

Stromatolites in Shark Bay, Australia, reveal fossil traces of the oldest forms of life. Photo by Mark Boyle.

The Origins: Life Shapes the Planet

4.6 TO 3.5 BILLION YEARS AGO:
LIFE BEGAN IN THE PRIMORDIAL OCEAN
Our planet was born approximately 4.6 billion years ago in a gas-rich atmosphere, when huge blocks of rocks and icy meteorites collided and aggregated. The melted rocks slowly cooled. From an atmosphere charged with water vapor and heavy rains, a primordial ocean formed. Did life begin in a small muddy pond, as Darwin thought, or near volcanic hydrothermal vents at the bottom of the ocean? Perhaps life arrived from outer space, contained in blocks of ice that fed the nascent ocean? Someday we may know.

In any case, evidence reveals that plankton lived in the primordial ocean some 3.5 billion years ago. These were likely primitive forms of bacteria and archaea able to survive in oxygen-free environments, deriving their energy from the metals, gases, and heat emanating from the planet. This process continues today in hydrothermal vents and hot springs. Microorganisms then began to transform the planet. They left their mark as stratified rocks, or stromatolites, that reflect the biochemical activity of bacteria billions of years ago. Mineralization was the work of films of cyanobacteria that created thick deposits. Cyanobacteria are champions of photosynthesis, a process that releases oxygen as a byproduct. Thus cyanobacteria were responsible for making the atmosphere and ocean breathable for other organisms. These early phytoplanktonic life forms initiated the planktonic food chain.

3.5 TO 2.4 BILLION YEARS AGO:
THE GREAT OXIDATION OF OUR "RED PLANET"
From the beginning of life in the ocean, cyanobacteria have been major photosynthesizers. For nearly a billion years they produced a considerable amount of oxygen, transforming the original methane and carbon dioxide–rich atmosphere. Cycles of photosynthetic activity by bacterial films caused extensive oxidation on primitive Earth. Huge banded strata of rusted iron were deposited on the ocean floor. Seen from space, 2.4 billion years ago our planet might have appeared as red as Mars appears today. Bacterial activity altered the composition of the primitive atmosphere, reaching about 10 percent of the current oxygen concentration. An ozone layer developed on the surface of the Earth, protecting life from ultraviolet rays. This was catastrophic for bacteria and archaea that could not tolerate the new atmospheric conditions. Many of the survivors were sheltered in the deep ocean and other locations we might consider

inhospitable, where they continue to live and evolve to this day. We call such organisms "extremophiles" for the extreme environments in which they thrive.

2.4 TO 1.4 BILLION YEARS AGO:
LIFE TRANSFORMS THE PLANET
INTO "SNOWBALL EARTH"
Under the effect of oxygenic photosynthesis, the planet experienced a period of almost total glaciation. Methane, a powerful heat-trapping greenhouse gas, was an abundant component of the primordial atmosphere. But oxygen in large quantities began to oxidize methane, converting it into CO_2. The decrease in methane contributed to the glaciation of the entire planet. This ice-covered condition, or "snowball Earth," lasted for tens of millions of years. Over time, volcanoes spewed enormous quantities of CO_2 that, together with methane produced by archaea deep in the ocean, gradually caused the planet to warm and the vast ice to retreat.

ONE BILLION YEARS AGO:
THE EMERGENCE OF PROTISTS
During the tumultuous times of early life on Earth, ancestral archaea and bacteria evolved in extraordinary ways. Some bacteria and archaea engulfed and domesticated other bacteria and archaea. Thus eukaryotes were born—internalized bacteria/archaea evolved into DNA-containing nuclei and other organelles, such as mitochondria and chloroplasts that generate energy. The resulting single-cell organisms with nuclei and organelles are considered to be the first unicellular eukaryotes and the ancestors of protists. Even today bacteria, archaea, and protists continue to exchange organelles, genes, and proteins, acquiring new functions and metabolic pathways. Some protists house symbiotic algae, and others form communities of nucleated cells, a first step on the path to multicellularity. The very first multicellular organisms are believed to have appeared in the ocean more than a billion years ago, but as they did not have a mineral skeleton, they left no fossil record.

WHAT ABOUT VIRUSES?
Viruses and phages (the name given to bacteria-infecting viruses) are ubiquitous in terrestrial, aquatic, and marine ecosystems. Unlike bacteria, archaea, and eukaryotes, viruses are not cells and are not capable of self-reproduction. Rather, viruses are made of nucleic acids, often encased in protein and lipid structures. In order to reproduce, viral DNA enters cellular organisms and hijacks the host's DNA replication machinery to make more of itself. In addition to replicating themselves, viruses can take control of other genetic programs in the infected host. In some cases viruses are thought to govern the life cycle of their host, sometimes with striking consequences. For example, viral infection may be involved in regulating or terminating spectacular blooms of planktonic organisms. Scientists recently discovered types of giant viruses called gyrus, mimivirus, or megavirus that are themselves infected by smaller viruses. Viruses may have existed as long as, and may even have come before, bacteria, archaea, and eukaryotes. Some scientists consider them a fourth domain on the tree of life.

Explosions, Extinctions, and Evolution of Life in the Ocean

800 TO 500 MILLION YEARS AGO:
THE FIRST ANIMALS
Ancestral animals, which appeared in the ocean between 800 and 700 million years ago, left few fossil traces. When and how they appeared is a question pursued by biologists and paleontologists who analyze morphological characteristics of fossils and living organisms, as well as genes and genomes from representative species. Sponges, ctenophores, and cnidarians are descended from the oldest animal lineages. Cnidaria is the animal phylum that includes corals and sea anemones, as well as large gelatinous plankton such as siphonophores and jellyfish. All cnidarians have in common stinging cells called cnidocytes, a term derived from the Greek word for nettle. As a group, cnidarians exemplify the full range of possibilities in lifestyle: sexual reproduction and asexual propagation; symbiosis and carnivory; regeneration and, according to some biologists, even immortality. Ctenophores, or "comb jellies," superficially resemble jellyfish, but in fact form their own phylum (Ctenophora) and display distinct morphological and life history traits. Despite their fragile appearance, cnidarians and ctenophores have proven champions of adaptation. They have survived and evolved through five major periods of extinction. In our modern ocean basins, which are rapidly being depopulated of fish and mammals, these gelatinous predators seem to proliferate and eventually may become dominant.

500 TO 200 MILLION YEARS AGO:
EXPLOSIONS AND EXTINCTIONS OF LIFE
The history of life on planet Earth is marked by successive periods of explosions and extinctions of life. Even before the Cambrian period, more than 540 million years ago, almost all the phyla now present on Earth had already appeared. Then, over a relatively short period of evolutionary history, the "Cambrian explosion" saw mollusks, arthropods, echinoderms, and annelids spread and diversify throughout the ocean. The ancestors of vertebrates also appeared during the Cambrian. Animals, though originating in the ocean, gradually conquered the land. This development, the date of which is constantly being pushed back, happened more than 500 million years ago. It probably involved several types of animals, ancestors of nematode worms and arthropods that crawled out of the sea onto land. These creatures most likely fed on marine debris, bacterial films, or lichens and plants.

Among the ancient animal groups, cnidarians and ctenophores are characterized by radial symmetry. In contrast, mollusks, arthropods, echinoderms, and vertebrates, including their embryos and larvae, have bilateral symmetry. Their bodies possess anterior and posterior regions, demarking a head, tail, front, back, right, and left sides. Many of them build shells, exoskeletons, or internal skeletons. These shells and

Chronological History of Life and the Ocean on our Planet

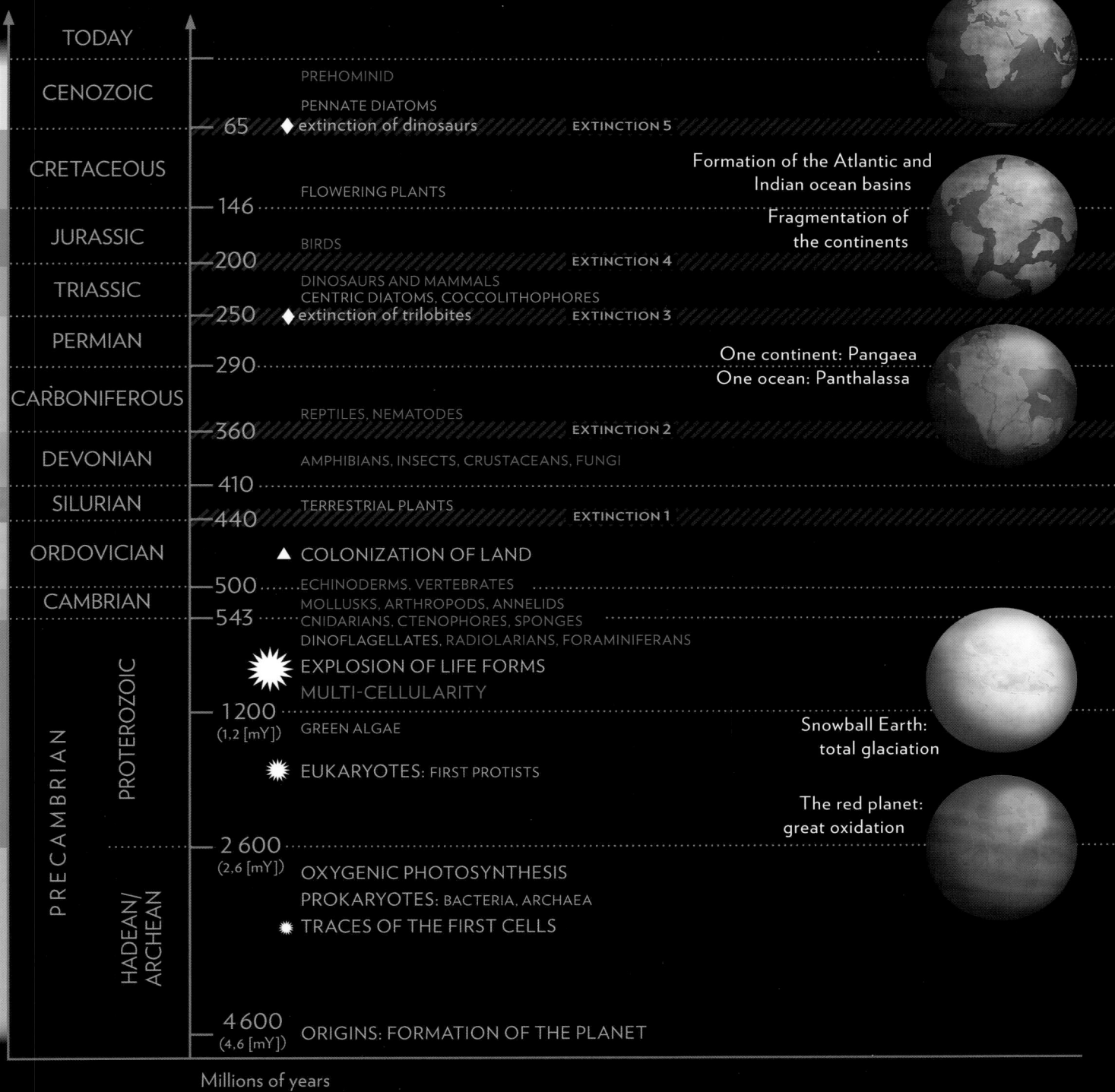

TODAY		
CENOZOIC	PREHOMINID	
	PENNATE DIATOMS	
— 65	◆ extinction of dinosaurs	EXTINCTION 5
CRETACEOUS		Formation of the Atlantic and Indian ocean basins
— 146	FLOWERING PLANTS	
JURASSIC		Fragmentation of the continents
— 200	BIRDS	EXTINCTION 4
TRIASSIC	DINOSAURS AND MAMMALS	
	CENTRIC DIATOMS, COCCOLITHOPHORES	
— 250	◆ extinction of trilobites	EXTINCTION 3
PERMIAN		
— 290		One continent: Pangaea
CARBONIFEROUS		One ocean: Panthalassa
	REPTILES, NEMATODES	
— 360		EXTINCTION 2
DEVONIAN	AMPHIBIANS, INSECTS, CRUSTACEANS, FUNGI	
— 410		
SILURIAN	TERRESTRIAL PLANTS	
— 440		EXTINCTION 1
ORDOVICIAN	▲ COLONIZATION OF LAND	
— 500	ECHINODERMS, VERTEBRATES	
CAMBRIAN	MOLLUSKS, ARTHROPODS, ANNELIDS	
— 543	CNIDARIANS, CTENOPHORES, SPONGES	
	DINOFLAGELLATES, RADIOLARIANS, FORAMINIFERANS	

EXPLOSION OF LIFE FORMS
MULTI-CELLULARITY

— 1200
(1,2 [mY])   GREEN ALGAE

EUKARYOTES: FIRST PROTISTS

Snowball Earth: total glaciation

The red planet: great oxidation

— 2 600
(2,6 [mY])

OXYGENIC PHOTOSYNTHESIS
PROKARYOTES: BACTERIA, ARCHAEA
TRACES OF THE FIRST CELLS

— 4 600
(4,6 [mY])   ORIGINS: FORMATION OF THE PLANET

PRECAMBRIAN — PROTEROZOIC — HADEAN/ARCHEAN

Millions of years

The history of planet Earth and the ocean began about 4.6 billion years ago. Our planet went through periods of oxidation and glaciation related to the emergence of life in the oceans, notably oxygenic photosynthesis by cyanobacteria. Represented here are the emergence of continents and ocean basins; five periods of major extinctions (450–440 / 375–360 / 250 / 200 / 65 million years ago); and major events in the history of life on Earth: appearance of the first eukaryotic cells, protists and multicellular organisms (during the Proterozoic), appearance of most animal phyla (Cambrian), colonization of land (Ordovician), disappearance of trilobites (end of Permian) and dinosaurs (end of Cretaceous). The vertical scale on the far left uses the color code adopted by the international community of geologists.

✴ Traces of the first cells (3.5 billion years ago [bY])	✺ Explosions of life (0.8 to 0.5 [bY])	◆ Disappearance of trilobites (0.25 [bY])
✸ Appearance of eukaryotes (2 to 1.5 [bY])	▲ Colonization of land (0.7 to 0.5 [bY])	◆ Disappearance of dinosaurs (0.065 [bY])

Tree of Life, from the Origins to the Present Day

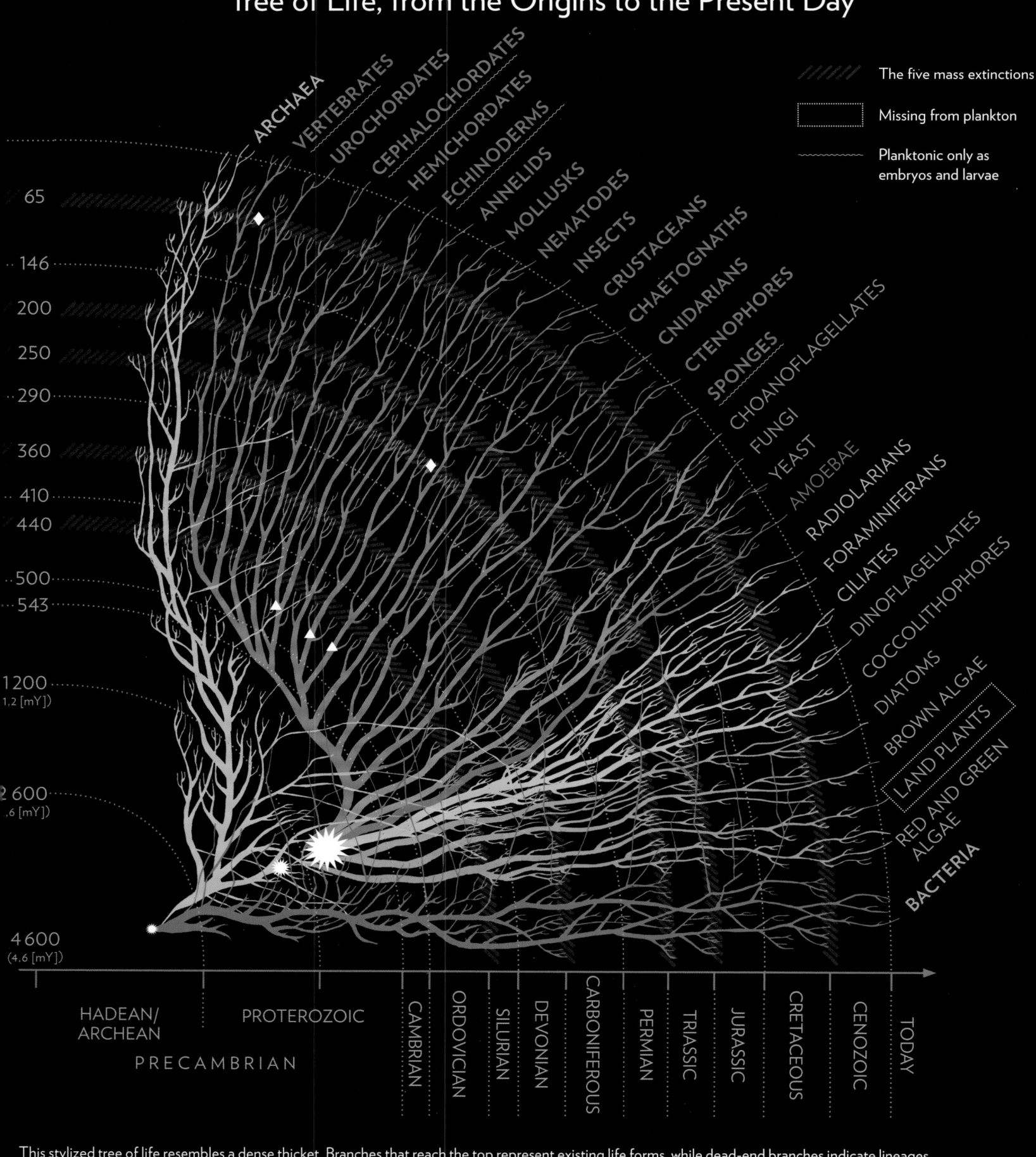

The five mass extinctions

Missing from plankton

Planktonic only as embryos and larvae

ARCHAEA · VERTEBRATES · UROCHORDATES · CEPHALOCHORDATES · HEMICHORDATES · ECHINODERMS · ANNELIDS · MOLLUSKS · NEMATODES · INSECTS · CRUSTACEANS · CHAETOGNATHS · CNIDARIANS · CTENOPHORES · SPONGES · CHOANOFLAGELLATES · FUNGI · YEAST · AMOEBAE · RADIOLARIANS · FORAMINIFERANS · CILIATES · DINOFLAGELLATES · COCCOLITHOPHORES · DIATOMS · BROWN ALGAE · LAND PLANTS · RED AND GREEN ALGAE · BACTERIA

65
146
200
250
290
360
410
440
500
543
1200
1,2 [mY]
2 600
,6 [mY]
4 600
(4,6 [mY])

HADEAN/ ARCHEAN · PROTEROZOIC · CAMBRIAN · ORDOVICIAN · SILURIAN · DEVONIAN · CARBONIFEROUS · PERMIAN · TRIASSIC · JURASSIC · CRETACEOUS · CENOZOIC · TODAY

PRECAMBRIAN

This stylized tree of life resembles a dense thicket. Branches that reach the top represent existing life forms, while dead-end branches indicate lineages that have died off, particularly during the five major extinction periods. Transverse branches symbolize the transfer of organelles and genes between bacteria, archaea and eukaryotes. Precise affiliations and origins of some of the branches remain unknown or controversial, so parts of this schema are hypothetical. The clade occupied by animals (in orange) is depicted larger than in actuality.

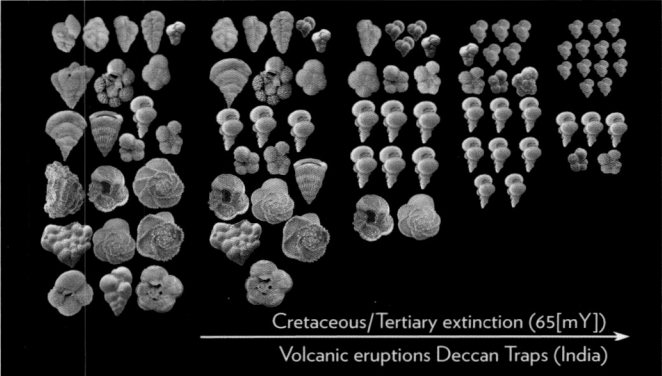

Cretaceous/Tertiary extinction (65[mY])
Volcanic eruptions Deccan Traps (India)

It is assumed that the dinosaurs disappeared 65 million years ago, after a giant meteorite hit the Earth. Recent analyses of fossil foraminifera by Gerda Keller and colleagues at Princeton University show that before the meteorite's fall, massive volcanic eruptions in India had already largely destroyed some species of foraminifera and caused a general decrease in their size, as seen here from left to right.

skeletons may be found by the millions as fossils in sediments, remnants of the great mass extinctions.

Widespread disappearances of species due to cataclysms of terrestrial origins (volcanoes and glaciers) or alien origins (meteorites) punctuate the history of our planet. Mass extinctions are probably most famous for killing the dinosaurs roughly 65 million years ago. But over the ages, extinctions also devastated planktonic organisms, as evidenced by the abundant microfossils left behind by foraminifera and diatoms. One of the most remarkable fossil records belongs to the thousands of species of trilobites that populated the ocean from the early Cambrian until the late Permian, roughly 245 million years ago. Trilobite fossils have been found from low-lying deserts like Death Valley to high mountains in the Himalaya, indicating that both areas are made from sedimentary rocks once under the sea. This extreme difference in fossil location gives us a sense of the magnitude of tectonic movement and of the enormous time scale on which geological and evolutionary changes take place. Today, many species of fish and marine mammals are endangered by overfishing and pollution, and many more organisms are threatened by rising atmospheric temperatures, leading to increased oceanic acidification and habitat destruction. Some fear we might be living through the beginnings of a sixth major extinction period, and human activities seem to be accelerating its progress.

Periods of extinction have profoundly shaped the tree of life. With each one, the tree has lost branches, but as some species survived and diversified, new branches emerged. Just as terrestrial mammals profited from the demise of the dinosaurs in order to get bigger and differentiate, certain surviving planktonic species spread to new ecological niches abandoned by their previous occupants. Centric diatom species, which gradually appeared 200 million years ago, survived the mass extinction at the end of the Cretaceous (the one that wiped out most dinosaurs, 65 million years ago). Soon other diatom species appeared, such the pennates, which were able to glide along surfaces. Since then, pennate and centric diatoms have diversified, becoming especially abundant in the cold, silica-rich waters of the polar regions.

200 MILLION YEARS AGO TO THE PRESENT DAY: NEW OCEAN BASINS, CONTINENTS, AND SPECIES

Continents, ocean basins, and seas have appeared and disappeared over hundreds of millions of years due to the movement of tectonic plates on the Earth's surface. The plates are continuously moving, albeit very slowly. The western Mediterranean, for example, was created 35 million years ago from the fragmentation of the southern European border, and will one day disappear as the African plate continues moving toward Europe.

The ocean basins and currents we know today began to form about 200 million years ago. At that time, there existed only one supercontinent, Pangaea, and a vast solitary ocean, Panthalassa. These took shape as the result of the slow agglomeration of the Earth's landmasses. Pangaea began to break up into several pieces, giving rise to our present continents, and the Panthalassa ocean divided among them. The present-day Atlantic basin arose 180 million years ago when the American and Eurasian tectonic plates gradually split. Large currents and gyres slowly developed, defining various marine provinces. Each province has its own dynamics and its share of resident species representing almost the entire tree of life. It is indeed remarkable to note that almost all the major phyla, except terrestrial plants, are part of the plankton as adults or larvae.

Taxonomy and Phylogeny: Hierarchical Categorizations

In the eighteenth century, Swedish naturalist Carl Linnaeus devised a system for classifying, grouping, and naming living organisms. His categories and criteria have changed over the years, as biologists and paleontologists discover and describe new species, alive or fossilized, and learn more about organismal physiology and relatedness. But the essence of the Linnaean approach remains the taxonomic standard today. Organisms are classified by hierarchy: domain, kingdom, phylum, class, order, family, genus, and species. The three major domains, the broadest groupings, are Bacteria, Archaea, and Eukarya, and some argue for a fourth to include viruses. Kingdoms consist of such familiar distinctions as plants, animals, and fungi. Within each kingdom—for example, the animals—phyla represent major groups such as arthropods, mollusks, or vertebrates. And so on down the line to genera, each genus containing related species.

The total number of living eukaryotic species (organisms whose cells possess nuclei and organelles) on the planet is estimated at around 10 million. Most of these remain unidentified, though a large fraction is thought to be terrestrial insects. Known marine eukaryotes are estimated at 226,000, but at least one million more species are thought to spend part or all their lives in the plankton. Clearly we still have much to discover! The notion

Representations of the Tree of Life

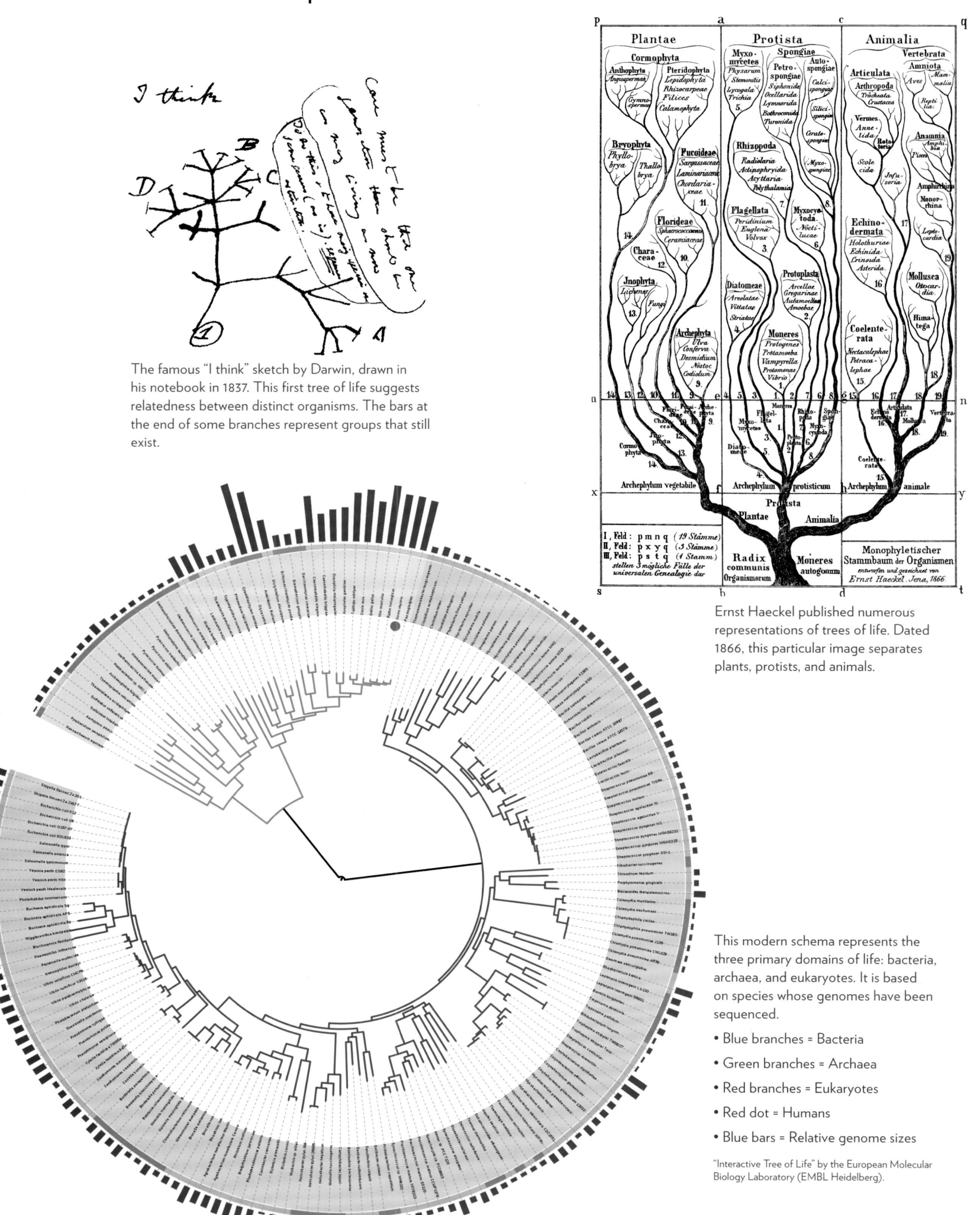

The famous "I think" sketch by Darwin, drawn in his notebook in 1837. This first tree of life suggests relatedness between distinct organisms. The bars at the end of some branches represent groups that still exist.

Ernst Haeckel published numerous representations of trees of life. Dated 1866, this particular image separates plants, protists, and animals.

This modern schema represents the three primary domains of life: bacteria, archaea, and eukaryotes. It is based on species whose genomes have been sequenced.

• Blue branches = Bacteria

• Green branches = Archaea

• Red branches = Eukaryotes

• Red dot = Humans

• Blue bars = Relative genome sizes

"Interactive Tree of Life" by the European Molecular Biology Laboratory (EMBL Heidelberg).

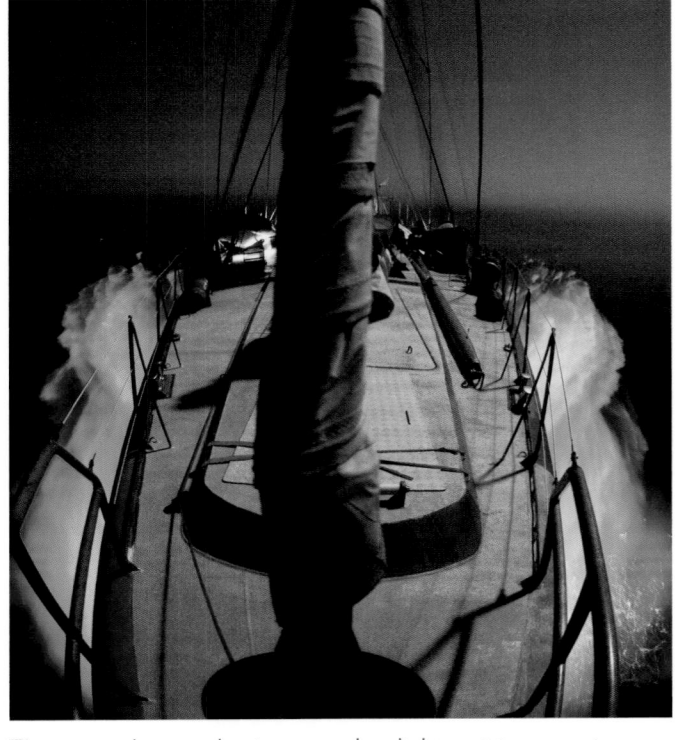

The sea can become luminescent when light-emitting organisms are particularly abundant. Here luminescence is due to a bloom of dinoflagellates, agitated by the schooner *Tara* in the Indian Ocean. Photo by Julien Girardot.

of species is increasingly blurred, however. A group of organisms is identified as a species if it has the capacity to produce fertile descendants. But we must also take into account "cryptic" species (those that look identical but are actually different), as well as the numerous forms of bacteria and protists arising anew from adaptations to their changing environments.

In 1837 Darwin made a drawing in his notebook that was the first representation of the "tree of life," what we now call a phylogenetic tree. Phylogeny is the study of relationships between organisms. The word is derived from *phulé*, meaning "tribe" in Greek. The phylogenetic tree in the drawing on page 13 looks more like a bush or a dense thicket, endeavoring to illustrate the major groups of living creatures over the history of Earth. Those branches extending all the way to the top represent lineages that have passed descendants to the present day. Branches that stop short indicate the many lineages that have died off over time, either being outcompeted or suffering extinction due to some environmental cause. Horizontal branches represent transfers of organelles and genes between bacteria, archaea, protists, and even some higher eukaryotes.

Phylogenetic classification was traditionally based on the study of morphological and life history traits of living beings and fossils. Today, computational analyses of gene and protein sequences provide a highly quantitative understanding of the common ancestors and evolutionary relationships of different groups. We can learn when and how different phyla, classes, and orders diverged, and estimate when two species split from their common form. For example, over more than 500 million years, arthropods have evolved into more than 50,000 species of crustaceans, and millions of species of insects. During roughly the same period, mollusks diversified into 100,000 species of gastropods (i.e., snails, slugs, pteropods) and more than 700 species of cephalopods (squid, octopus, and nautilus).

In this book, we use both the common names of organisms and the standard approach to taxonomic nomenclature. Take the example of modern humans: the genus and species name of all contemporary humans, regardless of their origin or skin color, is *Homo sapiens*, from the Latin "wise man." *Homo sapiens*, along with *Homo erectus*, *Homo neanderthalensis*, and other missing cousins, constitute the family of hominids whose most ancient fossils (first discovered in Africa) date back 13 million years. Hominids belong to the order of primates that appeared 55 million years ago. Primates are part of the class of mammals that began to dominate the land about 100 million years ago when movements of the tectonic plates started to separate the American and African continents. Mammals themselves are part of the chordate phylum that appeared more than 530 million years ago.

Classification of planktonic organisms follows the same logic. The jellyfish *Pelagia noctiluca* is a stinging species that drifts along the Mediterranean and Atlantic coasts. Different species of *Pelagia* jellyfish belong to the family Pelagidae, which is in the class Scyphozoa, and the phylum Cnidaria. The phylum Cnidaria, like the phylum Chordata (including us *Homo sapiens*) is part of the kingdom Animalia (also known as Metazoa) in the domain Eukarya. Man and jellyfish therefore share a common ancestor and our lineages inherited the genes of this primitive animal.

Sometimes different species look so similar that only a specialist such as taxonomist or systematist can identify them after dissection and molecular analyses. In this book, if we have a doubt about the identity of a species photographed, we adopt the general nomenclature "*Genus* sp." (e.g., *Homo* sp. for humans, or *Pelagia* sp. for *Pelagia* jellyfish).

Organisms of All Sizes, with Different Roles and Behavior

It is not easy to collect and study a drifting ecosystem consisting of a vast multitude of organisms ranging in size from less than 1 micron to tens of meters, over a 10-million-fold difference. The smallest beings are viruses, and then bacteria and archaea. The largest are threadlike colonial cnidarians (siphonophores such as *Praya dubia*) that can reach more than 50 meters when extending their fishing filaments.

Within the same class of organisms, sizes can vary greatly. For example, the giant "lion's mane" jellyfish, *Cyanea capillata*, can be more than 2 meters in diameter while the small jellyfish *Clytia hemispherica* rarely exceeds 2 centimeters. And *Clytia* is a giant compared with the numerous species of microscopic jellyfish. Although composed of a single cell, some radiolarian protists measure one to several millimeters in size and are bigger than many multicellular animal embryos or larvae. At the other end of the spectrum, many protists are barely more than a micron across and only

Left: Zooplankton collected in a net during the *Tara Oceans Expedition*. Top right: Close-up of a Bongo net with 180-micron (0.18 mm) mesh. Above right: 20-micron mesh, and 120-micron mesh, with the tip of a match to show relative size. Photos at left and top right by Anna Deniaud Garcia.

Collection and observation of zooplankton during the *Tara Oceans Expedition*, between the coast of Ecuador and the Galapagos Islands in May 2011.
Photos by Christoph Gerigk.

slightly bigger than most bacteria. This is also the size of *Prochlorococcus* sp., the most abundant unicellular organism on the planet. In contrast, filamentous bacteria often can be seen with the naked eye in plankton nets.

While each species has adapted to its niche and plays a particular role within the planktonic ecosystem, altogether different organisms frequently perform similar functions and represent "functional types." Take for example the photosynthetic protists: green algae, coccolithophores, dinoflagellates, and diatoms. All produce oxygen and absorb CO_2 to make living matter. These phytoplanktonic organisms are the "primary producers," like plants in terrestrial ecosystems. They are eaten by other protists such as radiolarians and foraminifera and by zooplankton and animal larvae. Copepods and their fellow crustaceans are all grazers of algae and protists. Salps, appendicularians, and pyrosomes filter large volumes of water and actively feed on bacteria and algae. Crustaceans, mollusks, and fish are higher on the food chain, eating the countless copepods, larvae, and other plankton.

Very different pelagic organisms have evolved similar strategies not just for feeding, but for growth and shelter. Gastropod mollusks mobilize calcium in the sea to make their shells and skeletons, as do some protists (foraminifera and coccolithophores), echinoderms like sea urchins and sea stars, as well as corals. Calcium is abundant in the ocean, whereas silica and strontium are present only in low concentrations. This limits the growth of organisms that need silica or strontium oxides to build their shells and skeletons, namely diatoms, radiolarians, and silicoflagellates.

Many plankton are transparent, gelatinous animals made 95 percent of water. The gelatinous envelopes of cnidarians, ctenophores, mollusks, and tunicates consist of a loose network of protein and sugar macromolecules. Tunicates build their outer layer, or tunic, using the carbohydrate polymer cellulose, a trait possibly coopted from a photosynthetic organism by gene transfer. These casings are sometimes eaten by fish or turtles, but most often are recycled by specialized crustaceans called amphipods. Different species of amphipods usually have a favorite host or prey (jellyfish, siphonophore, salp, or pyrosome) that they parasitize and eat. Some, like *Phronima*, appropriate the casing to make shelters for their young.

Collecting and Identifying Plankton, Then and Now

Plankton have been part of human culture for ages. We see historical evidence on antique vases depicting jellyfish and continue to find plankton in various guises, such as fish larva paste ("poutine") on restaurant menus along the French Riviera today. In the late 1700s and early 1800s the first microscopes revealed the invisible world of animalcules, tiny animals or protists we today call plankton. In the first decade of the nineteenth century the zoologist François Péron and illustrator Charles-

Alexandre Lesueur described the gelatinous organisms in the Bay of Villefranche-sur-Mer. Sailing on the *Beagle* from 1831 to 1836, Darwin dragged a small mesh net to collect microscopic organisms. Around 1887, the German zoologist Victor Hensen referred to beings that drift with the currents as "plankton," and the name stuck.

Inspired by Darwin's theory of natural selection and its implications for the evolution of organisms, zoologists such as Ernst Haeckel and Carl Vogt explored the coasts and initiated expeditions to collect, describe, and classify new species. Research was organized by universities and marine stations emerging in Europe, the Americas, and Japan. Situated in Concarneau, Naples, Plymouth, Roscoff, Banyuls, Villefranche-sur-Mer, La Jolla, Woods Hole, Pacific Grove, and Misaki, these field stations exposed a hidden marine world. The first oceanographic expeditions, including the voyage of HMS *Challenger* (1872–1876), developed collection techniques and described many organisms. This tradition of marine exploration continues today through the deployment of oceanographic research ships, satellite observations, recordings by automated buoys, and remote-controlled robots that roam the ocean depths.

Nowadays, images from submersible cameras, automated microscope analysis, and deciphered gene sequences and relationships (the sciences of molecular biology and genomics) complement one another. We are beginning to analyze the whole planktonic ecosystem and the countless associations and interactions between organisms. Using these new approaches, 140 years after the HMS *Challenger*, the *Tara Oceans Expedition* traversed the world ocean in three years, collecting organisms in 35,000 samples from more than 200 selected sites. The organisms range from tiny viruses to fish larvae and large gelatinous animals. The huge amount of amassed data will be used to model oceanic phenomena, with the goal of understanding and predicting the alteration of climate and ecosystems and the evolution of life in the ocean.

Planktonic organisms reveal an incredible diversity of forms and behavior. The photos in the book were taken with a variety of cameras, objective lenses, magnifiers, and microscopes during the *Tara Oceans Expedition* in the Mediterranean Sea, and in the Indian, Pacific, Atlantic, Antarctic, and Arctic Ocean basins. Some collecting and photography was done in marine stations in Europe, the USA, and Japan, with the help of local biologists and individuals. In addition to generating still images, one major objective of this work was to capture the unique and unfamiliar movements and habits of these creatures.

The *Plankton Chronicles* videos and photos combine science and art. They are directly accessible via the pages of this book thanks to Q codes. Connect with our *Plankton Chronicles* website and explore more deeply the fascinating world of plankton.

Plankton Chronicles website
www.planktonchronicles.org
An interactive website with
photos, texts and videos

In 1862 Ernst Haeckel published *Die Radiolarien*, an atlas that remains a scientific and artistic reference today. 1–2: *Lipthoptera mulleri*; 3–6: *Astrolithium dicopum, A. bifidum,* and *A. crutiatum*; 7–8: *Diploconus fasces.*

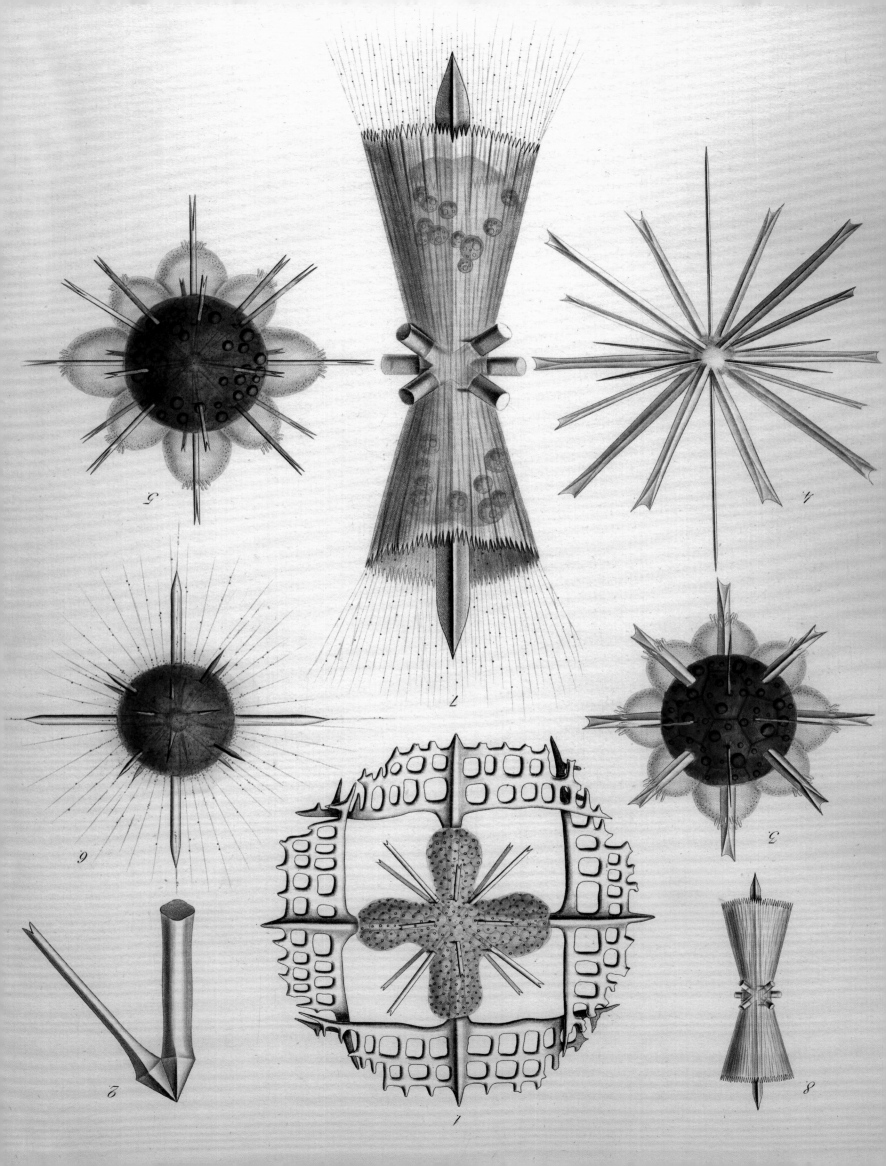

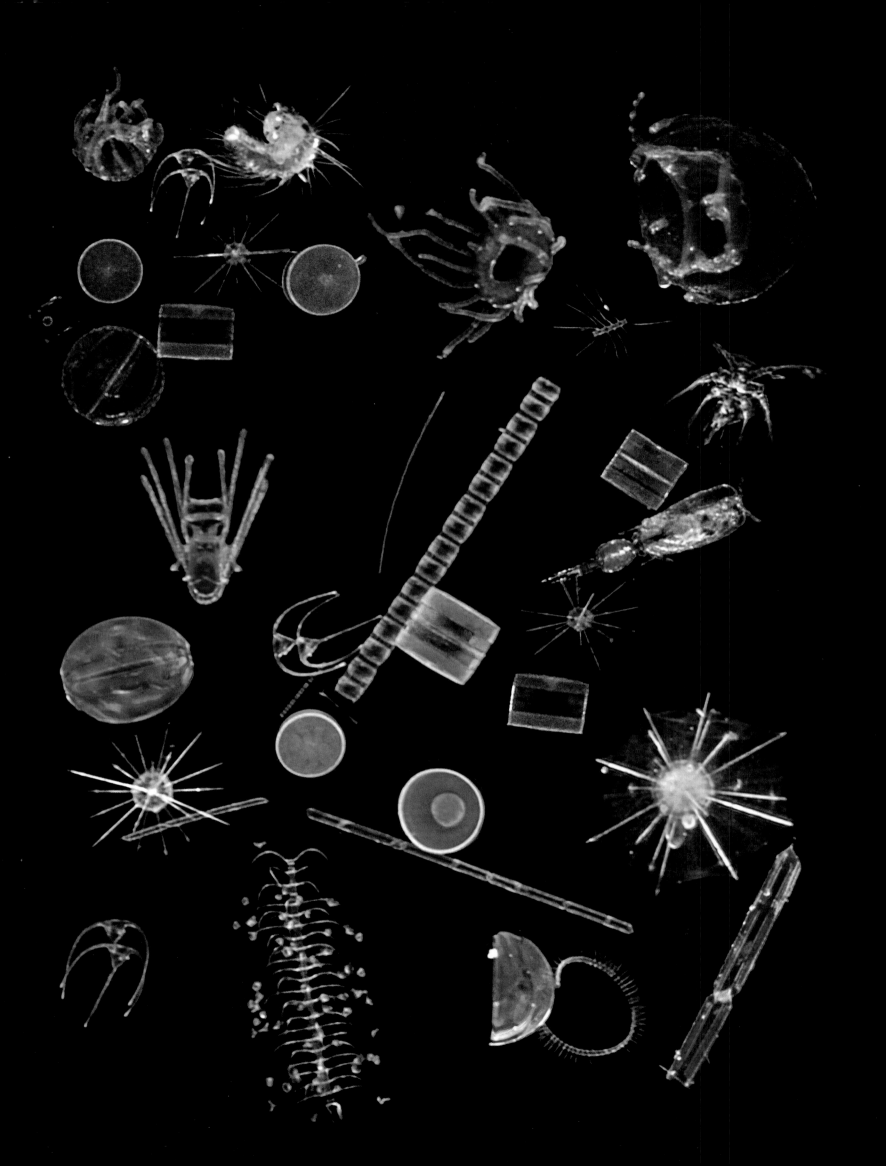

PLANKTON
of the WORLD

Any body of water, whether fresh, salty, or brackish, tropical or frozen, acidic or alkaline, has its share of uni- and multicellular planktonic organisms. Aquatic and marine ecosystems—rivers, lakes, open ocean, coastlines and estuaries, and glaciers—each have their own characteristics that change with the daily tides, periodic seasons, and fluctuating currents. The dynamics of resident plankton populations remain enigmatic. Are the different types of organisms in an ecosystem always present everywhere in the water column, ready to fill the ecological niche if conditions are favorable? What are the relationships between the composition of the ecosystem and its environment—temperature, salinity, oxygen, acidity, minerals, and nutrients? What is the role of large predators that come and go, and why are symbiosis and parasitism so prevalent?

Universities, institutes, and marine stations are mobilized for research to find answers to many of these questions. Oceanographic buoys and unmanned vehicles monitor and roam the ocean; satellites continuously scan the seas tracking blooms and fish stocks. In recent years, analysis of genes and microscopic images has begun to reveal the identity and extent of organisms constituting each ecosystem and how they function together. Researchers engage in vast participatory networks such as the Census of Marine Life and MAREDAT. Starting with the HMS *Challenger* over a century ago, and continuing with recent efforts like the *Global Ocean Sampling* and the *Tara Oceans Expedition*, scientists have collected tens of thousands of samples worldwide. Teams work to analyze the plankton, mapping out ecosystems and environmental parameters. Researchers model and simulate biodiversity and dynamics of water masses throughout the seas and ocean.

For this book, we collected, identified, photographed, and filmed plankton around the world. We harvested organisms in nets and bottles aboard the schooner *Tara* in the Mediterranean Sea and the Indian, Pacific, Atlantic, Antarctic, and Arctic Ocean basins. We accompanied colleagues and friends in the bays of Villefranche-sur-Mer, Toulon, Roscoff, Shimoda, and Sugashima, and in the wetlands of South Carolina. In this chapter, we describe our explorations and show the diversity of organisms found in these waters.

These computer-generated images give an idea of the dominant types of phytoplankton during the seasons of our collections in various regions of the world.
Computer simulation: ECCO2 and Darwin Project, MIT

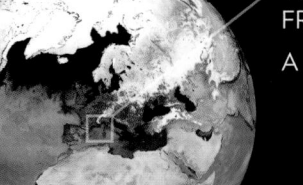

VILLEFRANCHE-SUR-MER, FRANCE

A bay famous for its plankton

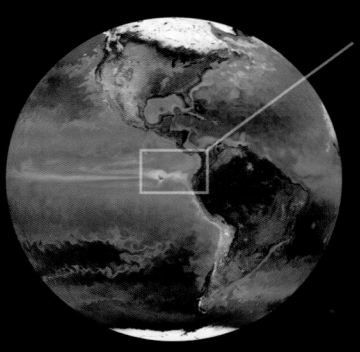

BETWEEN ECUADOR AND GALAPAGOS

Tara Oceans Expedition

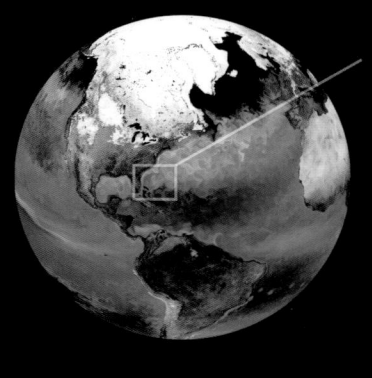

SOUTH CAROLINA, UNITED STATES

Salt marsh estuaries

THE IZU PENINSULA AND SHIMODA, JAPAN

Autumn plankton

- In red and yellow: Diatoms and other larger phytoplankton.

- In green and blue: Cyanobacteria *Prochlorococcus* sp., *Synechococcus* sp., and other tiny phytoplankton.

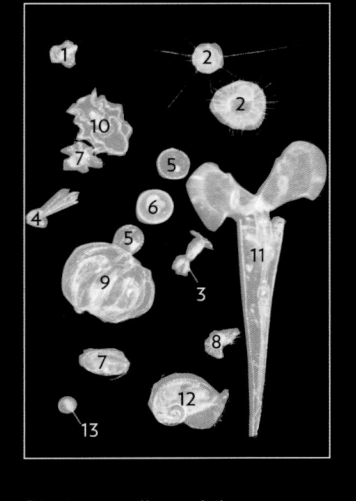

Cyanobacteria (green)
are the dominant
phytoplankton in winter.

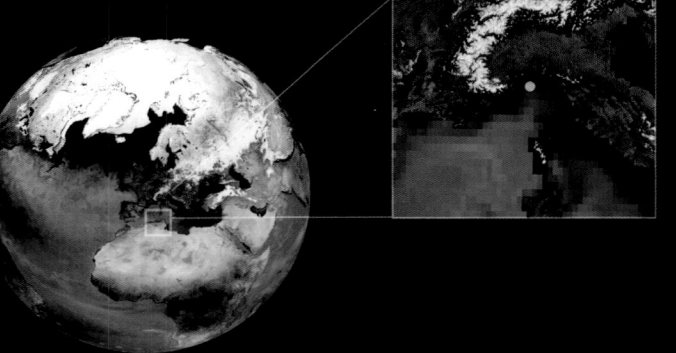

VILLEFRANCHE-SUR-MER, FRANCE
A bay famous for its plankton

The waters are often turquoise and azure in this region of the Mediterranean. The French Riviera stretches from the Italian border and the foothills of the Southern Alps past Villefranche-sur-Mer, Nice, Cannes, and all the way to Saint-Tropez. The proximity of the Ligurian current, flowing east to west, influences the distribution of surface plankton near the coast and in the bay of Villefranche-sur-Mer. As seasons and climate change, populations of plankton change too. Some organisms rise from the depths and remain trapped in the bay.

Boats from the marine station of Villefranche-sur-Mer (established in 1882) collect plankton at a reference site called Point B, at the mouth of the bay. This site is located at the end of a deep canyon that acts as a nursery for some planktonic species. Every day the boats bring in organisms to be identified, studied, and stored.

In the early 1800s, the zoologist François Péron, in collaboration with the illustrator Charles-Alexandre Lesueur, was the first to describe the pelagic organisms in the bay of Villefranche-sur-Mer. Jean Baptiste Vérany and Carl Vogt rediscovered and pursued studies of the gelatinous plankton in the bay in the 1850s. Thirty years later, Hermann Fol (who had recently discovered fertilization) joined a young French professor, Jules Barrois, to found the first laboratory. They were encouraged by Darwin and Vogt. Then Alexis Korotneff and his Russian colleagues joined them to host prestigious biologists at the Zoological Station of Villefranche-sur-Mer.

The research station has become a leading center for the study of embryos, larvae, and other planktonic organisms. More than 200 researchers, teachers, and students work together in state-of-the-art laboratories, now known as the Observatoire Océanologique de Villefranche-sur-Mer, under the responsibility of the Centre National de la Recherche Scientifique (CNRS) and the Université Pierre et Marie Curie (UPMC). In this distinguished place, Sharif Mirshak and Noé Sardet (of Parafilms, Montreal, Canada) and I filmed and photographed plankton for the *Plankton Chronicles* project.

Plankton collected during winter in the bay of Villefranche-sur-Mer, using a 0.2-mm mesh net. The longest organism here is the pteropod mollusk, measuring about 7 mm.

11. Mollusk larvae
2. Protists: radiolarians
3. Crustacean: copepod with eggs
4. Annelid larva
5. Protist: dinoflagellate
6. Protist: foraminiferan
7, 8. Crustacean larvae
9. Comb jelly larvae
10. Echinoderm larvae
11. Mollusk: pteropod
12. Mollusk: heteropod
13. Green algae

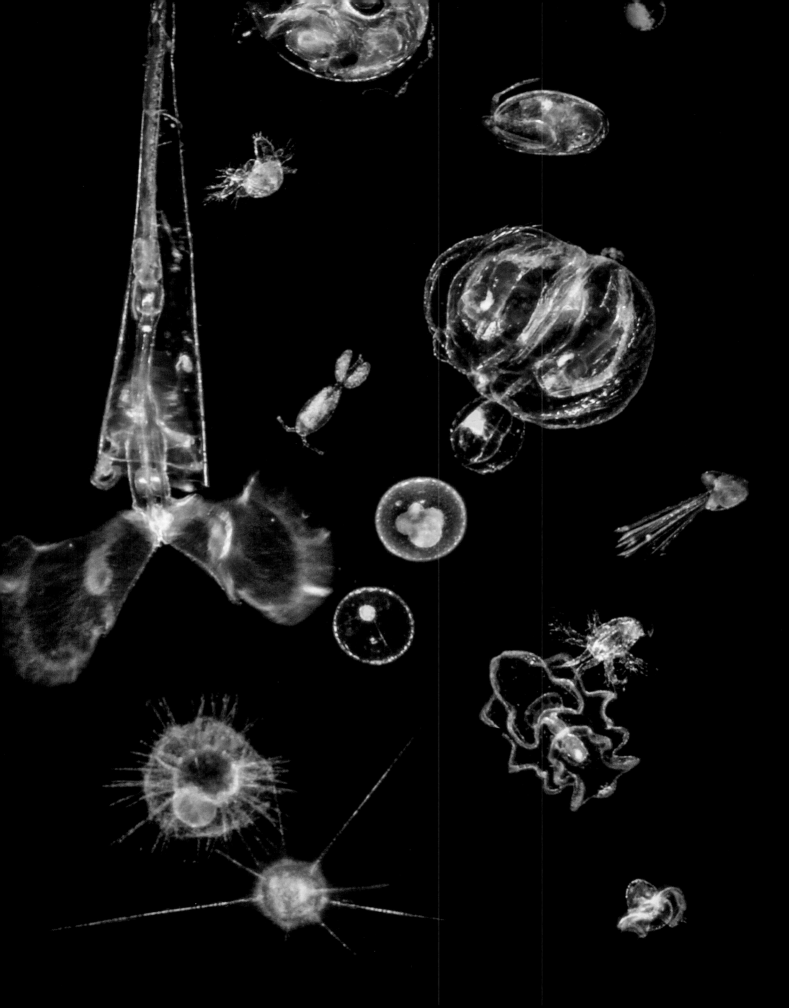

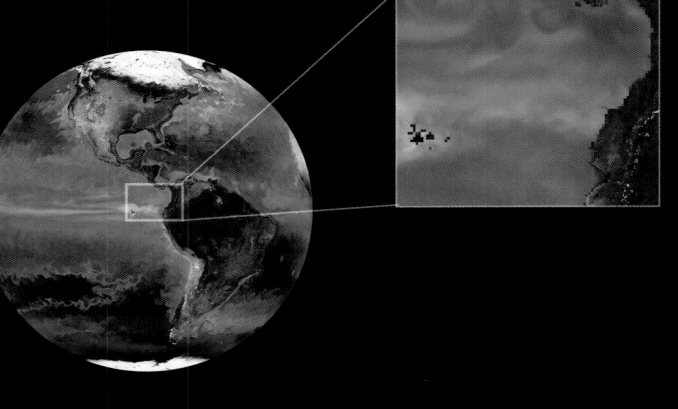

Cyanobacteria (green) and large phytoplantonic protists (yellow) are dominant in the spring.

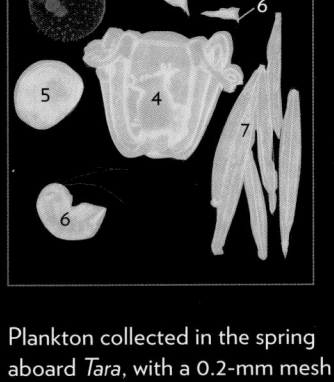

BETWEEN ECUADOR AND GALAPAGOS
Tara Oceans Expedition

n May 2011, in the middle of a thirty-month journey around the world, the *Tara Oceans Expedition* left the Guayas River delta to navigate between the coasts of Ecuador and the Galapagos Islands. Plankton are particularly abundant in these waters where warm equatorial currents mix with cold currents from the south.

The waters are rich in organisms of all kinds. Extraordinary protists, especially foraminifera and radiolarians, together with countless larvae showed up in our surface nets and in the bottles pulled up from the depths. But the large predators especially seized our attention. The nets were charged with fragments of iridescent ctenophores, especially Venus girdles and *Beroe*, as well as salps, photographed in situ by our divers.

One evening collecting in these waters our excitement reached a peak. We had begun with long-line fishing of large squids. Then the plankton nets brought up some strange creatures: planktonic sea cucumbers and gelatinous organisms in the form of long socks. Chief scientist Gaby Gorsky was thrilled when Stephane Pesant and Sophie Marinesque lined up the sock-like organisms on the collection table. These were tunicates called pyrosomes, each comprised of thousands of individuals living together as a colony,

Plankton collected in the spring aboard *Tara*, with a 0.2-mm mesh net. The jellyfish measures about 5 mm.

1. Appendicularian
2. Crustaceans: copepods
3. Crustacean: copepod with eggs
4. Cnidaria: jellyfish
5. Mollusk larva
6. Crustaceans: amphipod and copepod
7. Chaetognaths

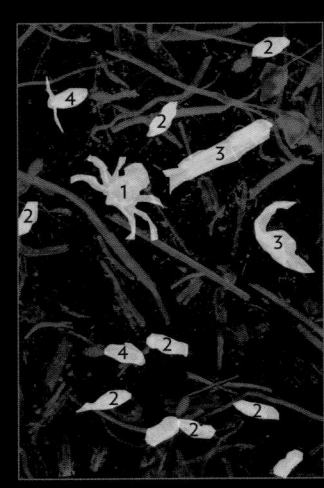

The dominant phytoplankton during the summer are the cyanobacteria *Synechococcus* (blue), and large phytoplanktonic protists (yellow).

SOUTH CAROLINA, UNITED STATES
Salt marsh estuaries

The vast salt marshes of South Carolina stretch from the Savannah River delta just below Hilton Head Island to Little River Inlet, just above Myrtle Beach. Tidal creeks wind their way through muddy banks and fields of *Spartina* grass, converging in bays and inlets sheltered from the Atlantic by sandy barrier islands. For thirty years, researchers at the Baruch Institute for Marine and Coastal Science and the University of South Carolina have been observing populations of planktonic organisms in Winyah Bay, near Georgetown.

It was a great pleasure to visit and work with Dennis Allen, director of the Institute. Dennis is a connoisseur of the marsh and its inhabitants and author of the illustrated book *Zooplankton of the Atlantic and Gulf Coasts.* During several hot summer days we collected organisms in a major salt marsh creek in the North Inlet estuary, as Dennis has done every week the past three decades.

In the laboratory, we observed the particular zooplankton living in the sediments and plant debris that give these waters their brown color. The plankton here are adapted to tidal changes in the brackish waters. As might be expected, larval forms of crustaceans and annelid worms were numerous, reflecting the omnipresence of adults that reside in the muddy wetlands. Among them were hordes of fiddler crab larvae, the predominant species in the plankton of the marsh creeks.

Plankton collected in the marsh creeks during summer, using a 0.36-mm mesh net. Copepods in the image are about 2 mm long.

1. Crab larva
2. Crustaceans: copepods
3. Shrimp larvae
4. Crustaceans: copepods with eggs

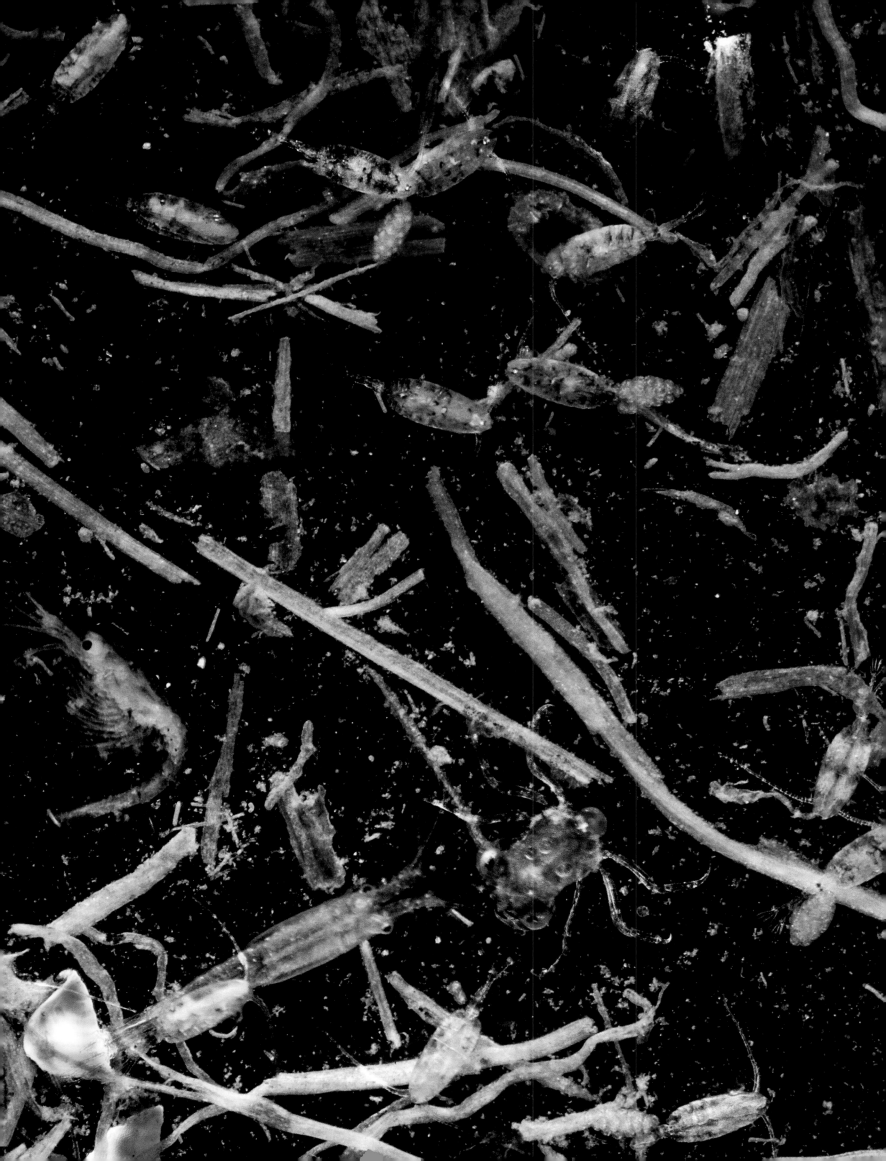

The dominant phytoplanktonic organisms in autumn are the cyanobacteria *Synechococcus* (blue).

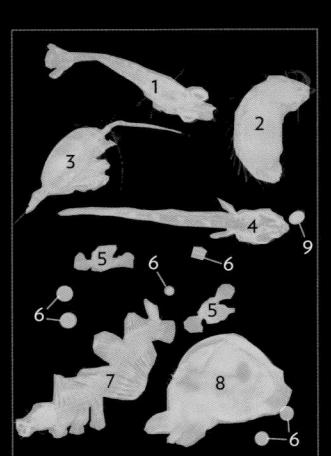

IZU PENINSULA AND SHIMODA, JAPAN
Autumn plankton

Shimoda is a small fishing town at the tip of the Izu Peninsula, three hours south of Tokyo. Nestled in a bay surrounded by green mountains with rounded peaks, the Marine Station of Tsukuba University has welcomed me for several years to work on marine eggs and embryos. In November 2012 under a threatening sky, my friend and colleague Kazuo Inaba and I embarked on the Tsukuba boat to collect plankton. In the bay of Shimoda, there was no bloom, but a great variety of organisms. Their beauty and diversity inspired me to film and photograph until dawn in the quiet of the laboratory. The next day, my hosts rallied around the collecting aquariums. Consulting illustrated books, researchers and students found the names and discussed the behavior of all these living creatures captured by my camera.

Plankton collected in Shimoda bay in autumn with a 0.2-mm mesh net. Organisms measure a maximum of 5 to 7 mm.

1. Crustacean: shrimp larvae
2. Polychaete annelid carrying eggs
3. Crustacean: copepod
4. Vertebrate: fish larva
5. Mollusks: pteropod larvae
6. Protists: diatoms
7. Polychaete annelid
8. Mollusk: pteropod
9. Crustacean: copepod eggs

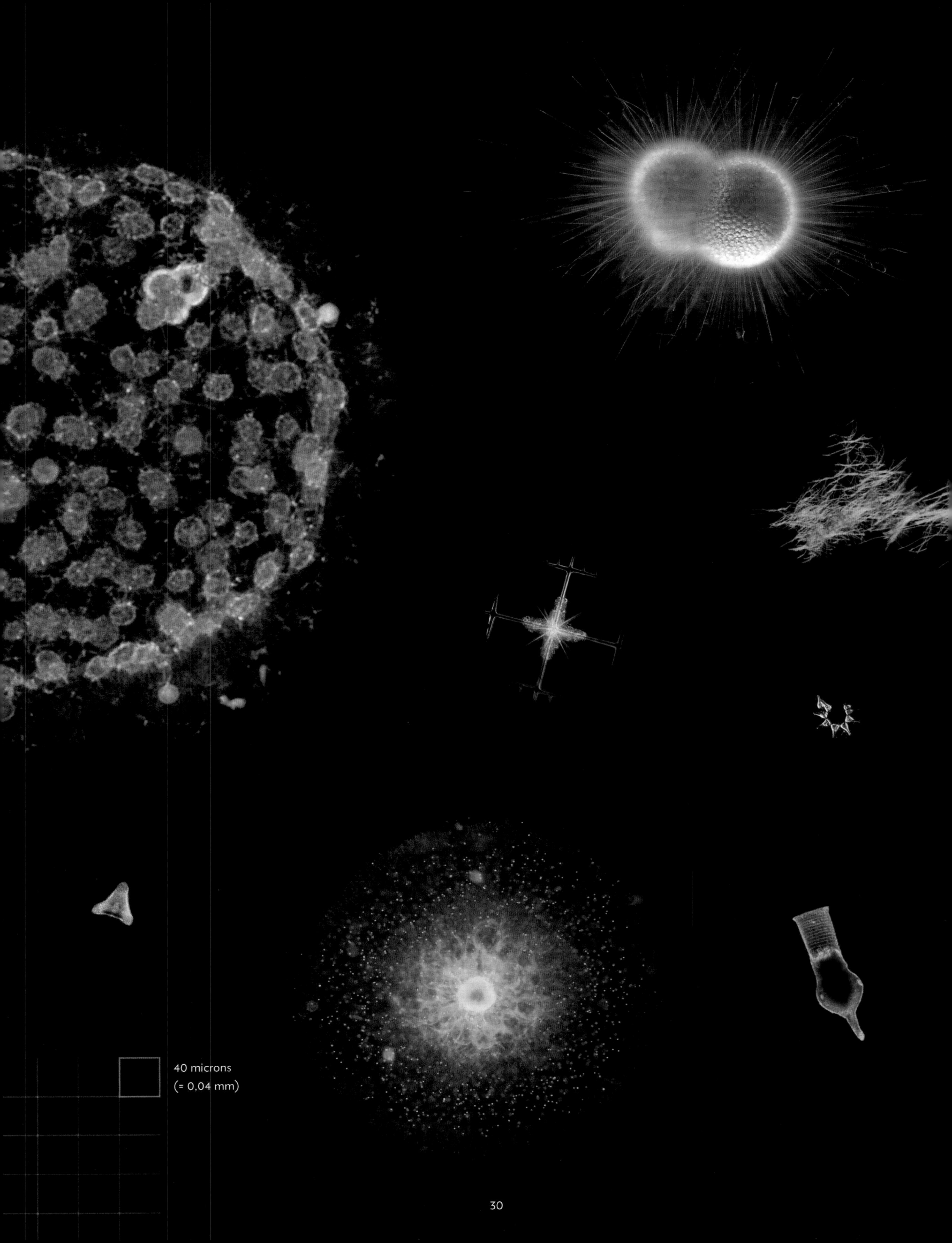

40 microns
(= 0,04 mm)

UNICELLULAR CREATURES

From the Origins of Life

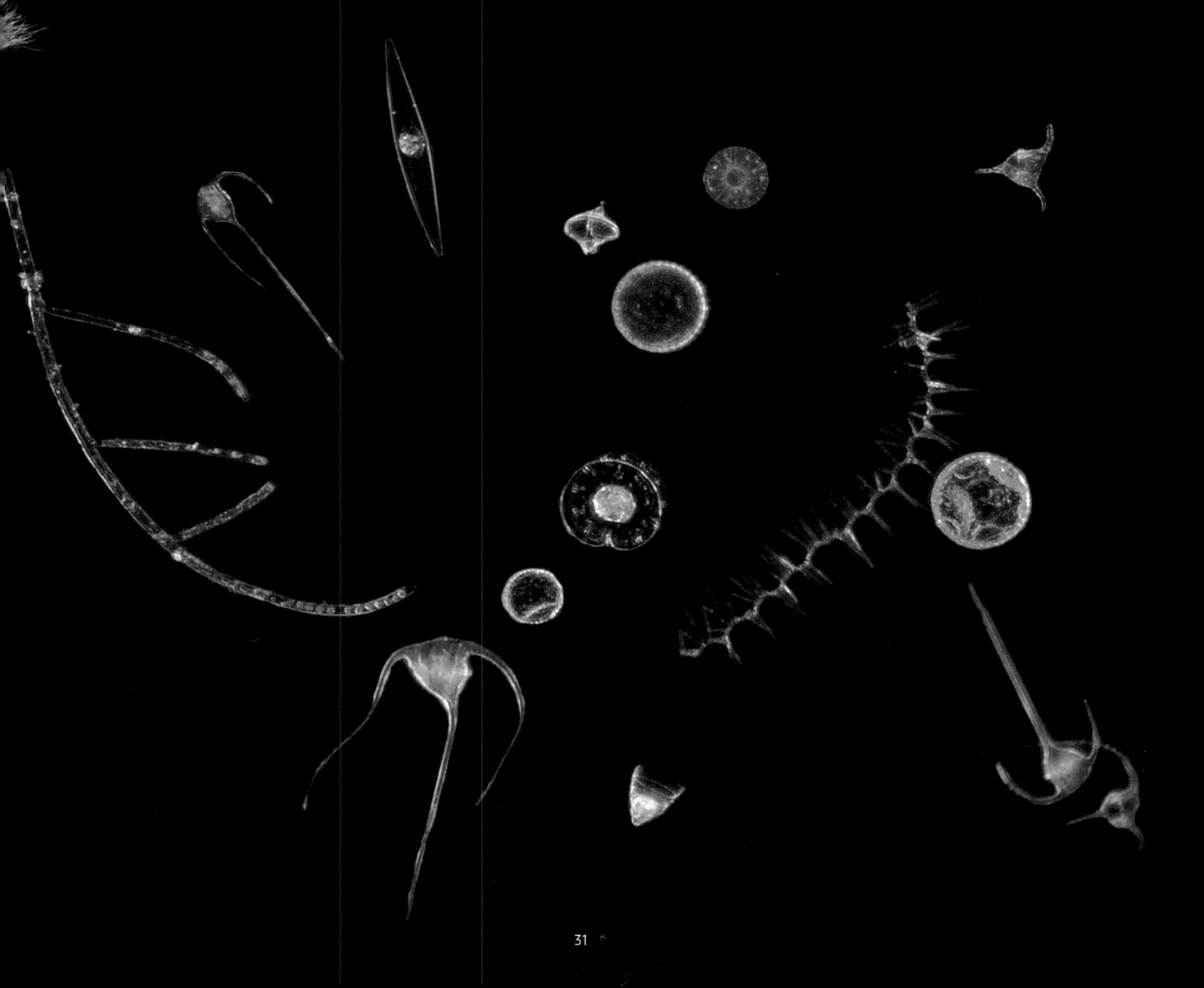

BACTERIA, ARCHAEA, AND VIRUSES

INVISIBLE BUT OMNIPRESENT

Bacteria and archaea, and their viruses known as phages, are found everywhere in the ocean and seas. Free-living, or as symbionts or parasites, they occupy the water column from the surface to bottom sediments. Bacteria and archaea—prokaryotes, cells lacking nuclei and other organelles—are especially abundant in the guts, feces, and cadavers of planktonic organisms. Their sizes vary greatly. An individual bacterium measures 1 or 2 microns at most; while in the form of films and filaments, bacteria can measure several millimeters or more.

Whether in oceanic "deserts" or in areas teeming with a diversity of organisms, bacteria may number from millions to billions in a single liter of water. Equipped with rotary flagella and hair-like appendages, bacteria and archaea constantly explore their microscopic universe, reacting to the presence of light, metals, nutrients, and chemical signals from other organisms.

Ancestral bacteria and archaea were probably the first living creatures to colonize the ocean. For more than two billion years, primitive single-celled organisms survived on energy from the oxidation of metals and light emitted by the sun. Eventually, symbioses of bacteria and archaea evolved into cells possessing nuclei and organelles, called eukaryotes. These were the ancestors of protists and all the plants and animals we know today.

Early bacteria and archaea shaped the Earth and its atmosphere. Cyanobacteria capable of photosynthesis produced oxygen on the young planet. Oxygen gradually accumulated in the atmosphere, thus establishing conditions more than 800 million years ago that were favorable for organisms whose existence depends on aerobic respiration. The planetary cycles of oxygen, as well as carbon and nitrogen, continue to depend on the activity of the enormous bacterial biomass.

Dividing rapidly and capable of exchanging genes, bacteria and archaea adapt quickly to new environments. They can utilize to their advantage various types of pollution such as nitrogen- and phosphorus-rich fertilizer runoff or gas and oil seeping from the ocean floor or spilled from tankers or drilling rigs. Bacteria even colonize the huge quantities of plastic particles and synthetic fibers that humans release into the sea.

Bacteria and archaea are in turn constantly infected and destroyed by phages in a widescale recycling of organic matter and genes. They are engulfed by protists and are consumed by countless filter feeders such as salps, doliolids, larvaceans, and pyrosomes.

Some planktonic bacteria are dangerous to people. Cholera bacteria (*Vibrio cholerae*) survive in plankton as spores associated with copepods and chaetognaths. During periods of bloom, these organisms spread cholera in human populations living along the coasts. Other planktonic bacteria have a much better reputation. Spirulina (*Arthrospira platensis*), a filamentous cyanobacterium, thrives naturally in tropical lakes. Easy to grow, Spirulina has become a dietary supplement, providing a valuable source of protein, essential amino acids, vitamins, and minerals for malnourished populations.

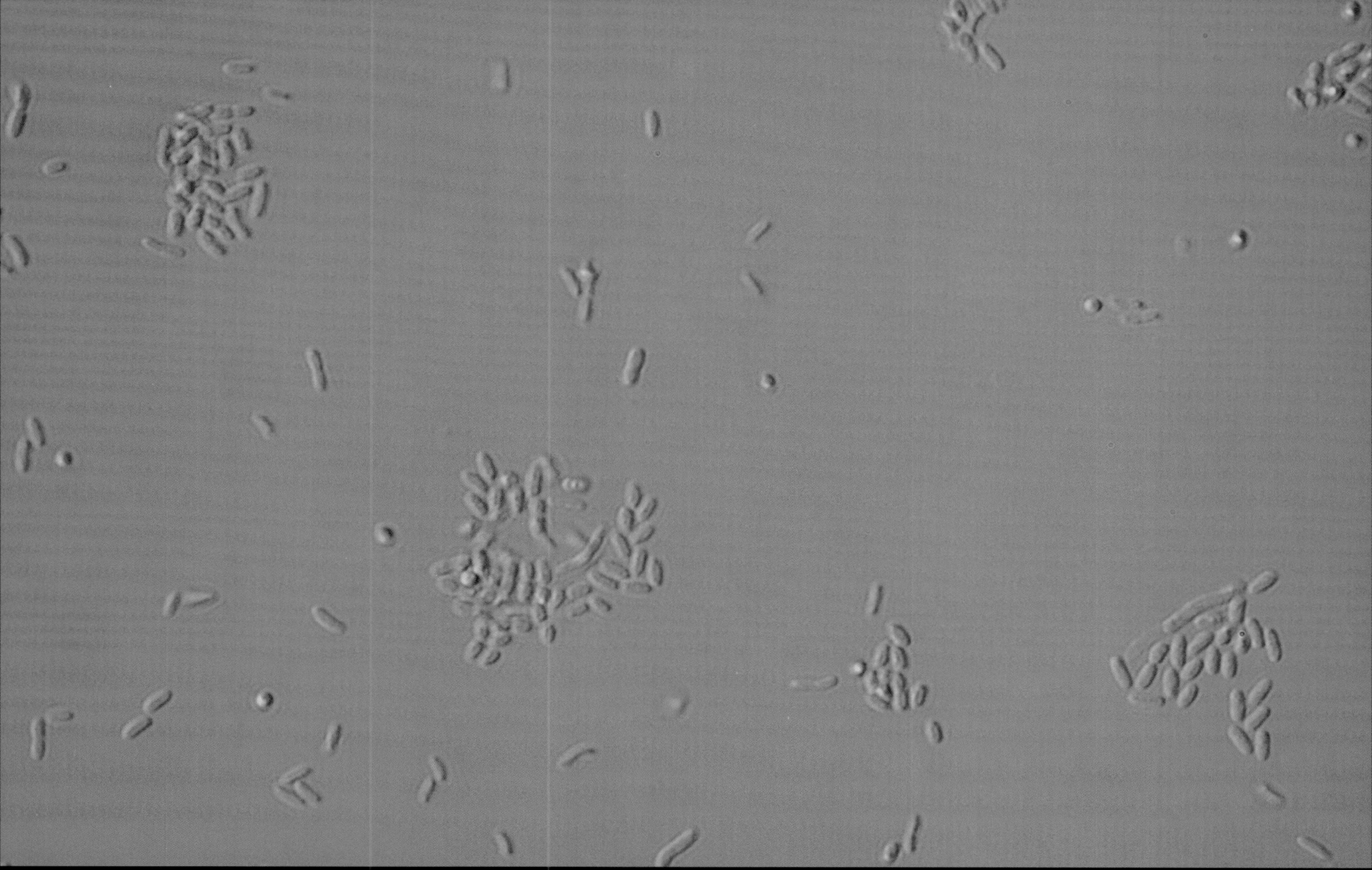

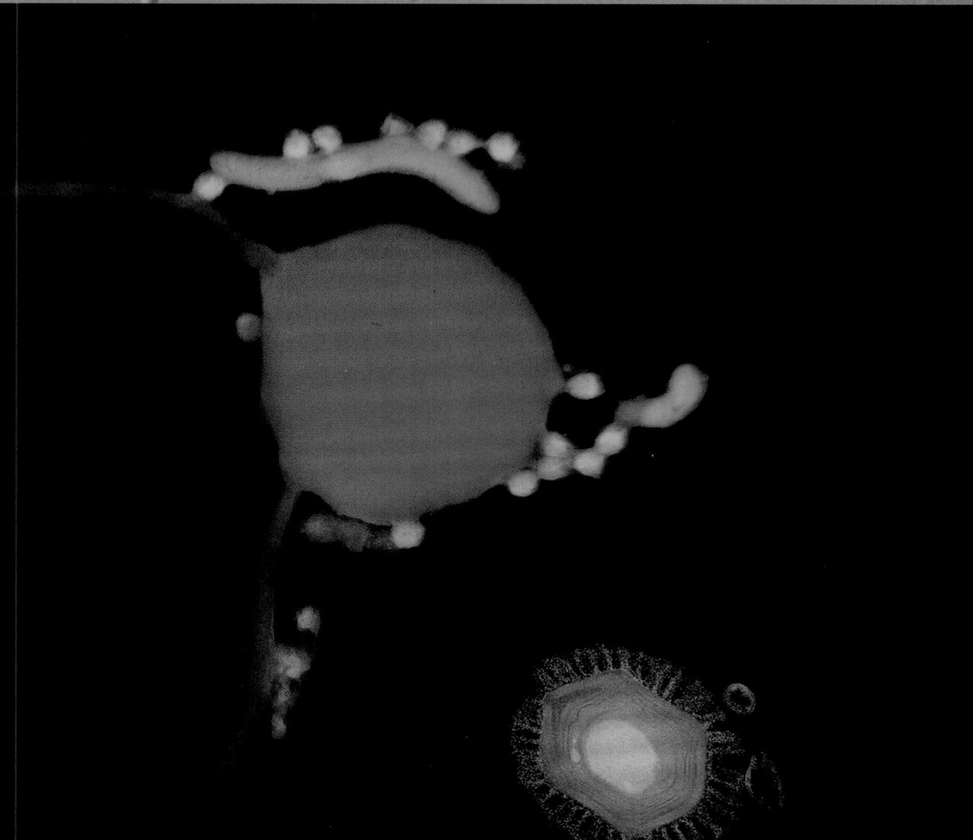

BACTERIA, VIRUS, AND GYRUS

Bacteria are often rod-shaped and rarely exceed a micron in size. In the bottom image, we colorized the bacteria blue and used pink for the phages and a giant virus called a gyrus (on the right). The bigger microalga with two flagella (in green) measures 2 microns in diameter.

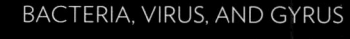

Above: Optical micrograph of live marine

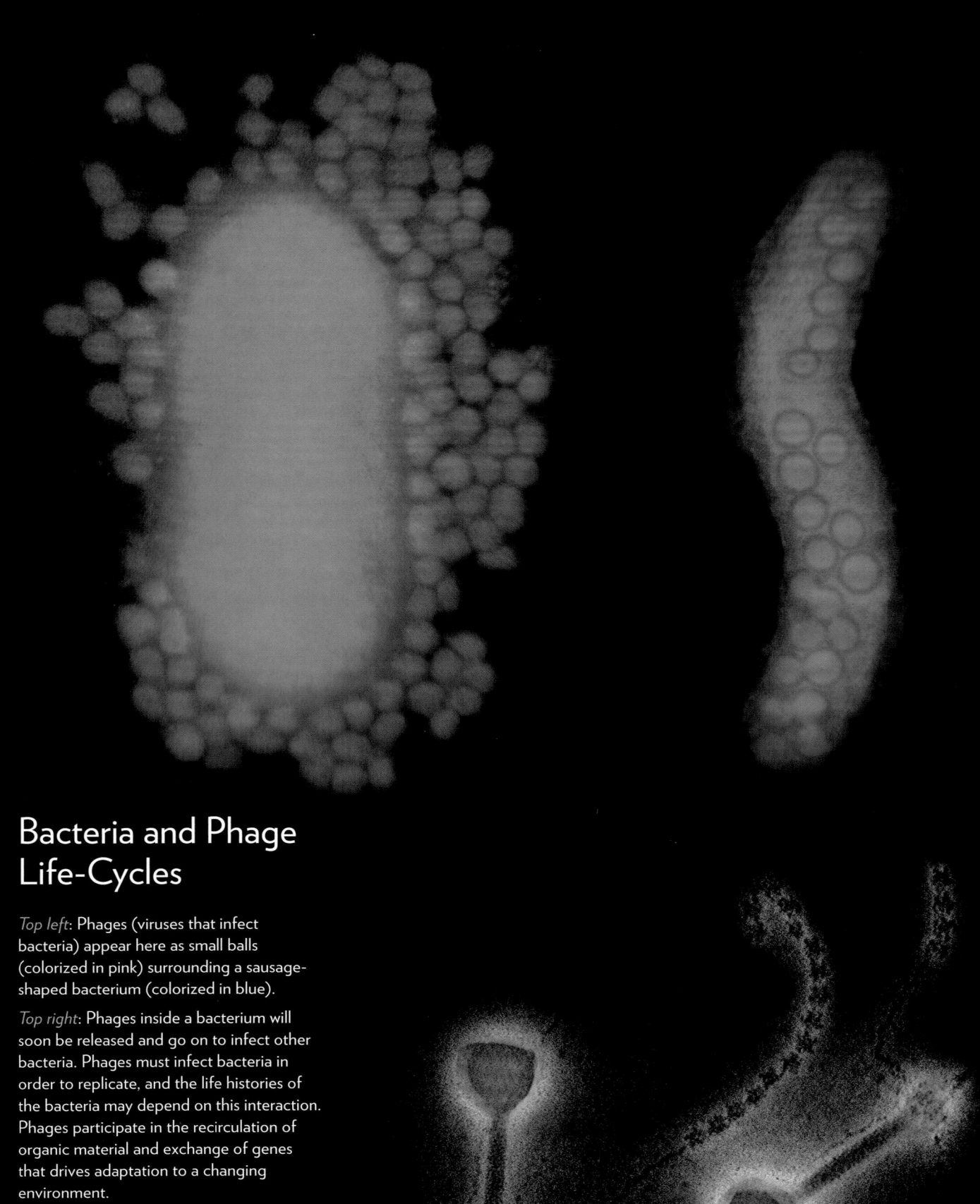

Bacteria and Phage Life-Cycles

Top left: Phages (viruses that infect bacteria) appear here as small balls (colorized in pink) surrounding a sausage-shaped bacterium (colorized in blue).

Top right: Phages inside a bacterium will soon be released and go on to infect other bacteria. Phages must infect bacteria in order to replicate, and the life histories of the bacteria may depend on this interaction. Phages participate in the recirculation of organic material and exchange of genes that drives adaptation to a changing environment.

ELECTRON MICROGRAPHS BY MARKUS WEINBAUER, CNRS, VILLEFRANCHE-SUR-MER.

Right: Different species of phages are characterized by the lengths and appearances of their tails. Phages use their tails like hypodermic syringes to inject genetic material into bacteria.

ELECTRON MICROGRAPHS BY MATTHEW SULLIVAN, JENNIFER BRUM, UNIVERSITY OF ARIZONA, USA.

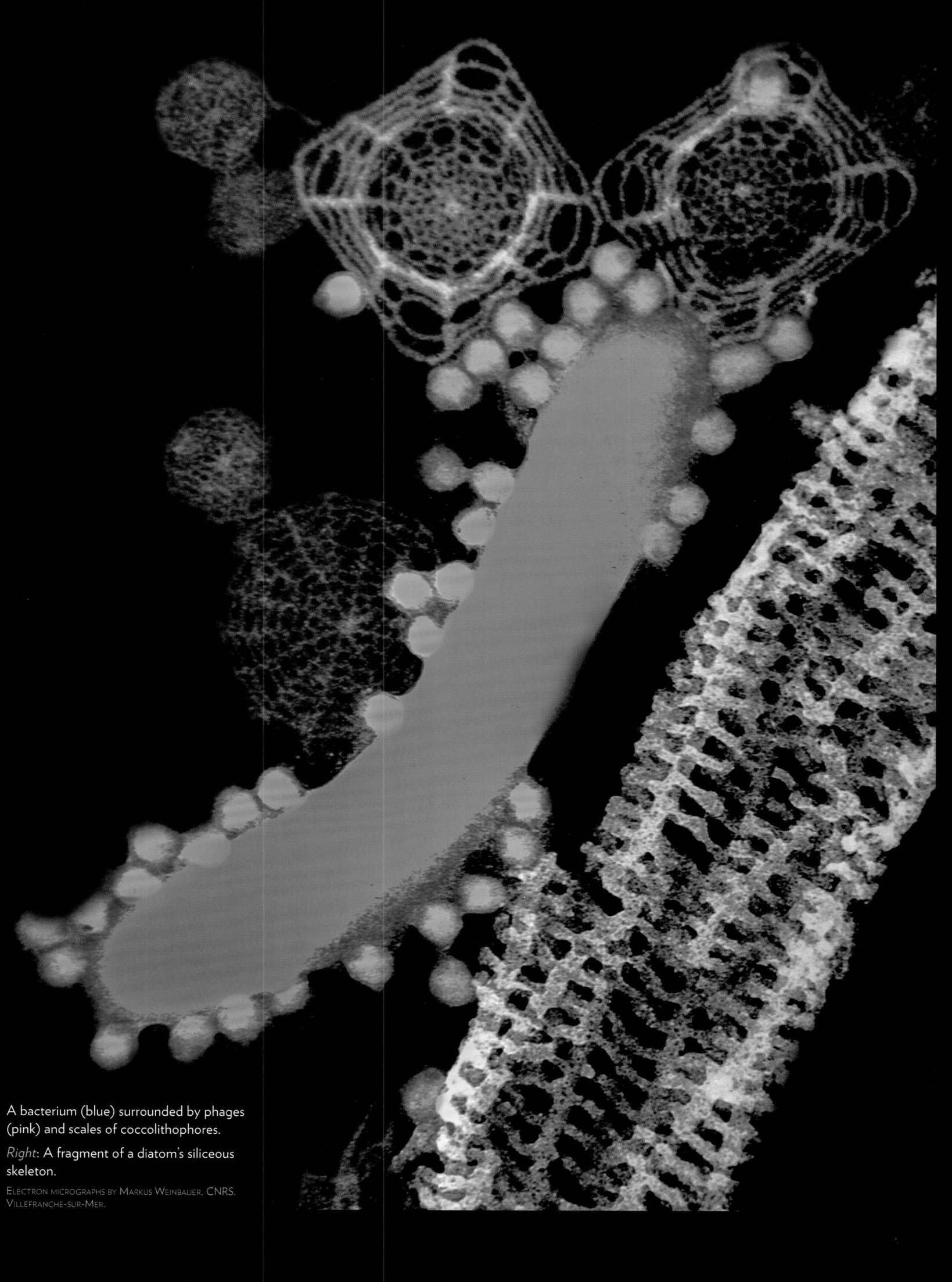

A bacterium (blue) surrounded by phages (pink) and scales of coccolithophores.

Right: A fragment of a diatom's siliceous skeleton.

Filamentous Bacteria

Trichodesmium sp. is a cyanobacterium that proliferates in warm regions, turning the water a golden color. Commonly called "sea sawdust," it forms bundles of entangled filaments covering large areas during blooms. This bacterium plays a major role in the global nitrogen cycle, "fixing" gaseous nitrogen from the air into bioavailable compounds like nitrites or nitrates. It is estimated that half the fixed nitrogen in the ocean is due to the activity of this cyanobacterium.

COLLECTED BETWEEN ECUADOR AND THE GALAPAGOS ISLANDS DURING THE *TARA OCEANS EXPEDITION.*

Top right: Another filamentous bacterium, *Roseofilum reptotaenium* is a cyanobacterium that slides along surfaces. Called "coral killer," it is the pathogenic agent responsible for the black band disease decimating coral reefs.

BIGELOW LABORATORY COLLECTION, BOOTH BAY MARINE LABORATORY, MAINE, USA.

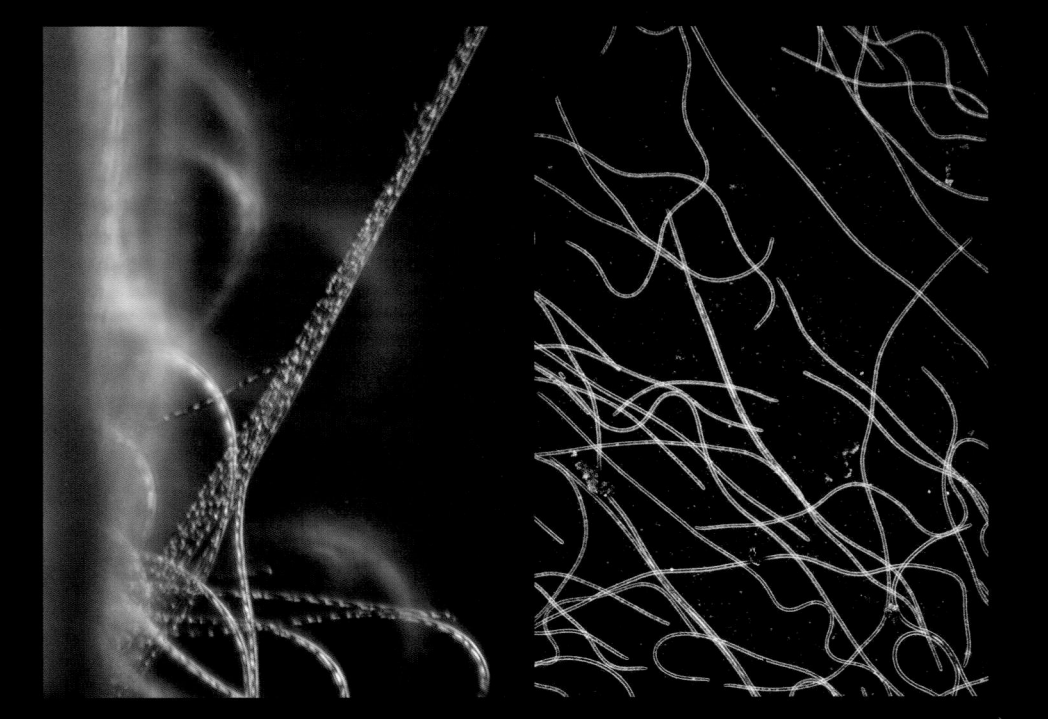

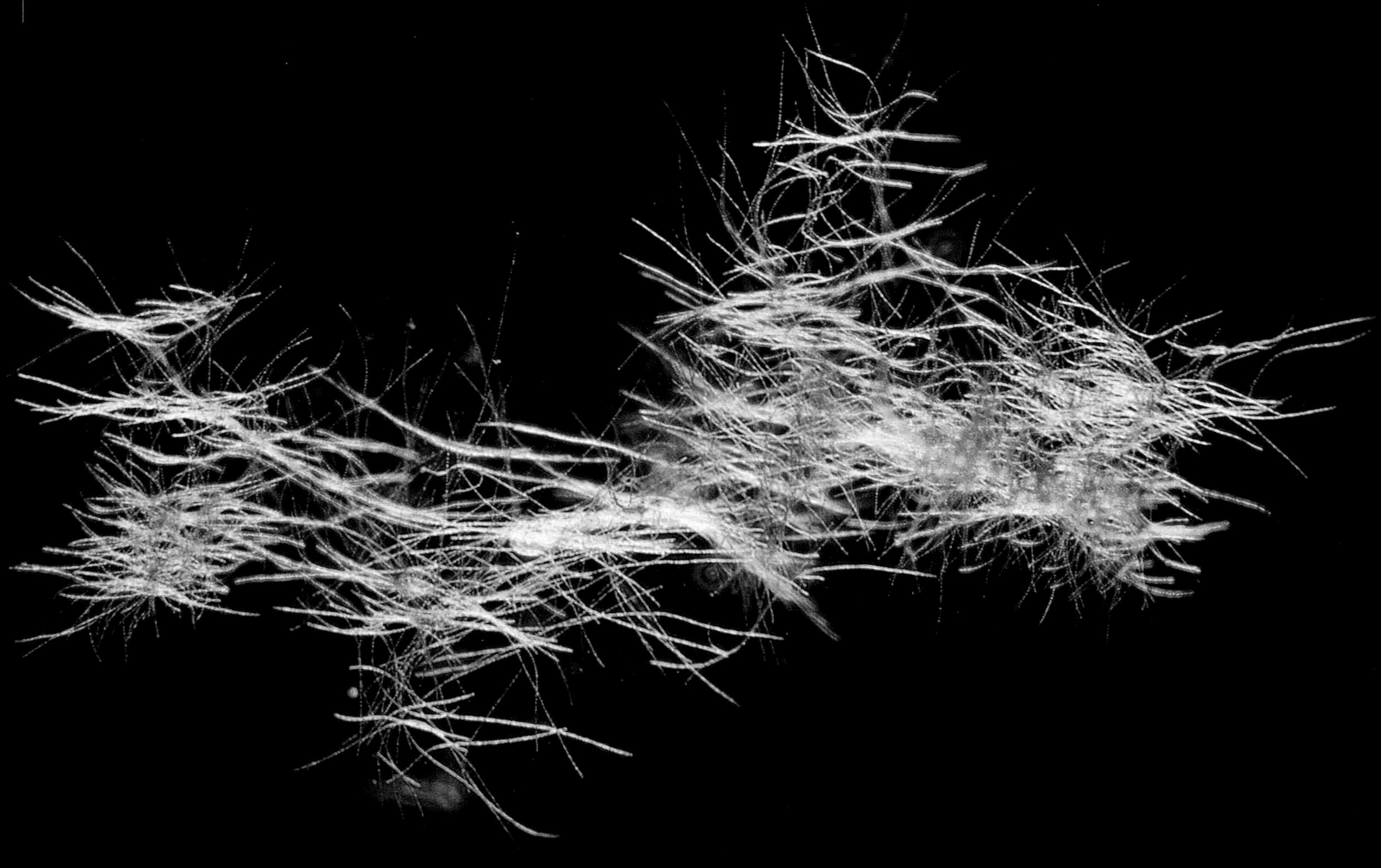

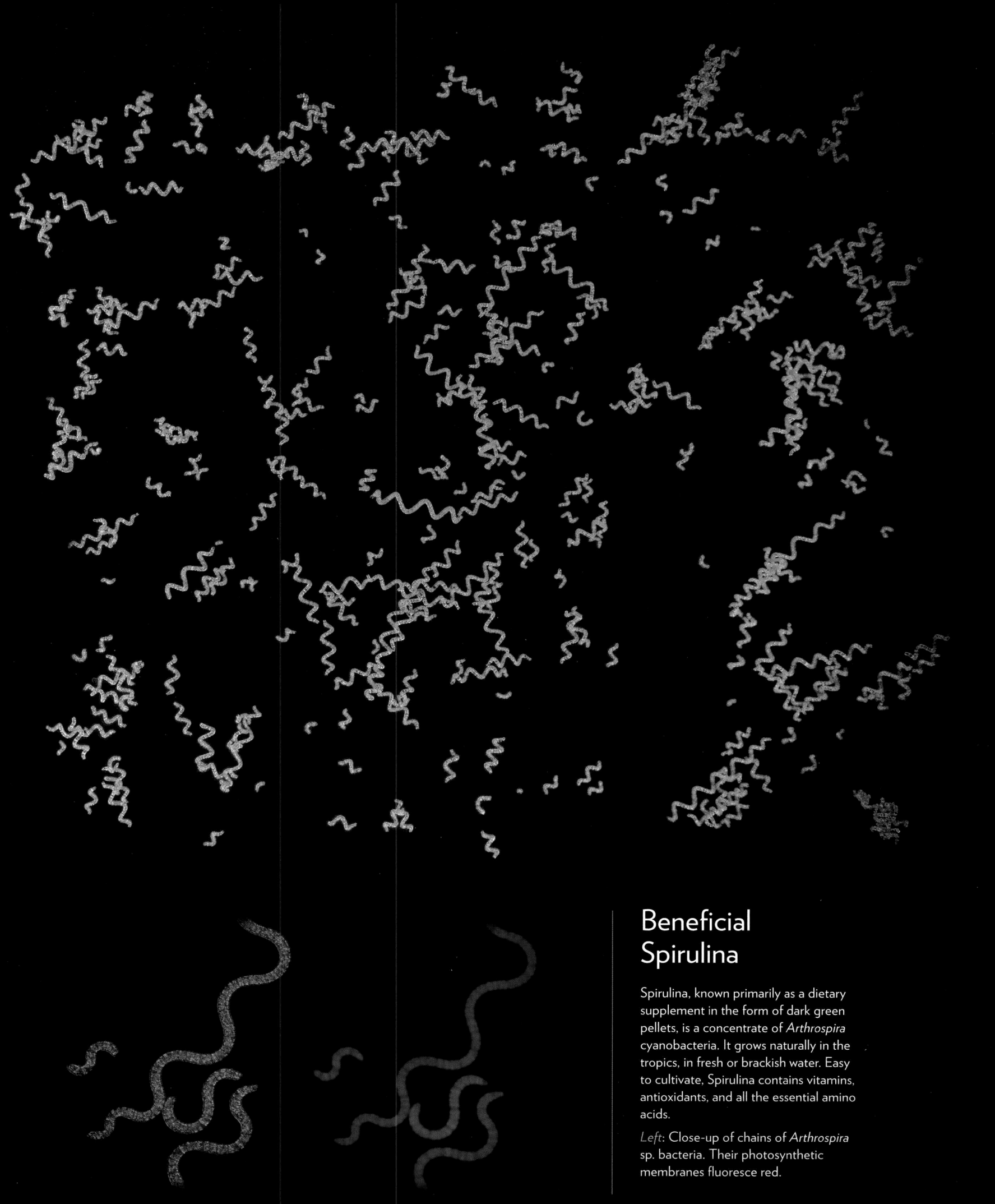

Beneficial Spirulina

Spirulina, known primarily as a dietary supplement in the form of dark green pellets, is a concentrate of *Arthrospira* cyanobacteria. It grows naturally in the tropics, in fresh or brackish water. Easy to cultivate, Spirulina contains vitamins, antioxidants, and all the essential amino acids.

Left: Close-up of chains of *Arthrospira* sp. bacteria. Their photosynthetic membranes fluoresce red.

UNICELLULAR PROTISTS

PRECURSORS OF PLANTS AND ANIMALS

Unlike bacteria that are devoid of nuclei and organelles, protists possess one or several nuclei, as well as organelles such as chloroplasts and mitochondria. Protists appeared more than a billion years ago from chimera of bacteria and archaea that gradually evolved into cells with internal organelles such as mitochondria and chloroplasts. Protists are part of the realm of eukaryotes—from the Greek *eu* ("true") and *karyon* ("nucleus"). Nuclei contain DNA packaged in the form of chromosomes, the carriers of genetic information.

Though protists are only single cells, they are our ancestors. More than 800 million years ago, certain unicellular organisms took the first steps toward multicellularity, a level of organization that characterizes all plants and animals. We don't know which protists first joined together to form colonies—possibly the volvocales (ancestors of green algae), or the choanoflagellates (see page 90), though multicellularity is thought to have evolved independently many times. Over a very long time, within a mass of identical protists, certain cells specialized, foreshadowing the tissues of the first plants and animals.

Most protists—diatoms, dinoflagellates, coccolithophores, radiolarians, ciliates, and foraminifera—can be seen only with the aid of a microscope. But a few species of foraminifera and radiolarians are large single cells visible with the naked eye. These giant protists are sometimes bigger than larvae and small planktonic animals made up of millions of tiny cells. Some diatoms and dinoflagellates form long chains of individual cells, and some radiolarians live in colonies of hundreds or thousands of individuals sharing a gelatinous home.

Protists display an incredible variety of structures and behavior. Thousands of species are already known, but new ones are discovered every year. Some species of diatoms, dinoflagellates, and coccolithophores get their energy from sunlight through photosynthesis, just like plants. These are part of the phytoplankton. Other protists get their energy by feeding on living organisms (just like animals do), in this case bacteria, larvae, and other protists. But most protists are capable of using multiple sources of energy and adapting to changes in the environment. To survive, many protists have become masters of symbiosis—cooperative living arrangements between different organisms that each contribute some of the necessities of life.

A MIXTURE OF PROTISTS

In the center, a colonial radiolarian consists of multiple, identical cells sharing a gelatinous envelope. It is surrounded by diatoms shaped like cylinders, dinoflagellates, and lobed foraminifera.

PLANKTON COLLECTED IN AUTUMN, USING A 100-MICRON MESH NET, IN TOBA BAY, JAPAN.

Plankton Chronicles website Protists 1

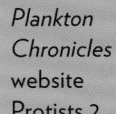

Plankton Chronicles website Protists 2

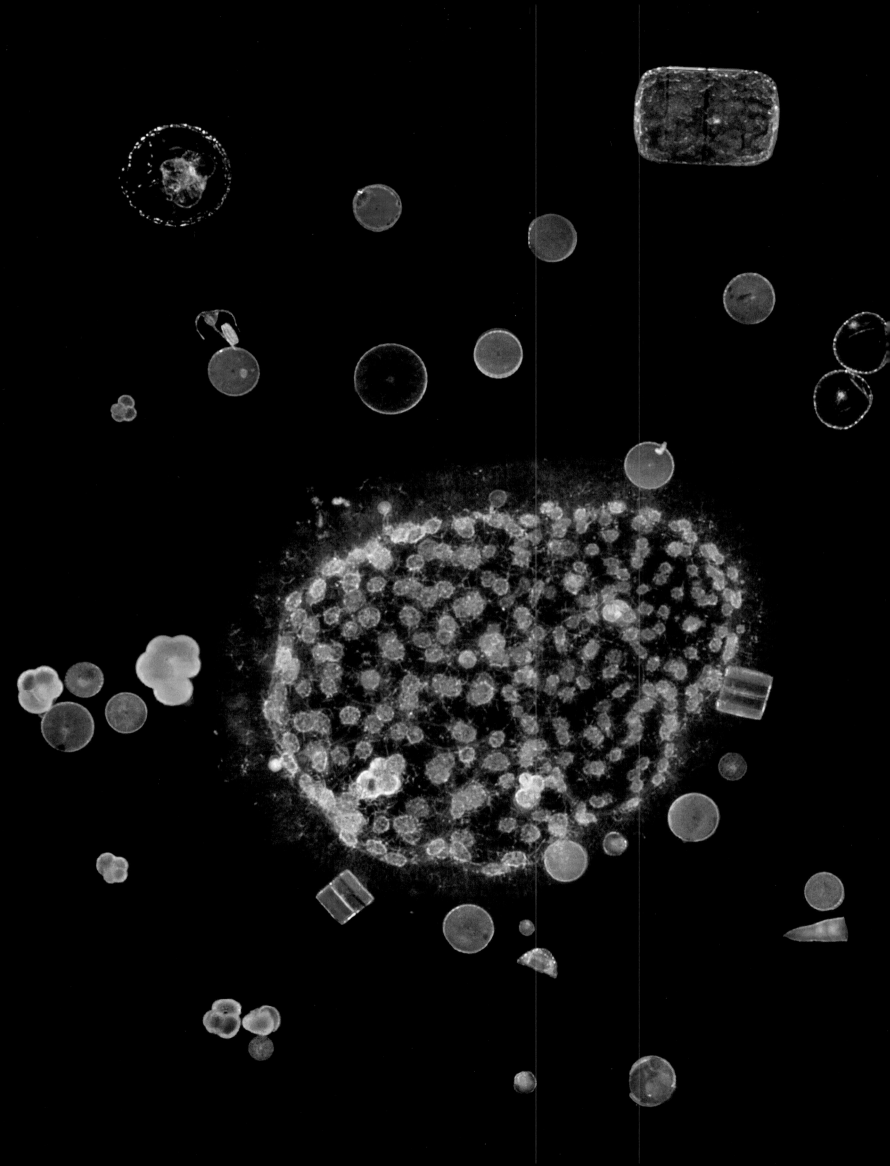

A Variety of Forms:
Who Is Who?

When looking through a binocular microscope at plankton collected with nets ranging in mesh size from 20 to 100 microns, we can sort different protists on the basis of size and form. Try to recognize who is who in this double page. There are 1 tintinnid, 4 foraminifera, 14 radiolarians, 12 dinoflagellates (including 3 couples), and 15 diatoms. The large greenish balls are colonies of green algae. Some multicellular organisms of comparable size are also present—a copepod, 3 jellyfish larvae, an annelid larva, an echinoderm larva, and a sac of copepod embryos.

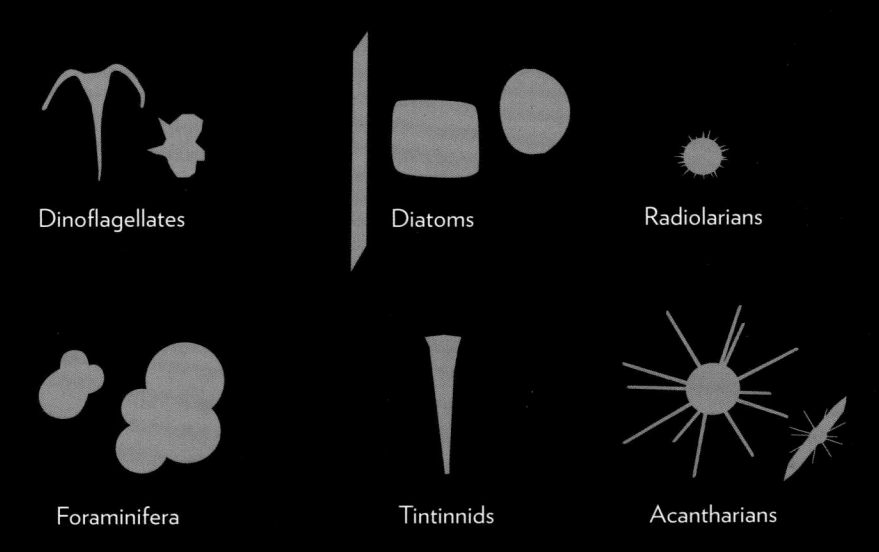

Dinoflagellates

Diatoms

Radiolarians

Foraminifera

Tintinnids

Acantharians

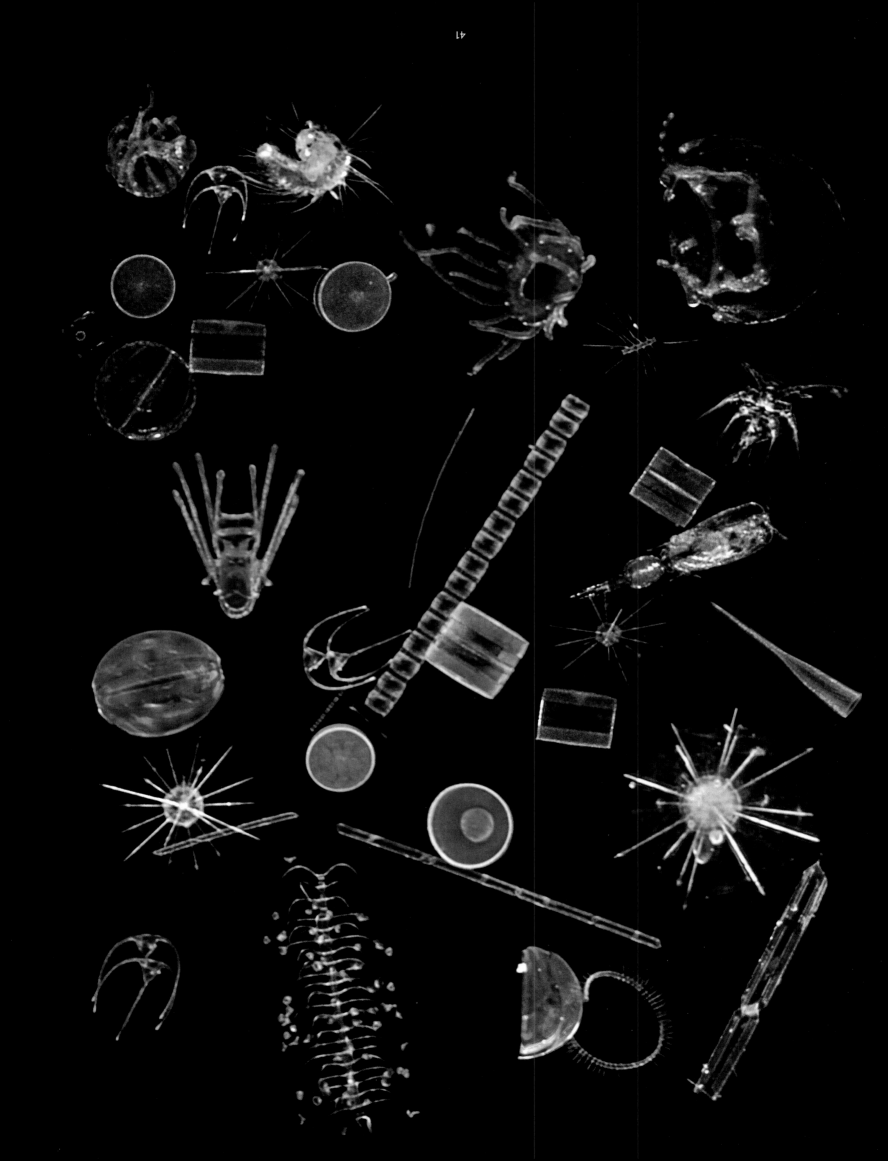

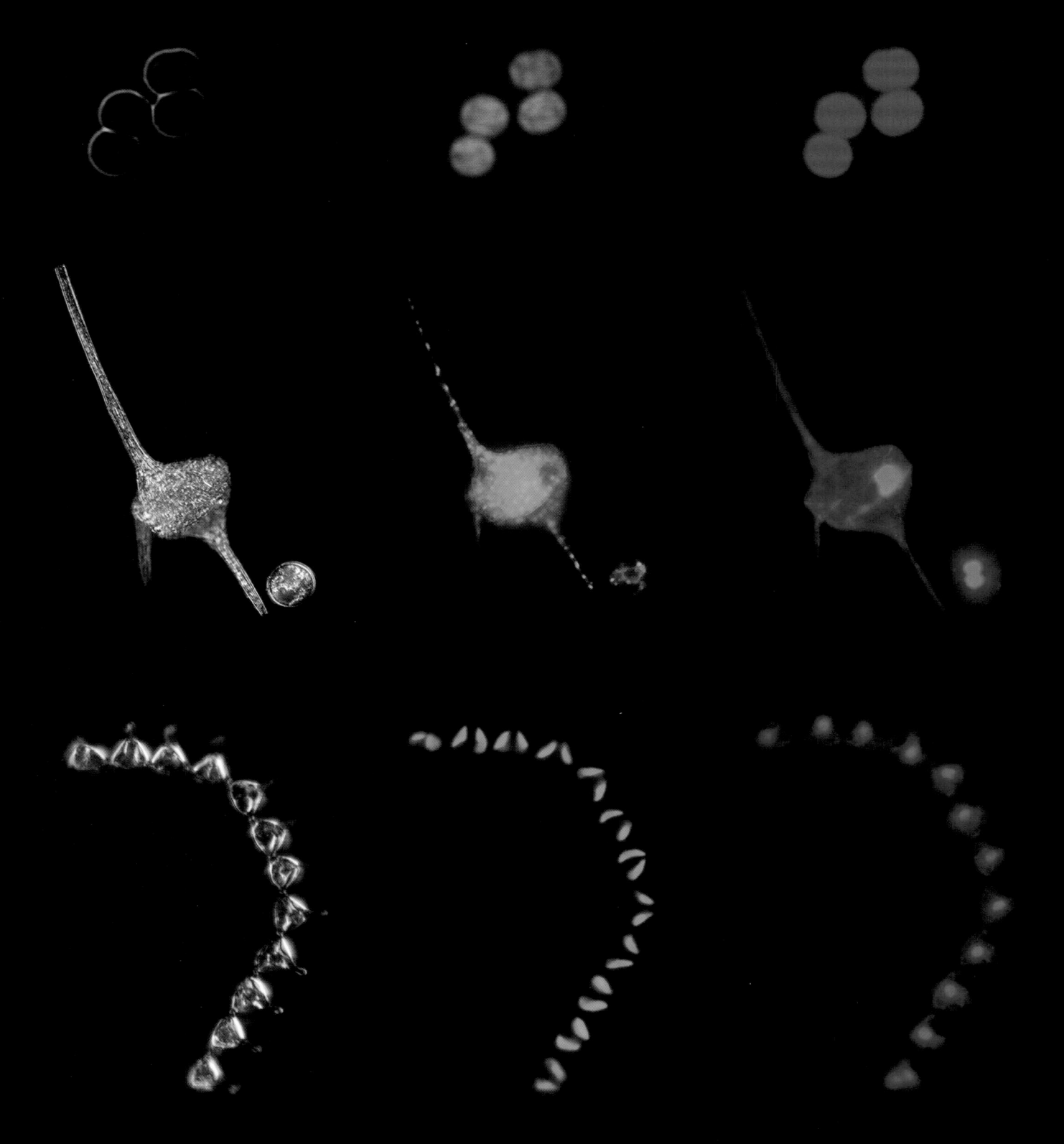

Photosynthesis and Chloroplasts

Photosynthesis is carried out by specialized membranes in bacteria, and by chloroplasts in the cytoplasm of protists and plants. Chloroplasts are organelles containing various pigment molecules (including chlorophyll) that fluoresce naturally.

Top: *Acaryochloris marina* bacteria do not have chloroplasts, but their membranes perform photosynthesis. They fluoresce green. DNA fluoresces blue when synthetic marker molecules are added to the cells. In this bacterium, photosynthetic membranes and DNA are distributed evenly throughout the cytoplasm.

Center: The dinoflagellate *Ceratium* sp. and the small, cylinder-shaped diatom *Coscinodiscus* sp. are filled with chloroplasts that appear as tiny fluorescent green dots. The nuclei of both protists show up as fluorescent blue when a fluorescent DNA-binding molecule is added.

Bottom: A chain of twelve *Asterionellopsis glacialis* diatoms. Each diatom has two big chloroplasts (fluorescent green) and a nucleus containing DNA (fluorescent blue).

Phytoplankton

Phytoplankton are organisms that derive their energy from sunlight through photosynthesis. Phytoplankton include cyanobacteria (prokaryotes, lacking nuclei) and single-celled protists (eukaryotes, possessing nuclei), including diatoms, dinoflagellates, and coccolithophores. Cyanobacteria employ special photosynthetic membranes in their cytoplasm. Phytoplanktonic protists, like the cells of plants, contain specialized organelles called chloroplasts. Chloroplasts harvest the energy from sunlight using chlorophyll pigments that confer their characteristic green, yellow, or red color to the photosynthetic cells of plants and phytoplankton. Chloroplasts fluoresce brightly when excited by certain wavelengths of light.

Photosynthesis

Specialized membranes and chloroplasts perform photosynthesis—the biochemical process that produces organic matter from carbon dioxide (CO_2) and water (H_2O). These biochemical reactions are essential for the growth and reproduction of phytoplankton and depend on the presence of salts and minerals—potassium, phosphates, nitrates, iron, silica. The photosynthetic conversion of CO_2 and H_2O into sugars and other organic molecules generates molecular oxygen (O_2). Phytoplankton produce an estimated half of all atmospheric oxygen. Terrestrial plants produce the other half. Phytoplankton—especially cyanobacteria and diatoms—absorb CO_2 from the atmosphere, playing an important role in carbon sequestration and climate regulation. Bacteria and protists comprise over 90 percent of the biomass in the ocean. They produce enormous quantities of the active molecules in water and the atmosphere. Coccolithophores, for example, are the source of dimethyl sulfide, an important factor in the formation of clouds.

Base of the Food Chain

Phytoplanktonic organisms live in the surface layers of the ocean penetrated by sunlight, known as the euphotic zone. The huge biomass formed by surface phytoplankton constitutes the very base of the food chain. Phytoplankton are consumed by protists such as dinoflagellates, ciliates, foraminifera, radiolarians, and groups of zooplanktonic animals and larvae. These organisms are themselves food for large predators—jellyfish, fish, birds, marine mammals, and humans.

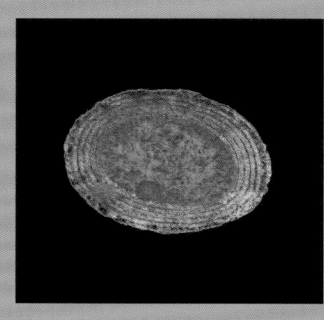

The diatom *Phaeodactylum tricornutum* dividing in two. The nuclei containing DNA are colorized blue, chloroplasts green, mitochondria red, and cytoplasm brown. The cell wall (yellow) made of silicate oxides duplicates during cell division. This diatom is a model species used in many laboratories. It measures about 5 microns in length.

ELECTRON MICROGRAPH BY ATSUKO TANAKA AND CHRIS BOWLER, CNRS, ÉCOLE NORMALE SUPÉRIEURE (ENS), PARIS.

The cyanobacterium *Prochlorococcus marinus* measures 1 micron. Photosynthetic membranes (green) surround the DNA (blue).

ELECTRON MICROGRAPH BY FRÉDÉRIC PARTENSKY, CNRS, ROSCOFF BIOLOGICAL STATION.

The Forests of the Ocean

Phytoplankton fulfill an essential role in their marine environment, comparable to the role forests play on land. Drawing their energy from the sun, phytoplanktonic bacteria and protists fix atmospheric carbon dioxide dissolved in the ocean to make organic material. They also release oxygen that then passes from the ocean into the atmosphere. Organic matter resulting from photosynthesis supports primary growth, but also nourishes the surrounding ecosystems. One important difference between phytoplankton and land plants is the rapid rate at which unicellular organisms renew themselves compared to slow-growing trees and plants. In this photo, the large spheres are the green algae *Halosphera* sp., and the iridescent cells are the diatoms *Rhizosolenia* sp.

These were the two predominant species of plankton collected with a 0.1-mm mesh net, during the winter in Roscoff, France.

PHAEOCYSTIS GLOBOSA

These unicellular algae measure up to 6 microns
and belong to the haptophytes. Abundant in all
ocean basins, they live as solitary flagellated cells
or assemble in spherical colonies. Sometimes they
appear in major proliferations, forming stinking scum
or foam. These algae emit 3-dimethylsulfopropionate
(DMSP), a precursor molecule of dimethyl sulfide
(DMS). This sulfurous compound regulates the
condensation of water droplets in the atmosphere,
and thus the formation of clouds and rain. These
microalgae are also involved in symbioses with protists
such as the acantharian *Lithoptera* sp. described in
pages 82–83.

NCMA SAMPLE COLLECTION, BIGELOW LABORATORY, BOOTH BAY, USA.

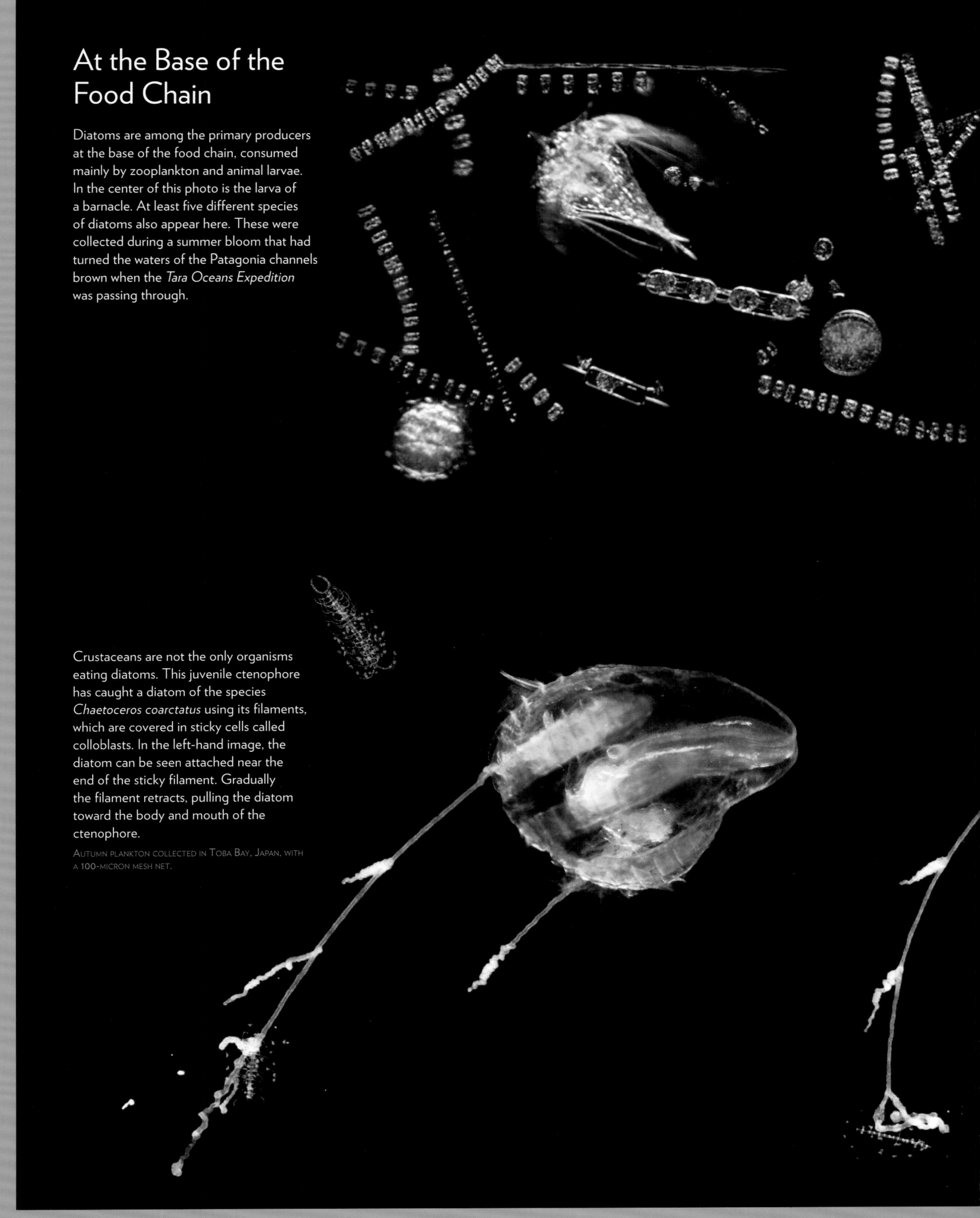

At the Base of the Food Chain

Diatoms are among the primary producers at the base of the food chain, consumed mainly by zooplankton and animal larvae. In the center of this photo is the larva of a barnacle. At least five different species of diatoms also appear here. These were collected during a summer bloom that had turned the waters of the Patagonia channels brown when the *Tara Oceans Expedition* was passing through.

Crustaceans are not the only organisms eating diatoms. This juvenile ctenophore has caught a diatom of the species *Chaetoceros coarctatus* using its filaments, which are covered in sticky cells called colloblasts. In the left-hand image, the diatom can be seen attached near the end of the sticky filament. Gradually the filament retracts, pulling the diatom toward the body and mouth of the ctenophore.

AUTUMN PLANKTON COLLECTED IN TOBA BAY, JAPAN, WITH A 100-MICRON MESH NET.

Opening its expandable mouth wide, this young ctenophore is about to swallow the diatom *Coscinodiscus* sp.

Autumn plankton in Toba Bay, Japan.

COCCOLITHOPHORES AND FORAMINIFERA
LIMESTONE ARCHITECTS

Coccolithophores and foraminifera are protists. These single-celled organisms produce skeletons made of calcium carbonate. Like other calcifying organisms such as corals and mollusks, only to a greater extent, coccolithophores and foraminifera are important regulators of CO_2 concentration in the ocean and in the atmosphere and are an essential component in the overall carbon cycle on Earth. Coccolithophores are sometimes so abundant they appear on satellite images, like the one on this page depicting a bloom of *Emiliania huxleyi*. Predominant in the Atlantic basin, *E. huxleyi* is increasingly adopted by laboratories to study the process of biomineralization and the adaptation of organisms to ocean acidification.

For hundreds of millions of years, the skeletons of calcifying cells have fallen to the ocean floor, forming thick sediments composed of microfossils. Over the ages these sediments compacted, rose up from the sea, and eroded, eventually forming chalk cliffs, such as the cliffs of Dover. Billions of microfossils are preserved in the calcareous stone of cathedrals and pyramids.

Coccolithophores are a type of photosynthetic algae known as haptophytes. They measure from 2 to 50 microns, have two flagella and a thin appendage, a haptonema, that gives its name to the whole group. Coccolithophores secrete delicate external scales called coccoliths that they modulate according to their life cycle and environmental conditions. Some species modify their coccoliths into elaborated appendages that may be used to avoid predation by copepods and other zooplankton.

IMAGE COURTESY OF NASA.

Foraminifera appeared during the Cambrian, the first geological period of the Paleozoic era (between 540 and 500 million years ago). They are five to one hundred times bigger than the biggest coccolithophores. Most of the 20,000 living species of foraminifera are benthic and cling to the bottoms, and some planktonic species are large enough to be collected by divers. Using cytoplasmic amoeboid extensions, foraminifera entwine and engulf all kinds of prey, including bacteria, small crustaceans, mollusks, and larvae.

The skeleton of a foraminiferan, called a test, develops inside the cell. It generally contains several chambers made from grains of sediment or calcium carbonate. The 38,000 fossil species of foraminifera, some hundreds of millions of years old, allow geologists to assign dates to fossil rocks. Thanks to these microfossils, we can locate and identify oil deposits and better understand the history of the Earth.

FORAMINIFERA AND COCCOLITHOPHORES

The intracellular skeleton (test) of the foraminiferan *Globigeronidoides ruber* (measuring about 400 microns) and the smaller extracellular shells of four coccolithophores. These extracellular shells, called coccospheres, are made of calcium carbonate scales, or coccoliths. From left to right: *Emiliania huxleyi*, a model species often used in laboratories; *Umbilicosphaera hulburtiana*; *Discosphaera tubifera*; and the larger *Scyphosphaera apsteinii* bearing several types of coccoliths.

SCANNING ELECTRON MICROGRAPHS: LAURENCE FROGET & MARIE JOSEPH CHRETIENNOT-DINET, CNRS PHOTOTHÈQUE, CEA; MARGAUX CARMICHAEL, STATION BIOLOGIQUE DE ROSCOFF; JEREMY YOUNG, UNIVERSITY COLLEGE, LONDON.

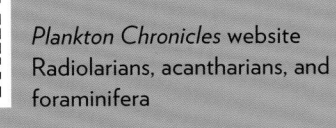

Plankton Chronicles website
Radiolarians, acantharians, and foraminifera

Coccospheres Made of Calcium Carbonate

Coccolithophores are flagellated cells that produce scales from calcium carbonate to form a protective shell, the coccosphere. Depending on environmental conditions and life cycle of the cell, the coccosphere varies in thickness and constitution. The scales, called coccoliths, are manufactured and calcified in vacuoles within the cell, then secreted. The flagellar region of the cell is covered by different types of coccoliths, revealing an asymmetry particularly evident in the two species shown at the bottom of this page: on the left *Rhabdosphaera clavigera*, and on the right, *Ophiaster formosus*. Above them, the coccosphere of *Discosphaera tubifera* exhibits a very delicate structure and arrangement of coccoliths.

Scanning electron micrographs by Jeremy Young, University College, London, and Margaux Carmichael, Roscoff Biological Station (bottom left image).

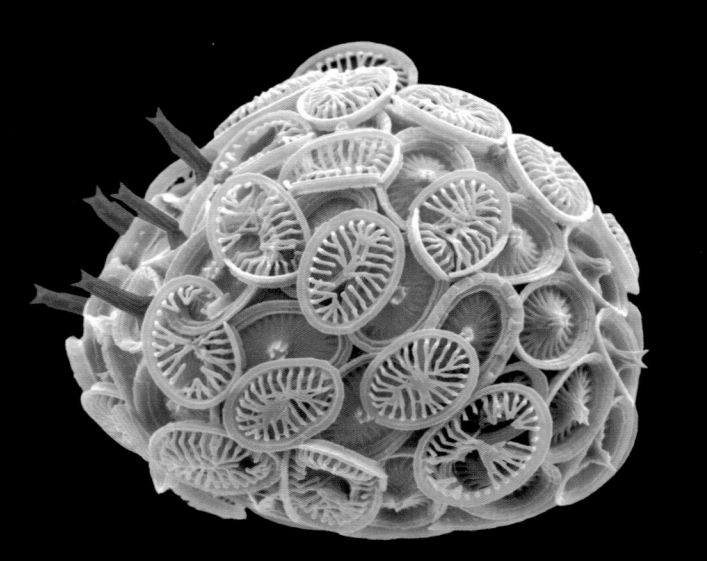

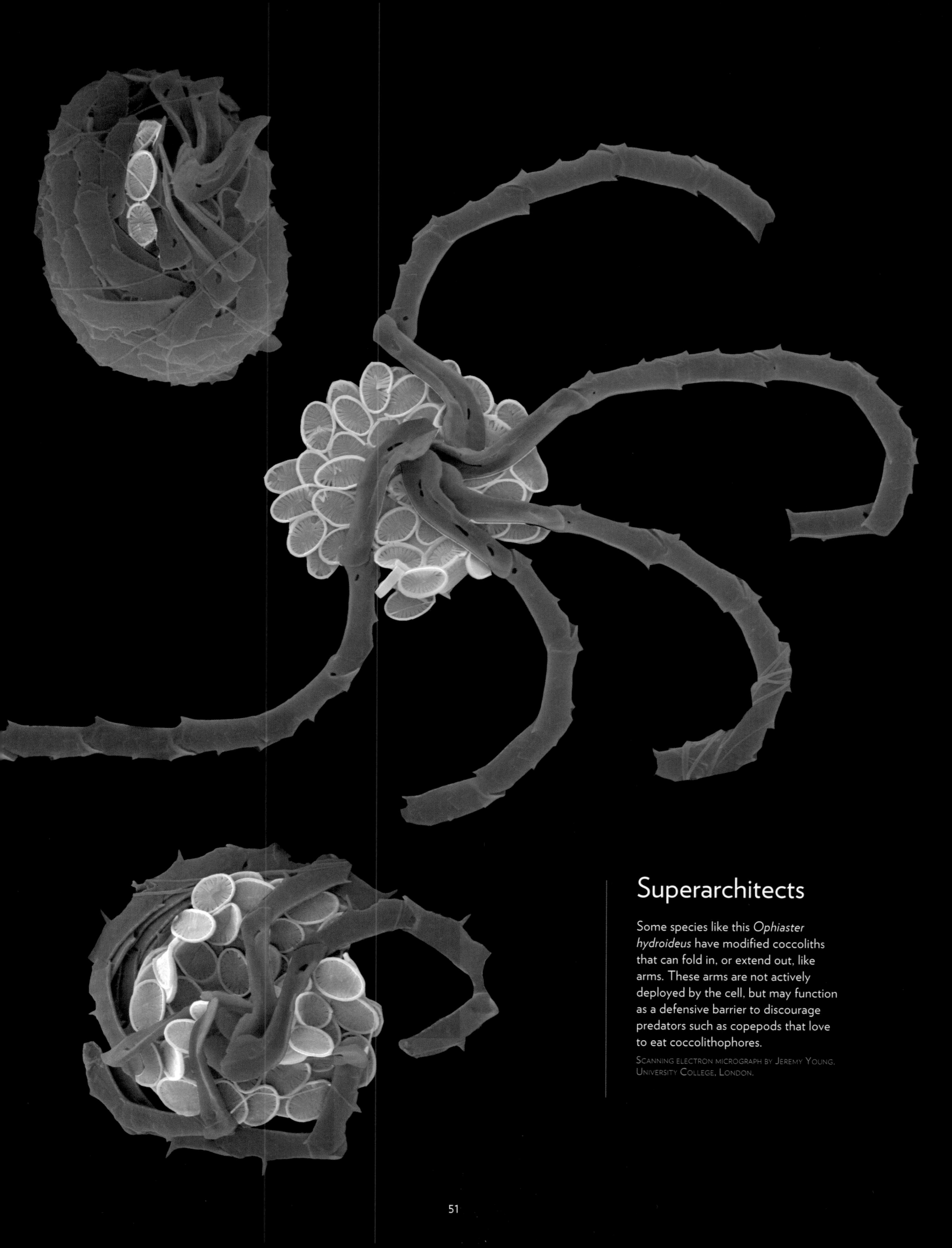

Superarchitects

Some species like this *Ophiaster hydroideus* have modified coccoliths that can fold in, or extend out, like arms. These arms are not actively deployed by the cell, but may function as a defensive barrier to discourage predators such as copepods that love to eat coccolithophores.

SCANNING ELECTRON MICROGRAPH BY JEREMY YOUNG,
UNIVERSITY COLLEGE, LONDON.

Hastigerinella digitata, a foraminiferan measuring
about 2 mm, photographed at a depth of 300
meters off the coast of Monterey, California. A
copepod shell is visible at its periphery.

PHOTO BY KAREN OSBORN, SMITHSONIAN NATIONAL MUSEUM OF
NATURAL HISTORY, WASHINGTON DC, USA.

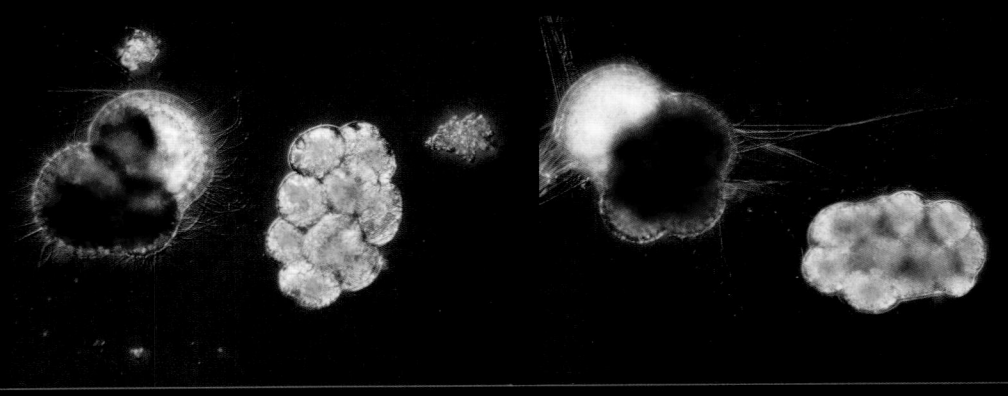

FORAMINIFERA AS PREDATORS

Foraminifera, like radiolarians, are rhizopods,
a name given to organisms that have amoeba-
like movements. They deploy their numerous
membrane extensions, called pseudopodia,
through pores in their calcareous tests. With
these temporary protrusions, they find, capture,
and envelope various prey. Foraminifera feed
on bacteria, other protists, and larvae. Here, a
globigerina (*Globigerinoides bulloides*) collected
in the bay of Villefranche-sur-Mer, is probing a
bunch of copepod eggs.

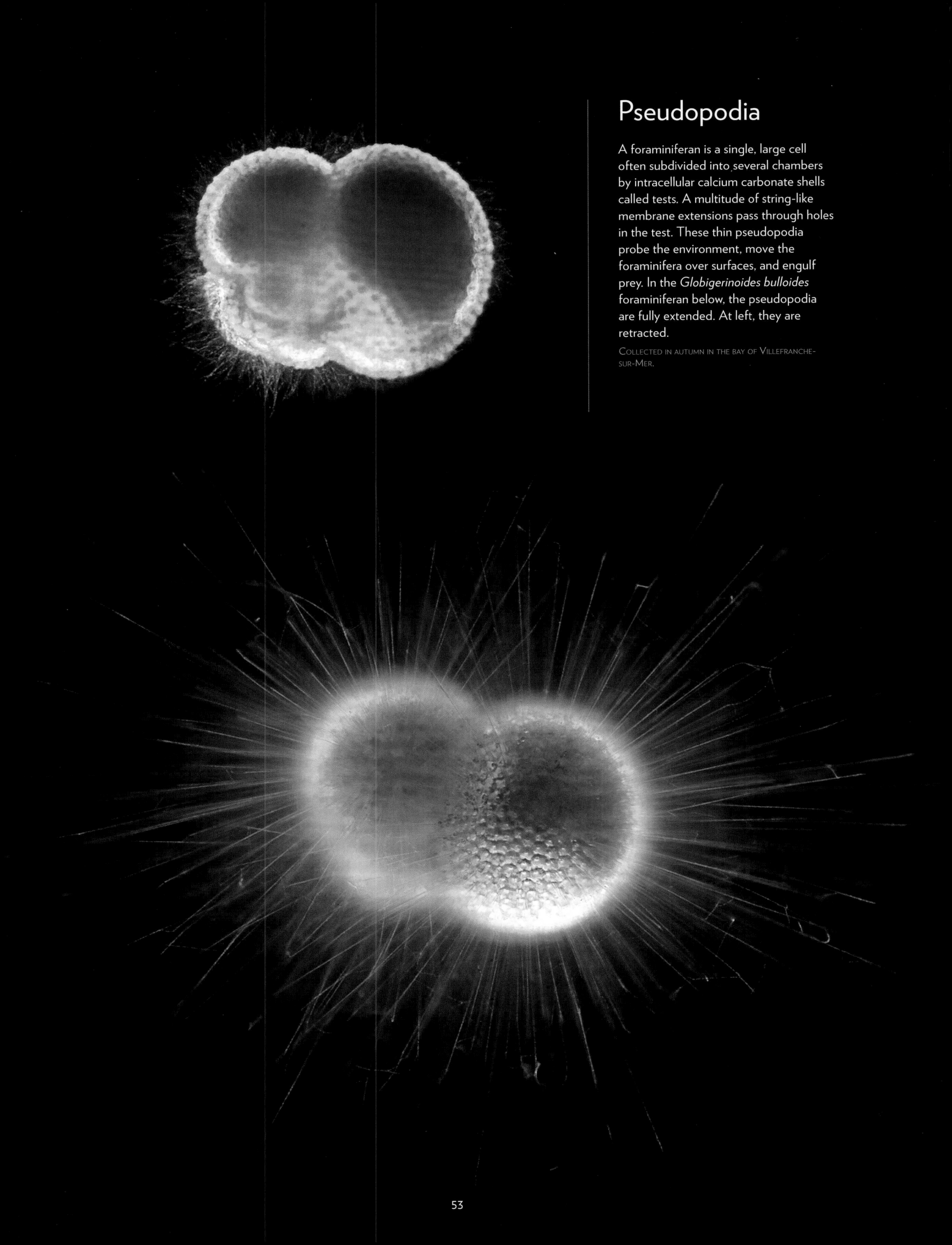

Pseudopodia

A foraminiferan is a single, large cell often subdivided into several chambers by intracellular calcium carbonate shells called tests. A multitude of string-like membrane extensions pass through holes in the test. These thin pseudopodia probe the environment, move the foraminifera over surfaces, and engulf prey. In the *Globigerinoides bulloides* foraminiferan below, the pseudopodia are fully extended. At left, they are retracted.

COLLECTED IN AUTUMN IN THE BAY OF VILLEFRANCHE-SUR-MER.

DIATOMS AND DINOFLAGELLATES

SILICATE OR CELLULOSE HOUSES

Of all the protists among the phytoplankton, dino-flagellates and diatoms are the most numerous. Diatoms alone generate an estimated quarter of the oxygen on our planet. Their fossils may date as far back as the Jurassic period, 200 to 150 million years ago, and are prevalent in the Cretaceous, the third and last period of the Mesozoic era, 145 million to 65.5 million years ago. Diatoms and dino-flagellates live as solitary cells and as groups of cells forming chains and arrays. Thousands of diatom species inhabit freshwater and marine environments. They thrive in icy waters and are particularly abundant in the Arctic and Antarctic regions.

Using siliceous salts dissolved in seawater, diatoms manufacture rigid envelopes, called frustules, composed of two shells that fit one inside the other. Many solitary or colonial species have long spines or bristles that keep them afloat in the currents. Dead diatoms eventually fall to the seabed, where they form sediment. Over millions of years they are compacted and compressed, creating layers of sedimentary rock, and in some places pockets of gas and oil. Deposits of diatom frustules form a friable siliceous sedimentary rock known as "diatomaceous earth" or diatomite, widely used in agriculture, but also in industry for making paints and abrasives. Diatoms even find their way into toothpaste.

In contrast to diatoms, which are rather static or glide slowly along surfaces, dinoflagellates move briskly using two flagella that undulate in rhythm. Dinoflagellates are most often photosynthetic, like plants, but some species also feed on bacteria or fellow protists, as do animals, while still others adopt both strategies. Some dinoflagellates can survive only as parasites. Dinoflagellate cells have an envelope, but unlike diatoms, their carapace is organic, made of cellulose. Most dinoflagellates produce and secrete their carapace as plates, which can be very ornate. Despite their hard shells, diatoms and dinoflagellates are at the base of the food chain, readily devoured by copepods and other zooplanktonic organisms.

When environmental conditions are favorable, certain diatoms and dinoflagellates proliferate, creating huge red, green, or yellow areas in the ocean. These blooms are visible from aircraft and monitored by satellites. Some blooms of diatoms and dinoflagellates—the notorious "red tides" for example—are toxic for other marine organisms and can decimate coastal ecosystems and aquafarms.

Plankton Chronicles
website Diatoms

DIATOM AND DINOFLAGELLATE BLOOM IN TOBA BAY

In autumn we took samples of plankton off the coast of Toba, Japan, using a 100-micron mesh net. All together the collected plankton had a beautiful pink color. Under the dissecting microscope we could see large cylindrical centric diatoms *Coscinodiscus* sp.; *Hemidiscus* sp. resembling transparent domes; chains of *Skeletonema* sp.; and the bright pink *Protoperidinium depressum*—dinoflagellates that ingest diatoms. At the center of this image is a yellow star-shaped radiolarian, and next to it, a pointed cirriped crustacean (barnacle) larva. Below is another zooplanktonic organism, an echinoderm larva in the shape of a spaceship.

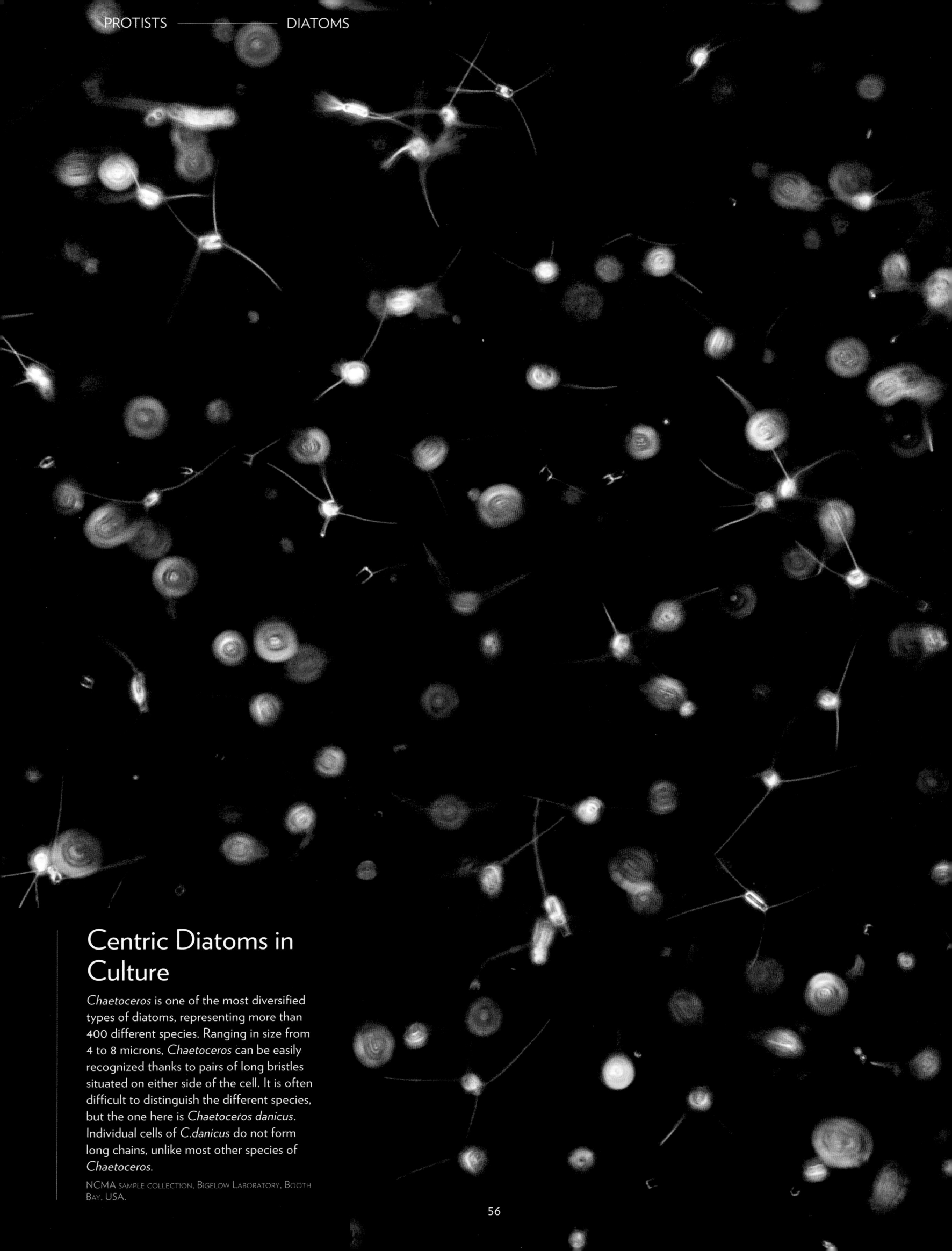

Centric Diatoms in Culture

Chaetoceros is one of the most diversified types of diatoms, representing more than 400 different species. Ranging in size from 4 to 8 microns, *Chaetoceros* can be easily recognized thanks to pairs of long bristles situated on either side of the cell. It is often difficult to distinguish the different species, but the one here is *Chaetoceros danicus*. Individual cells of *C.danicus* do not form long chains, unlike most other species of *Chaetoceros*.

NCMA SAMPLE COLLECTION, BIGELOW LABORATORY, BOOTH BAY, USA.

Toxic Pennate Diatoms

These pennate diatoms glide along
surfaces or against each other. Here we
mixed four different species of *Pseudo-
nitzschia*: solitary diatoms and species
forming chains of variable lengths. Some
species of *Pseudo-nitzschia* are highly
toxic, producing a neurotoxin called
domoic acid. This causes food poisoning
in humans who consume shellfish or fish
that have ingested *Pseudo-nitzschia* or
their zooplanktonic predators.

NCMA SAMPLE COLLECTION, BIGELOW LABORATORY,
BOOTH BAY, USA.

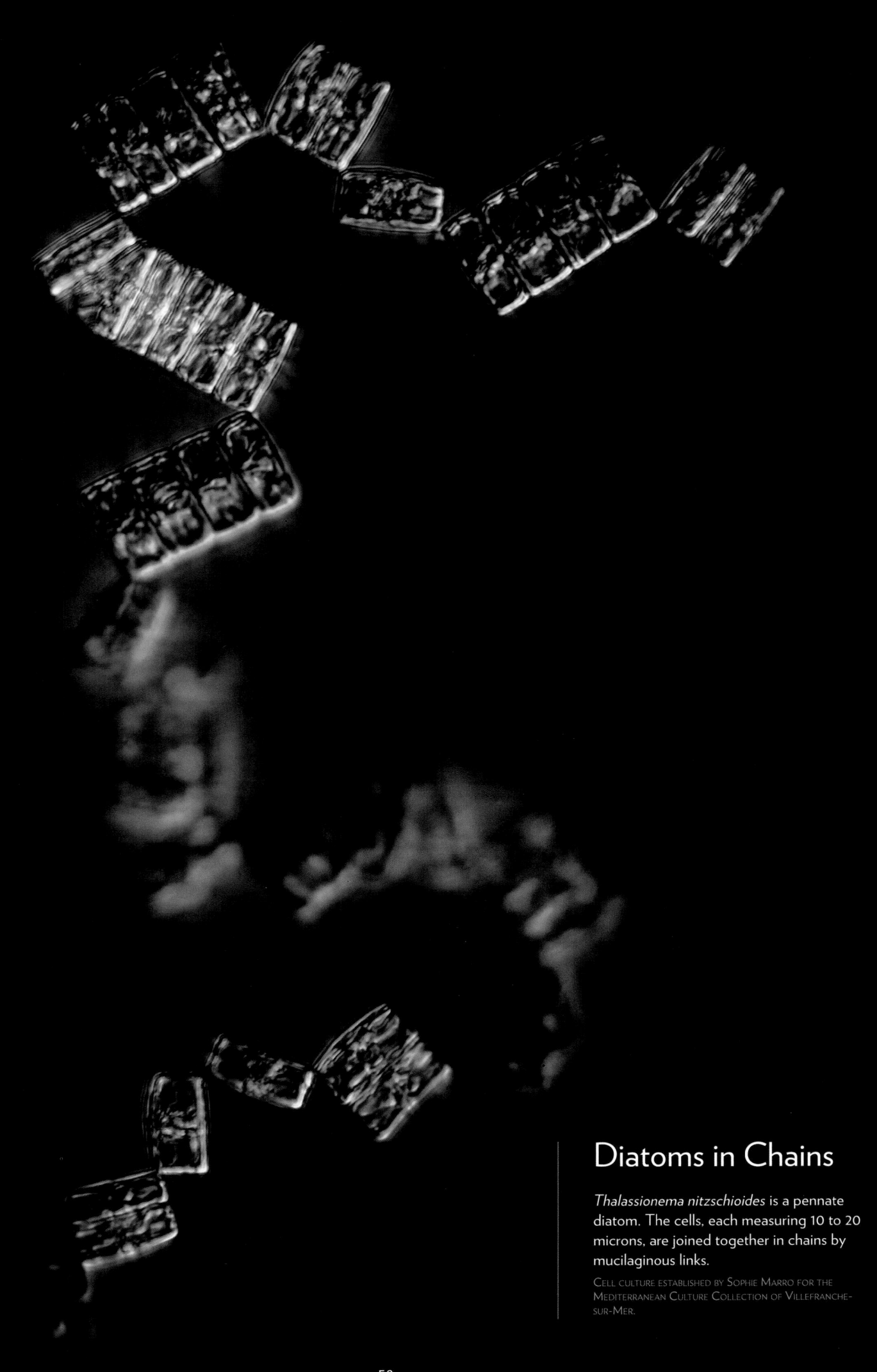

CELL CULTURE ESTABLISHED BY SOPHIE MARRO FOR THE
MEDITERRANEAN CULTURE COLLECTION OF VILLEFRANCHE-
SUR-MER.

Diatoms in Chains

Thalassionema nitzschioides is a pennate
diatom. The cells, each measuring 10 to 20
microns, are joined together in chains by
mucilaginous links.

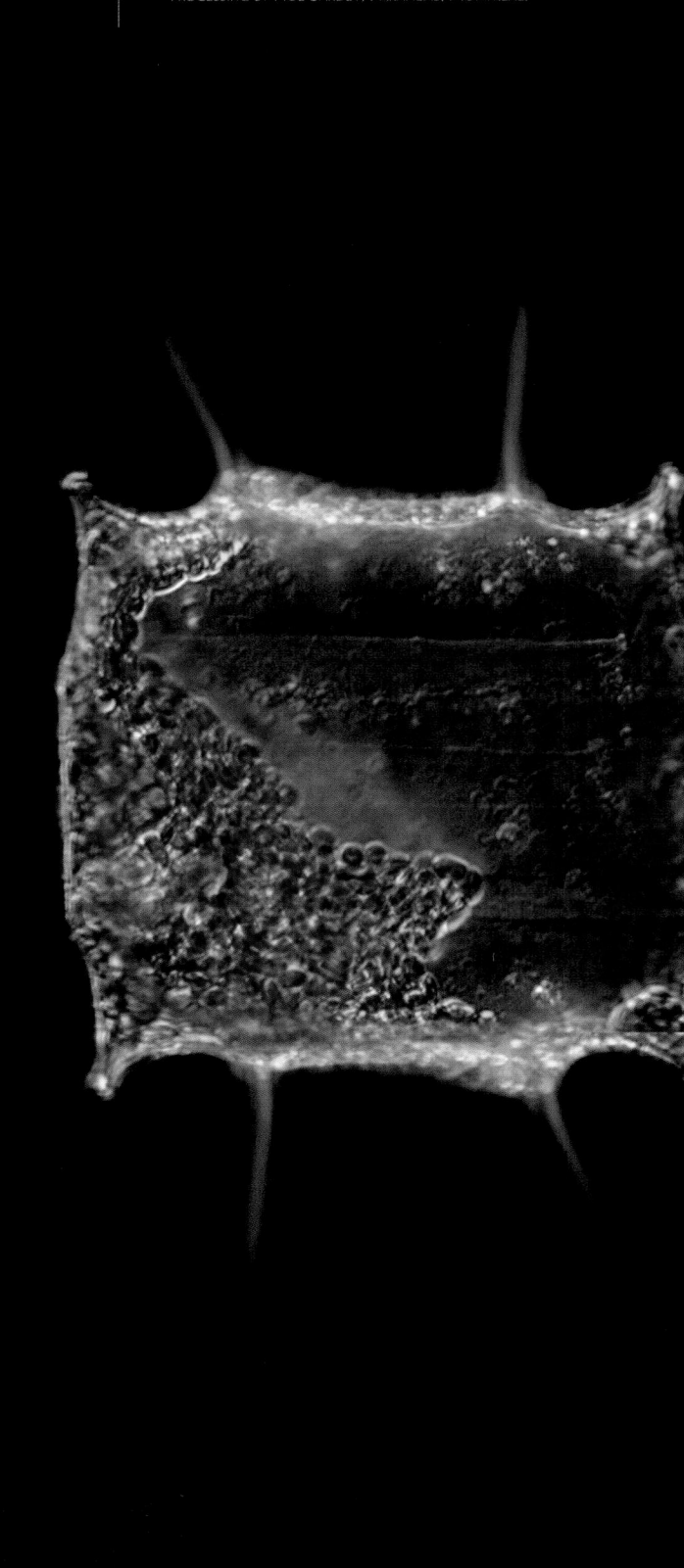

Two Solitary Diatoms

These centric diatoms measure 100 to 200 microns (0.1–0.2 mm). The small particles within the cells are chloroplasts.

On the left: *Odontella sinensis.*

On the right: *Odontella mobiliensis.*

COLLECTED IN ROSCOFF. PHOTOGRAPHY AND IMAGE PROCESSING BY NOÉ SARDET. PARAFILMS, MONTREAL.

IRIDESCENT GLASS SHELLS

Depending on the type and direction of illumination, the siliceous envelope of a diatom reflects light like a mirror and becomes iridescent. When the cell inside dies, the iridescence is at its brightest, as in the feather-shaped *Gyrosigma* diatom in the middle of this page. In contrast, the diatom at the bottom left (genus *Amphora*) is filled with chloroplasts.

Upper right: A small centric diatom of the genus *Actinoptychus*.

Lower right: A diatom of the genus *Cerataulina*.

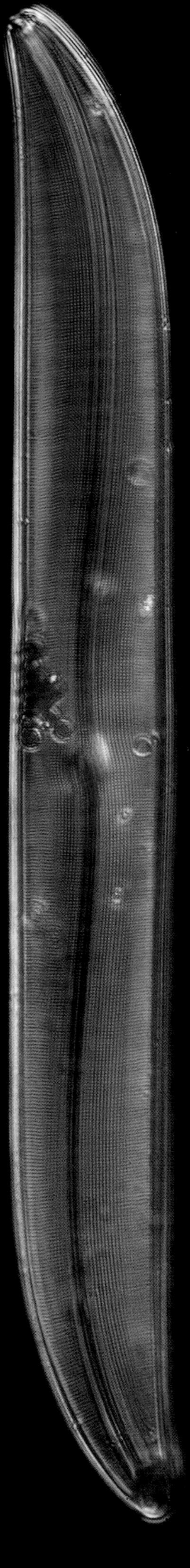

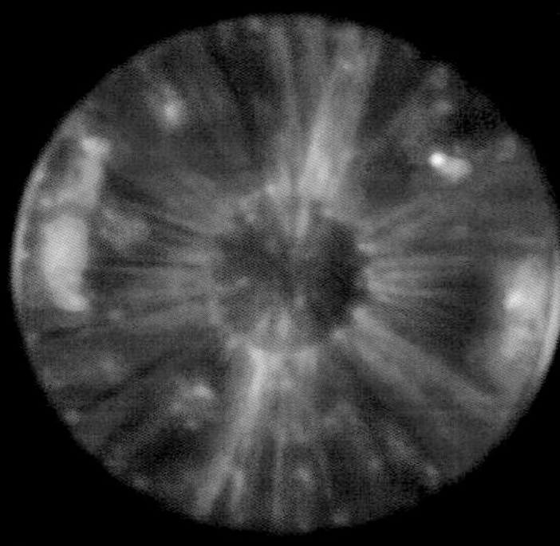

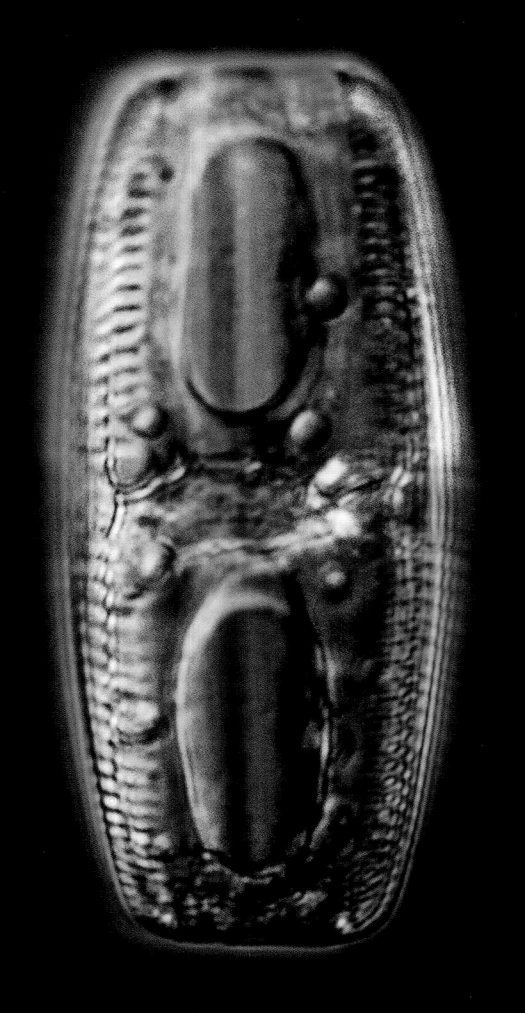

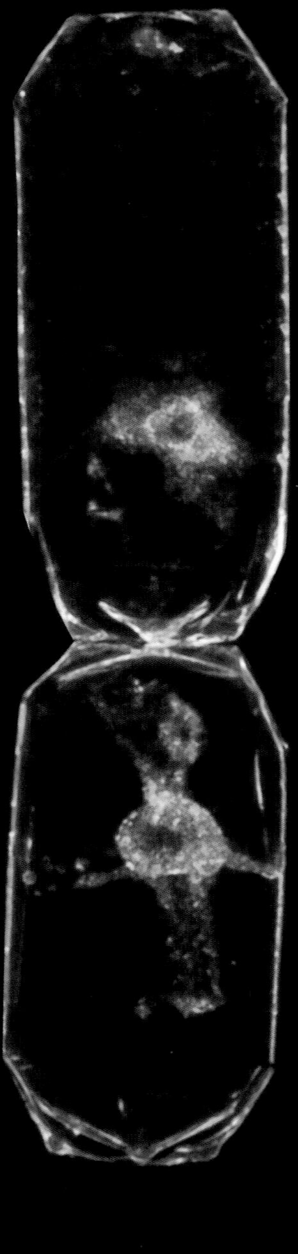

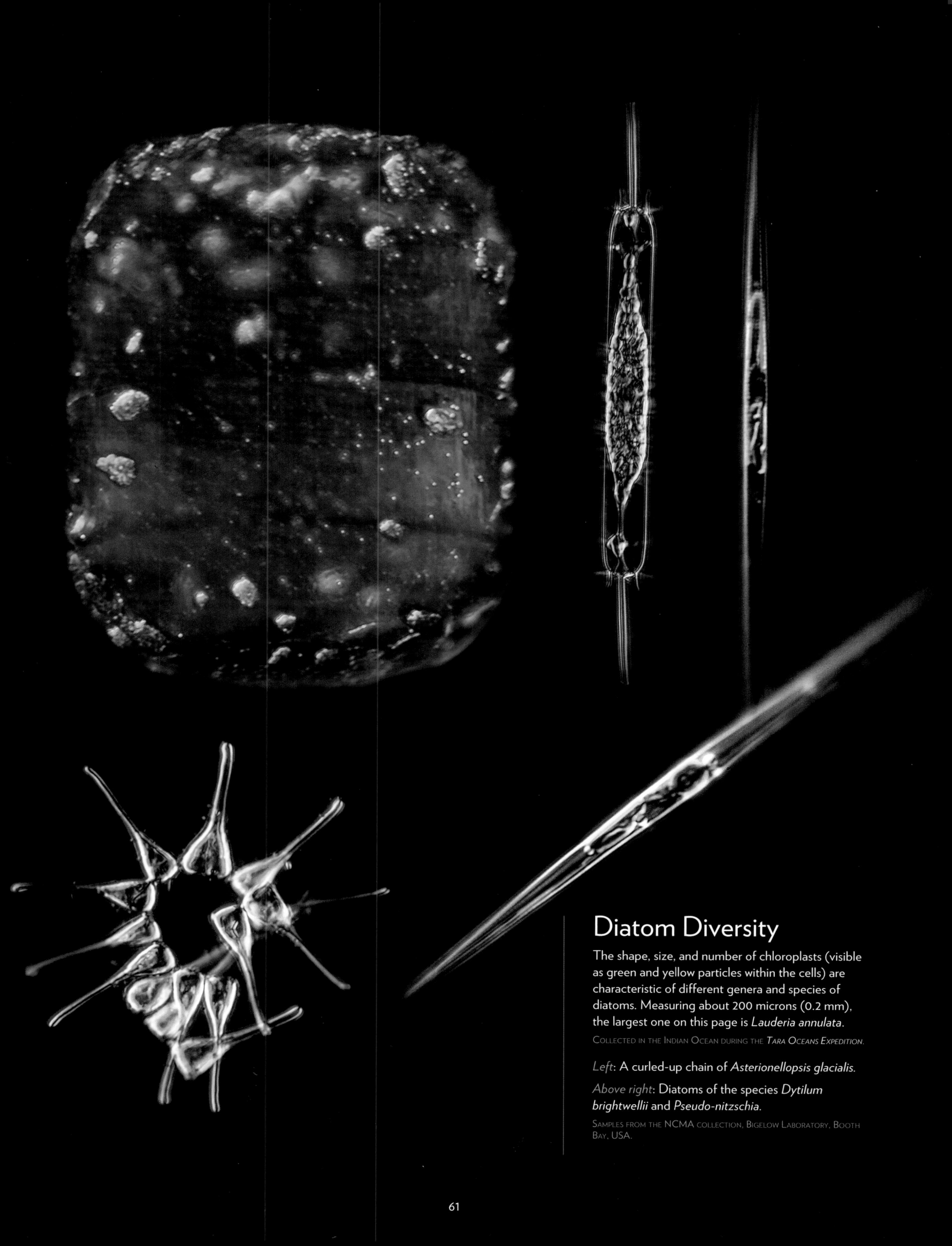

Diatom Diversity

The shape, size, and number of chloroplasts (visible as green and yellow particles within the cells) are characteristic of different genera and species of diatoms. Measuring about 200 microns (0.2 mm), the largest one on this page is *Lauderia annulata*.

<small>COLLECTED IN THE INDIAN OCEAN DURING THE *TARA OCEANS EXPEDITION*.</small>

Left: A curled-up chain of *Asterionellopsis glacialis*.

Above right: Diatoms of the species *Dytilum brightwellii* and *Pseudo-nitzschia*.

<small>SAMPLES FROM THE NCMA COLLECTION, BIGELOW LABORATORY, BOOTH BAY, USA.</small>

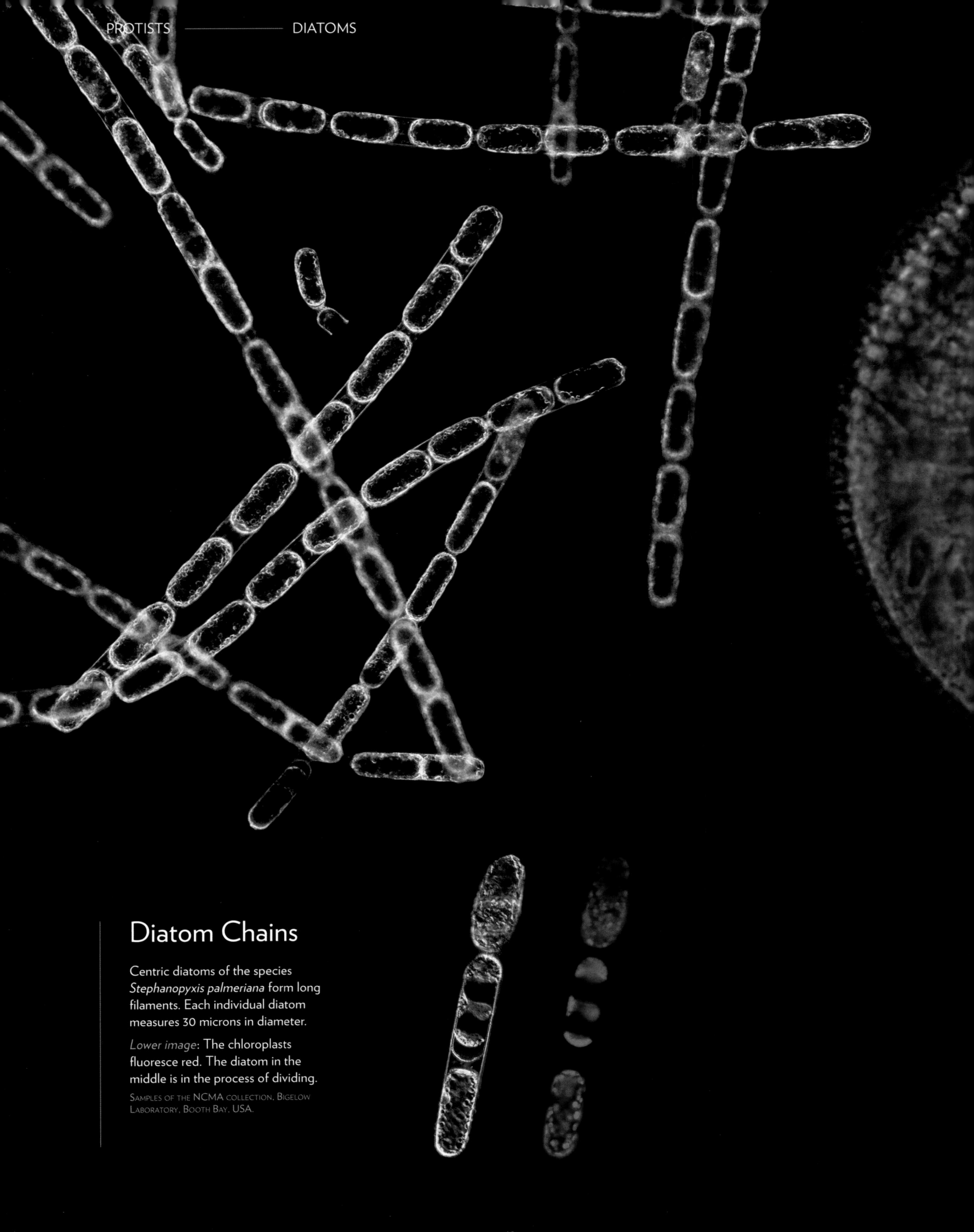

PROTISTS ———————————— DIATOMS

Diatom Chains

Centric diatoms of the species
Stephanopyxis palmeriana form long
filaments. Each individual diatom
measures 30 microns in diameter.

Lower image: The chloroplasts
fluoresce red. The diatom in the
middle is in the process of dividing.

SAMPLES OF THE NCMA COLLECTION, BIGELOW
LABORATORY, BOOTH BAY, USA.

Diatom Shells

The shell of diatoms, or more precisely, their extracellular envelope called a frustule, consists of two parts fitting one into the other. These shells are made of hydrated silicate oxides deposited onto a protein matrix synthesized inside the diatom cell. The ornate structures decorating the frustule are specific to each species. Different centric diatoms of the genus *Coscinodiscus* are displayed on this page.

RIGHT: SCANNING ELECTRON MICROGRAPHS BY NILS KROEGER, GEORGIA INSTITUTE OF TECHNOLOGY, USA, AND CHRIS BOWLER, ENS, PARIS.

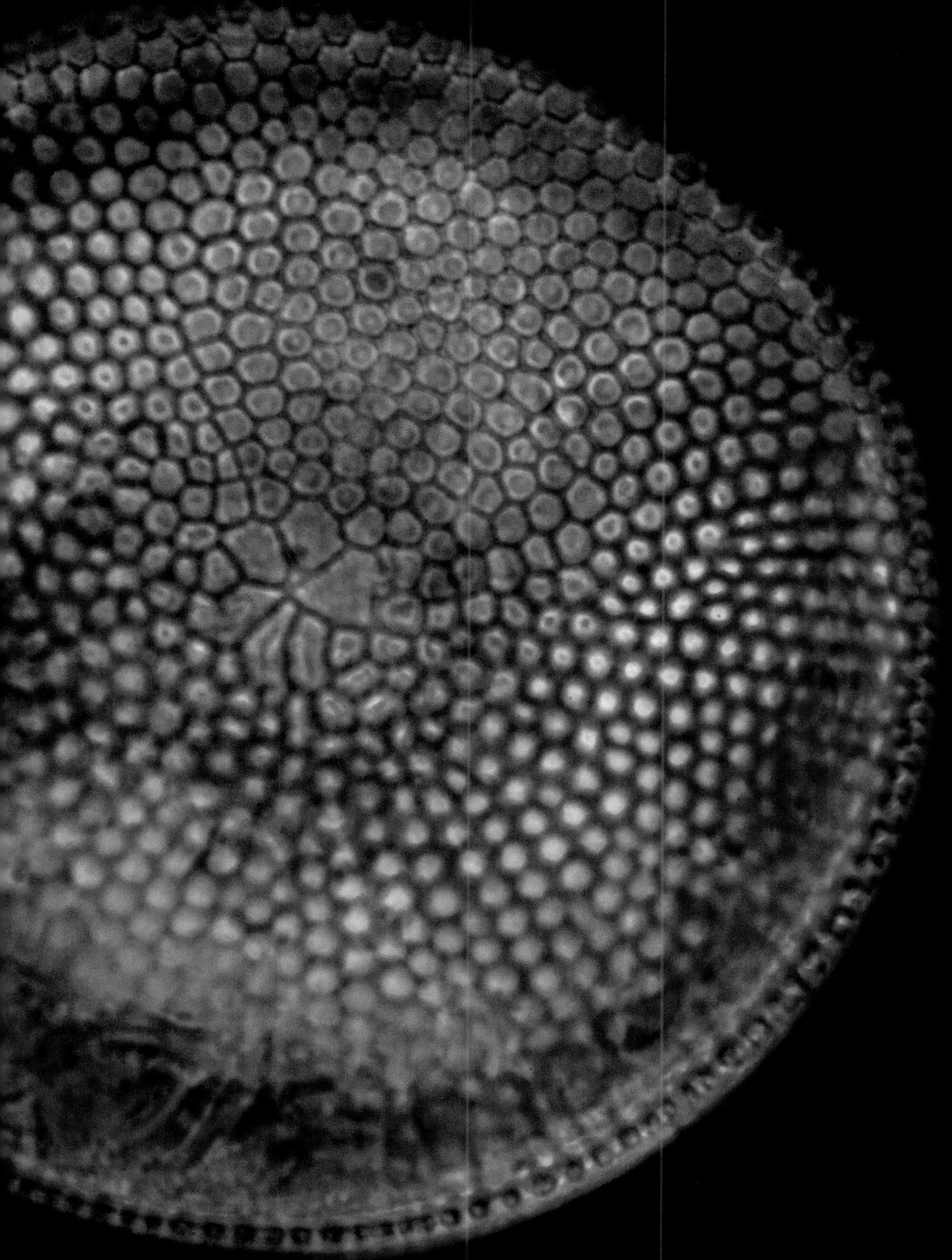

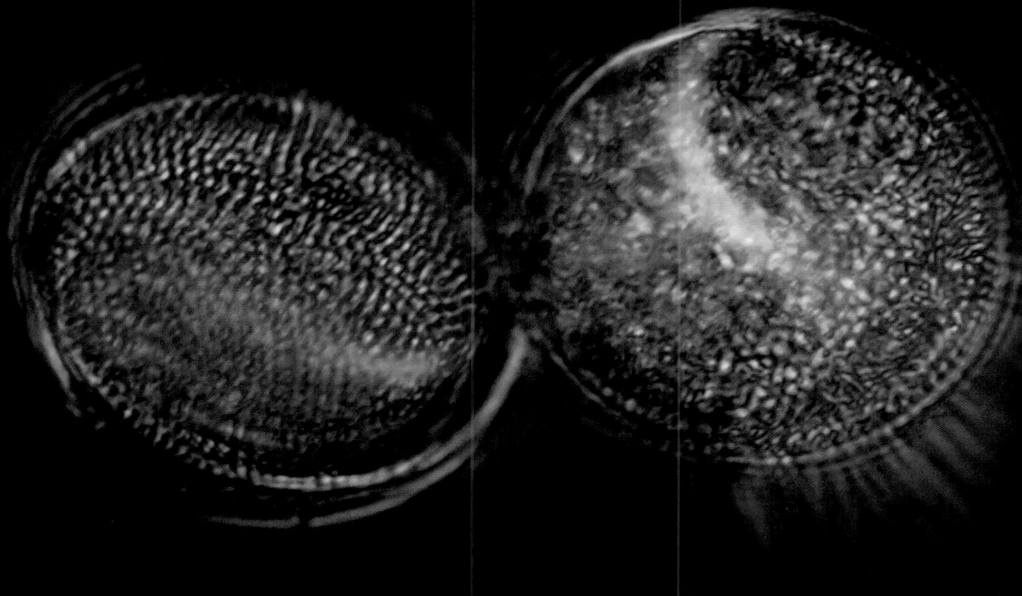

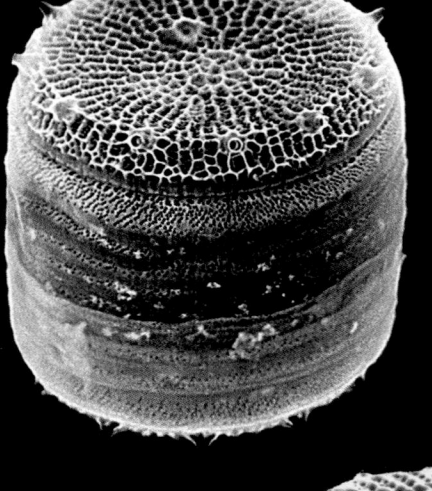

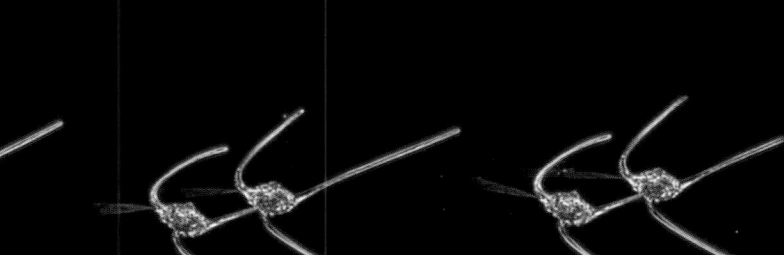

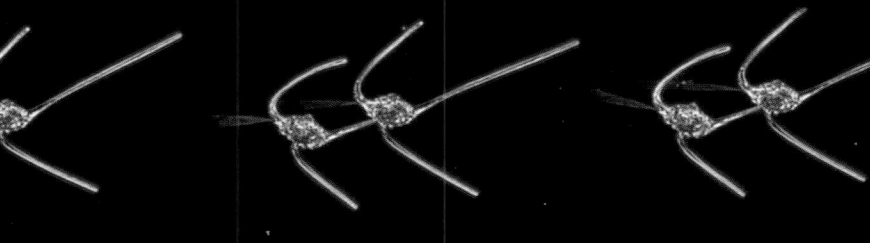

Dinoflagellate Motility

Top left: *Alexandrium tamarense*, a dinoflagellate that produces paralyzing toxins, often proliferates near the coasts, turning the water red. The cell measures 25 to 50 microns and has two flagella that propel it forward.

Samples from the NCMA collection, Bigelow Laboratory, Booth Bay, USA.

Left: *Ceratium hexacanthum* dinoflagellates remain in motion while undergoing cell division. They move by means of two flagella. Only the longitudinal flagellum is visible here, while the transverse flagellum is hidden in a groove circling the cell. The longitudinal flagellum propels the cell forward, while the transverse flagellum makes it whirl. This whirling motion is at the origin of the term *dinoflagellate*, from the Greek *dino* ("whirl") and Latin *flagellum* ("small whip").

Sample from the Mediterranean Culture Collection of Villefranche-sur-Mer.

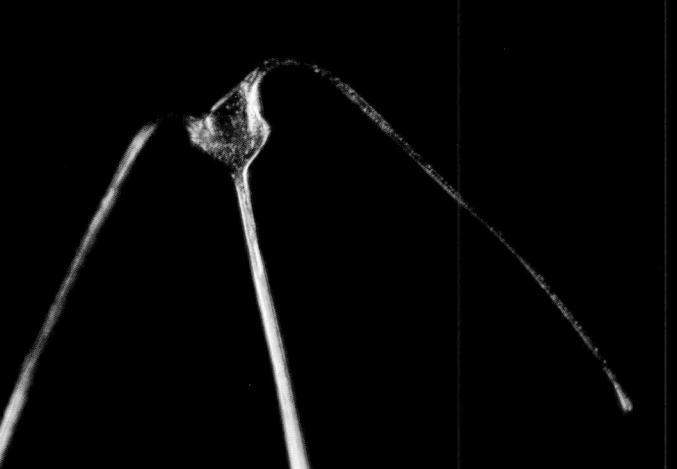

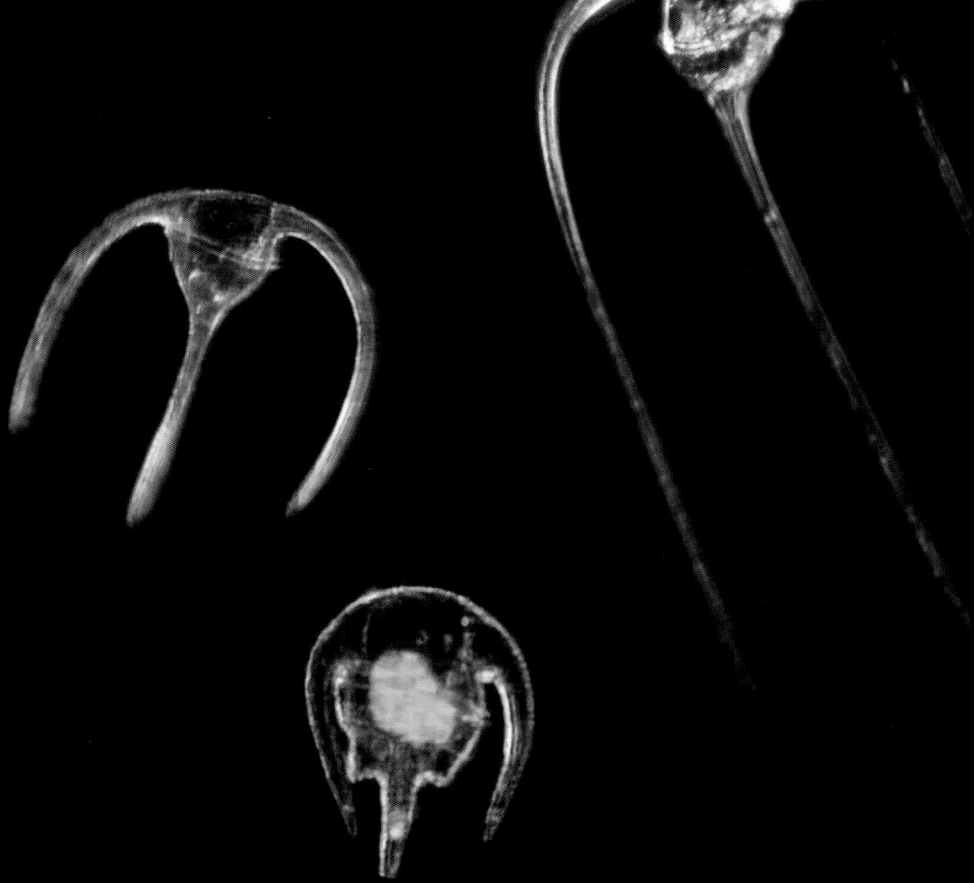

In early summer 2012, dinoflagellates were particularly abundant in plankton catches taken in the bay of Villefranche-sur-Mer using 20-micron mesh nets.

Opposite page: The majority of protists here are dinoflagellates of the genera *Protoperidinium* and *Dinophysis*, measuring between 30 and 60 microns.

This page, bottom image: Various *Ceratium* dinoflagellates measuring from one to several hundred microns. Their anchor-like shape is characteristic of the genus. From left to right: *Ceratium massiliense*, *C. symmetricum*, *C. limulus*, *C. longissimum*.

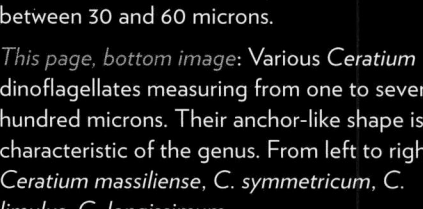

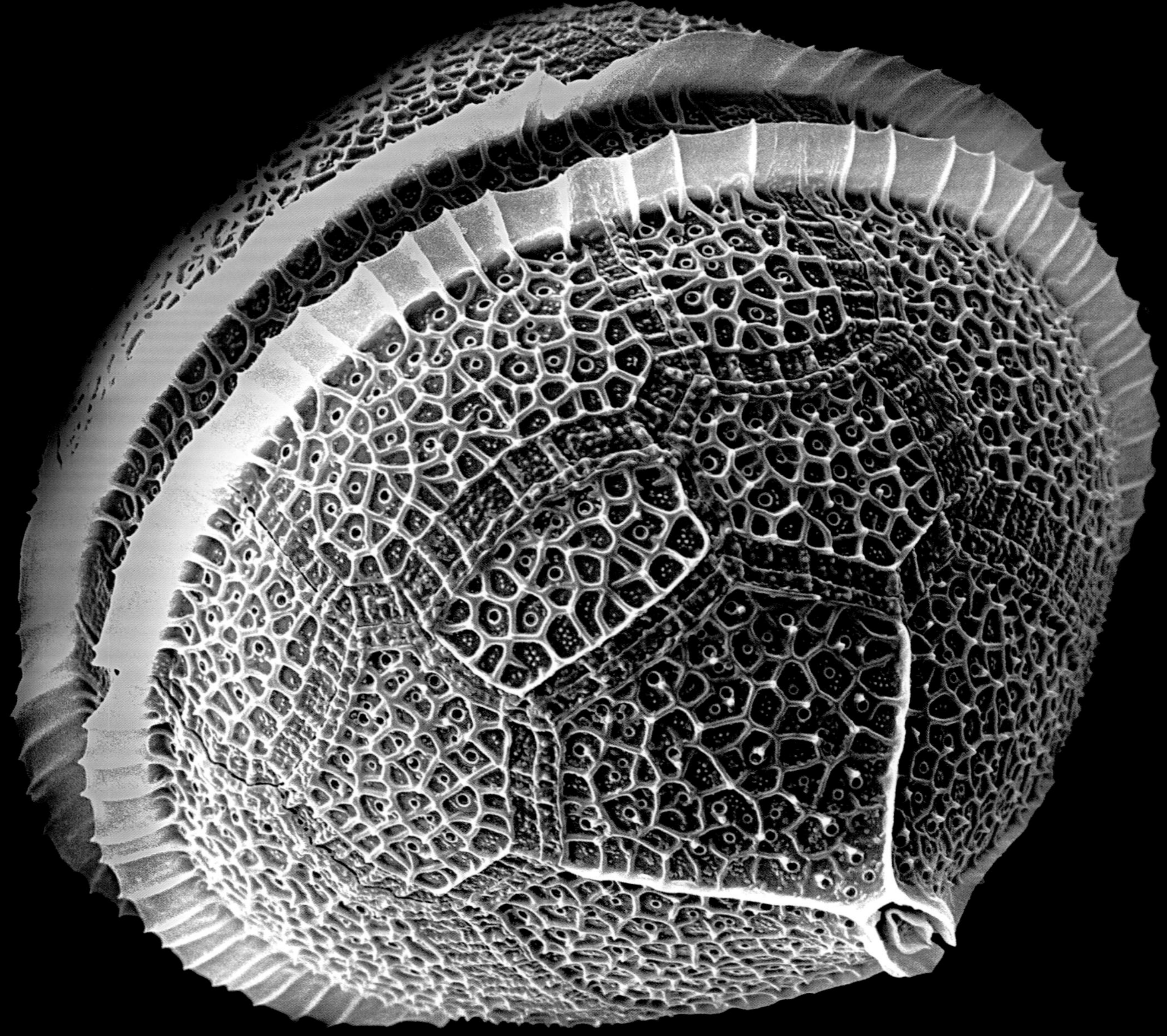

Cellulose Armor

Most dinoflagellates, like this *Protoperidinium* sp. (above), build envelopes made of delicately ornamented cellulose plates called thecas. Produced within flattened vesicles and secreted by the cell, the thecas are assembled on the cell surface.

ABOVE: SCANNING ELECTRON MICROGRAPH BY MAR-GAUX CARMICHAEL, ROSCOFF BIOLOGICAL STATION, FRANCE.

Right: The nucleus of the dinoflagellate *Ceratium candelabrum* fluoresces blue after addition of a dye that binds to DNA. Chloroplasts naturally fluoresce red.

CONFOCAL MICROSCOPY BY CHRISTIAN ROUVIÈRE AND CHRISTIAN SARDET, CNRS, OBSERVATOIRE OCÉANO-LOGIQUE DE VILLEFRANCHE-SUR-MER.

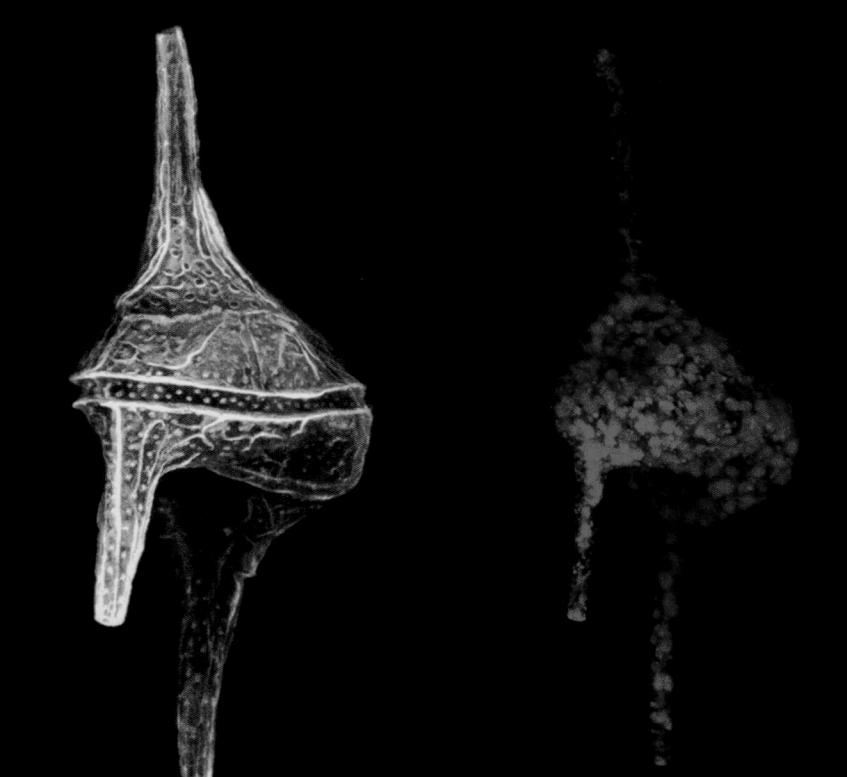

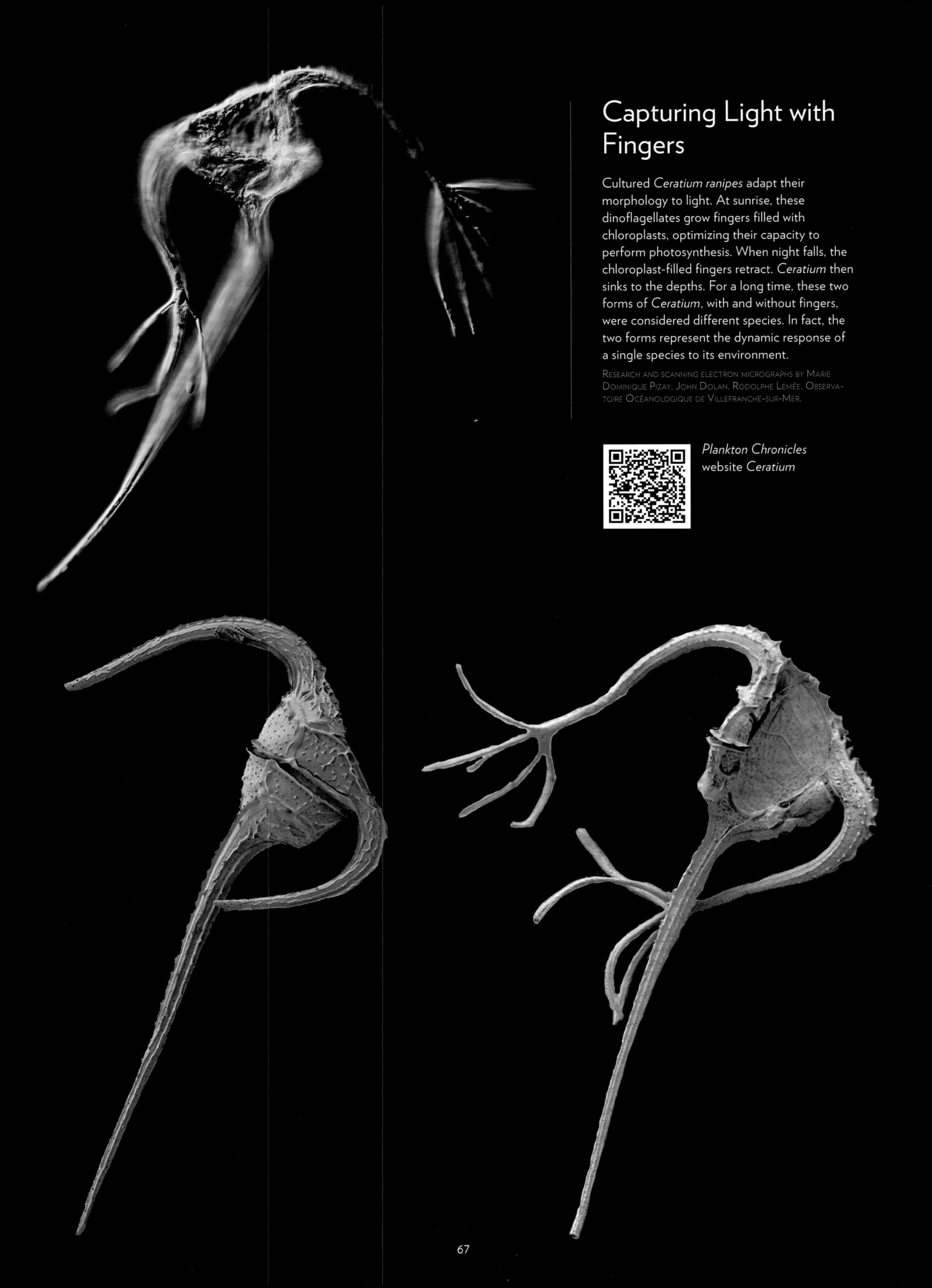

Capturing Light with Fingers

Cultured *Ceratium ranipes* adapt their morphology to light. At sunrise, these dinoflagellates grow fingers filled with chloroplasts, optimizing their capacity to perform photosynthesis. When night falls, the chloroplast-filled fingers retract. *Ceratium* then sinks to the depths. For a long time, these two forms of *Ceratium*, with and without fingers, were considered different species. In fact, the two forms represent the dynamic response of a single species to its environment.

RESEARCH AND SCANNING ELECTRON MICROGRAPHS BY MARIE DOMINIQUE PIZAY, JOHN DOLAN, RODOLPHE LEMÉE, OBSERVATOIRE OCÉANOLOGIQUE DE VILLEFRANCHE-SUR-MER.

Plankton Chronicles
website *Ceratium*

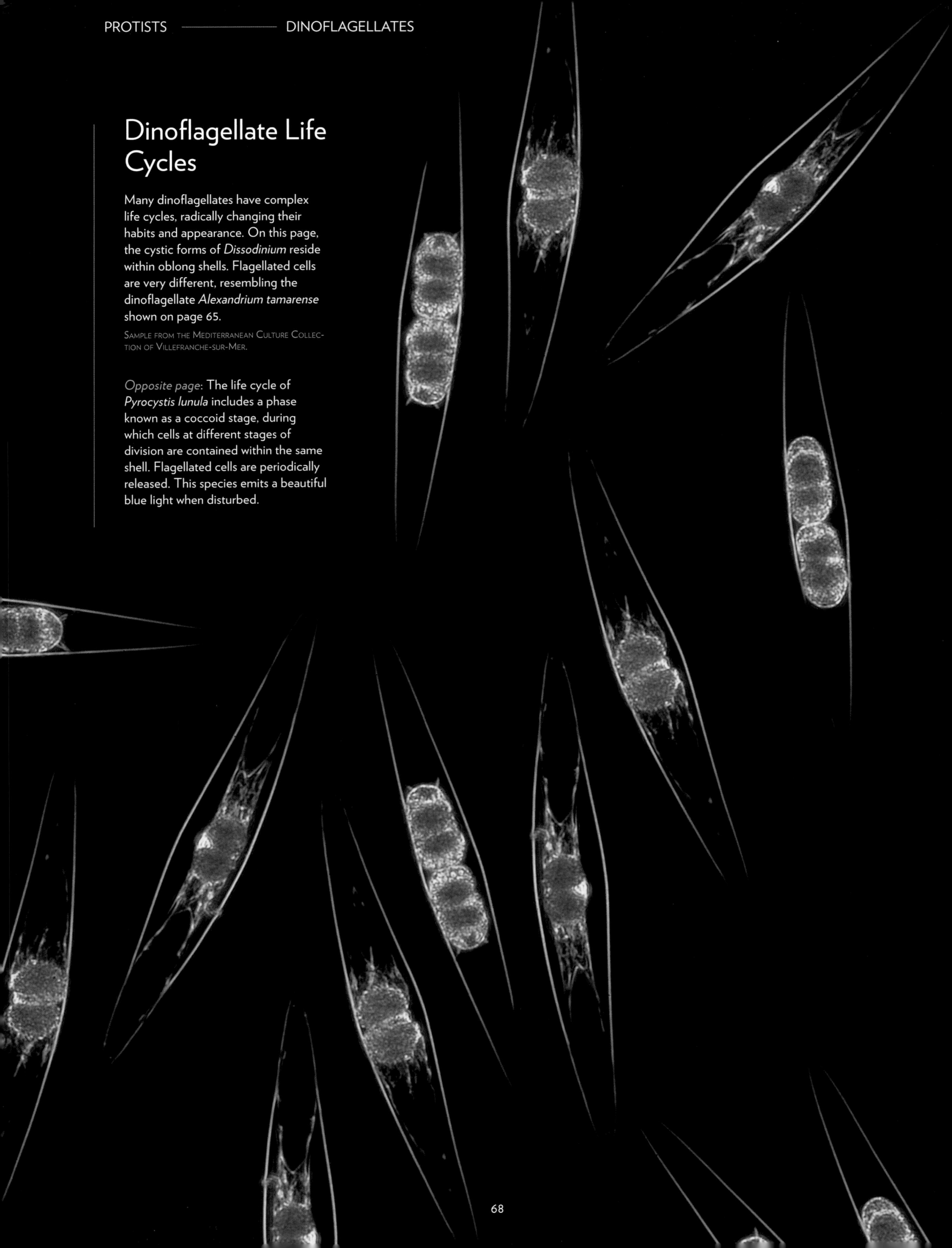

Dinoflagellate Life Cycles

Many dinoflagellates have complex life cycles, radically changing their habits and appearance. On this page, the cystic forms of *Dissodinium* reside within oblong shells. Flagellated cells are very different, resembling the dinoflagellate *Alexandrium tamarense* shown on page 65.

SAMPLE FROM THE MEDITERRANEAN CULTURE COLLECTION OF VILLEFRANCHE-SUR-MER.

Opposite page: The life cycle of *Pyrocystis lunula* includes a phase known as a coccoid stage, during which cells at different stages of division are contained within the same shell. Flagellated cells are periodically released. This species emits a beautiful blue light when disturbed.

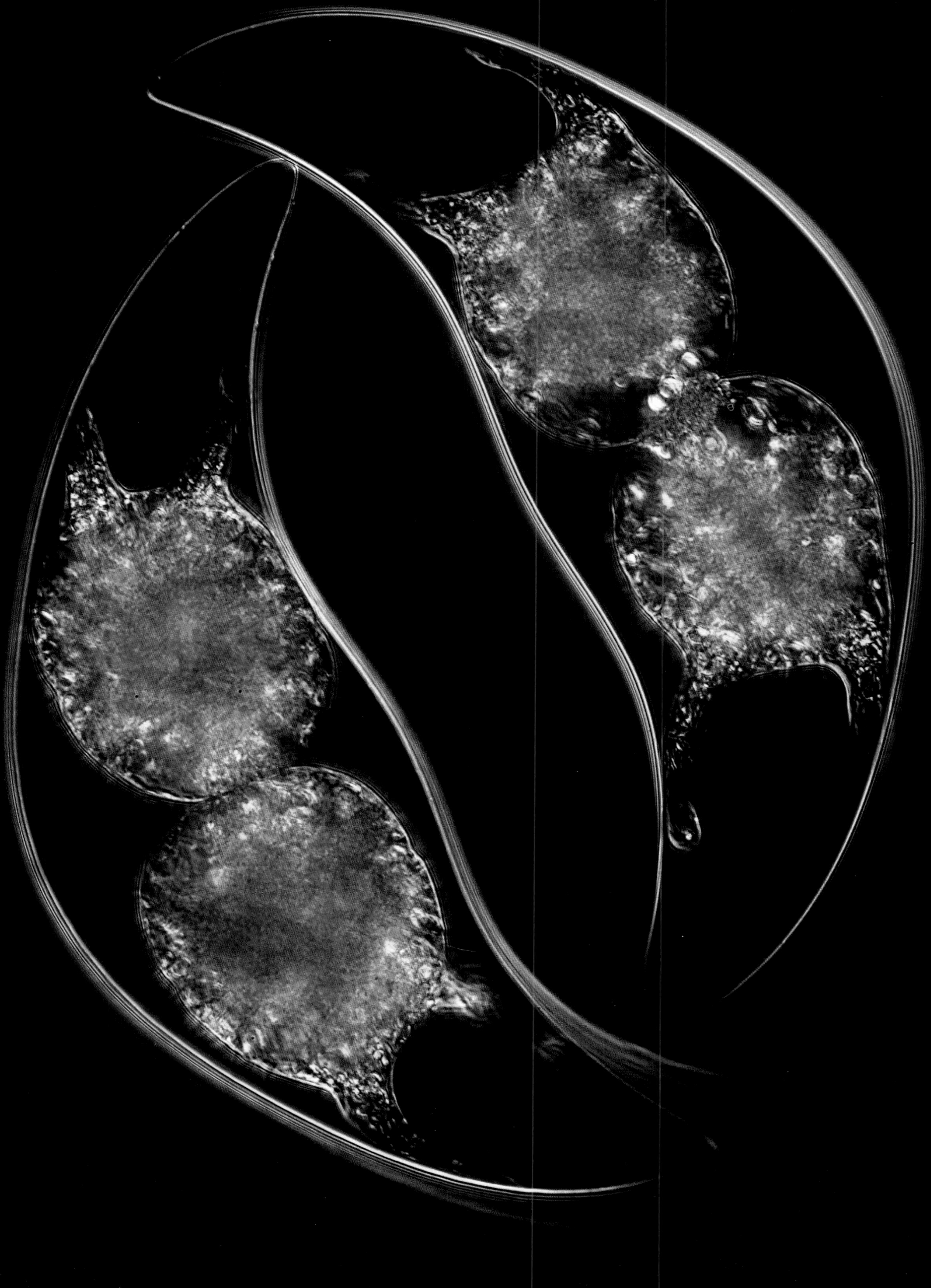

RADIOLARIANS: POLYCYSTINES AND ACANTHARIANS

SYMBIOSIS AT THE OCEAN SURFACE

Radiolarians are single-celled planktonic protists, classified in two main groups called polycystines and acantharians. Most radiolarians are microscopic, but some species are visible to the naked eye. The larger radiolarians became well known in the nineteenth century thanks to Ernst Haeckel's superb illustrations (see pages 19 and 85). Some polycystine radiolarians live as colonies forming gelatinous spheres easily seen by scuba divers. Most of the thousands of identified radiolarian species build sophisticated skeletons made of silica. Their fossils are found in sediments dating back over 500 million years, bearing witness to the evolution of the ocean and changes in climate. These fossils are often used to determine the origin of hydrocarbon deposits. The presence of radiolarian fossils in the Himalaya is strong evidence of the movement of tectonic plates.

Acantharians form remarkable needles and shield-like structures using strontium sulfate instead of silica. Like amoeba, acantharians and polycystines extend their membranes into thin extensions called pseudopodia, rhizopodia, and axopodia, structures used for exploring their environment and capturing and consuming prey—bacteria, other protists, or tiny animals.

In addition to feeding on other creatures, many species of polycystines and acantharians rely on long-term symbioses, much like coral. They harbor microalgae on their surfaces or inside their cytoplasm. The polycystines and acantharians provide shelter and a nutritive cellular environment while the microalgae capture energy via photosynthesis. They usually drift in the surface layers of the ocean to profit from a maximum of sunlight. Because radiolarians and their symbionts seemed a kind of vegetable/animal hybrid, our colleagues on the *Tara Oceans Expedition* jokingly called them "vegimals."

Another example of an organism that relies on symbiosis is *Collozoum*—a species of colonial radiolarian comprised of thousands of individuals sharing one gelatinous envelope full of tiny microalgae. Other species of common microalgae (haptophytes) live within the cytoplasm of the beautiful acantharian *Lithoptera*. This instance of cellular cooperation seems to date back to the Jurassic era, 200 to 150 million years ago when the ocean was very poor in nutrients. Planktonic symbioses have evolved to help these organisms survive in the hot spots and deserts of the world ocean.

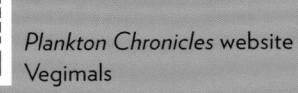

Plankton Chronicles website
Vegimals

POLYCYSTINE RADIOLARIANS FROM THE BAY OF VILLEFRANCHE-SUR-MER

Ten *Aulacantha scolymantha* radiolarians (*Phaeodaria*), about 1 mm in diameter, and two larger radiolarians *Thalassicolla pellucida* and *Thalassolampe margarodes* (see close-ups in the following pages). Bottom left, a spumellarian radiolarian, *Collozoum inerme*, consisting of many individual cells sharing a common gelatinous envelope.

PLANKTON COLLECTED IN AUTUMN, USING A 120-MICRON MESH NET.
PHOTO BY CHRISTIAN SARDET AND NOÉ SARDET, PARAFILMS, MONTREAL.

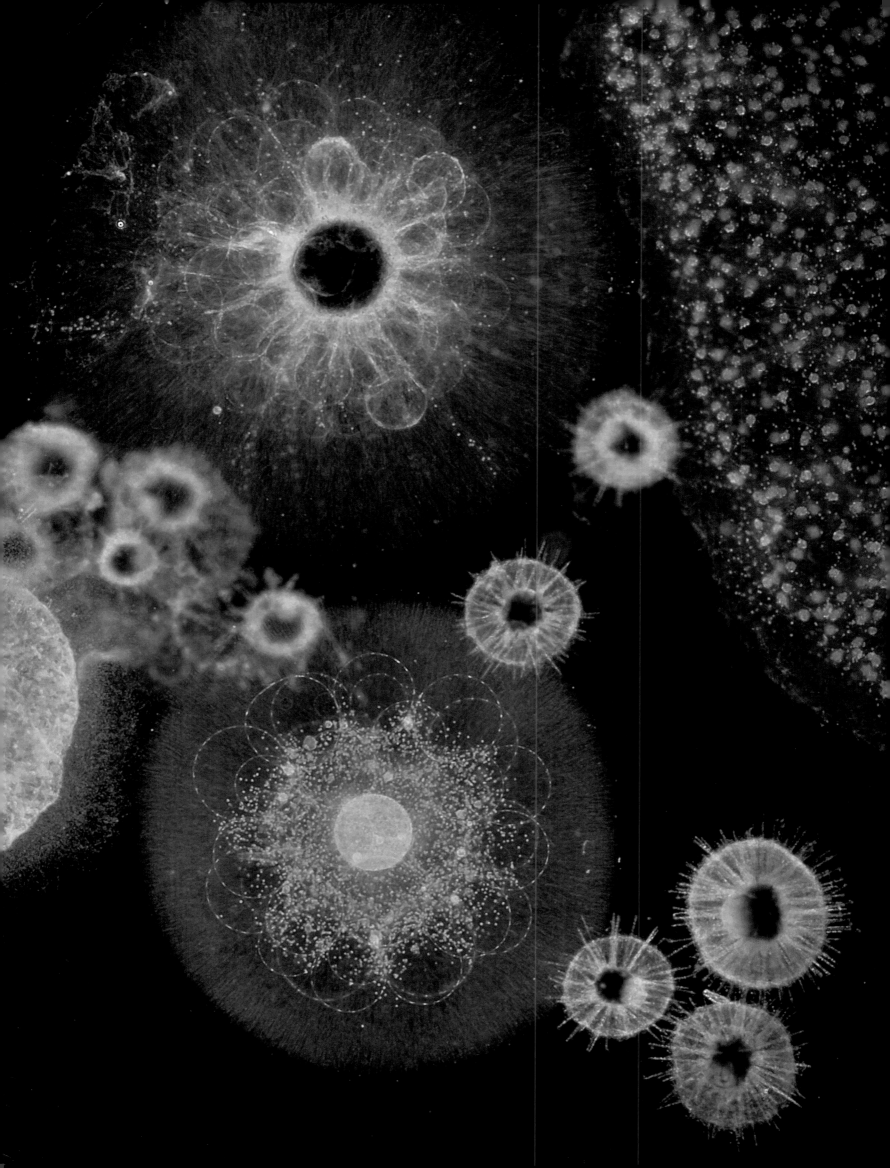

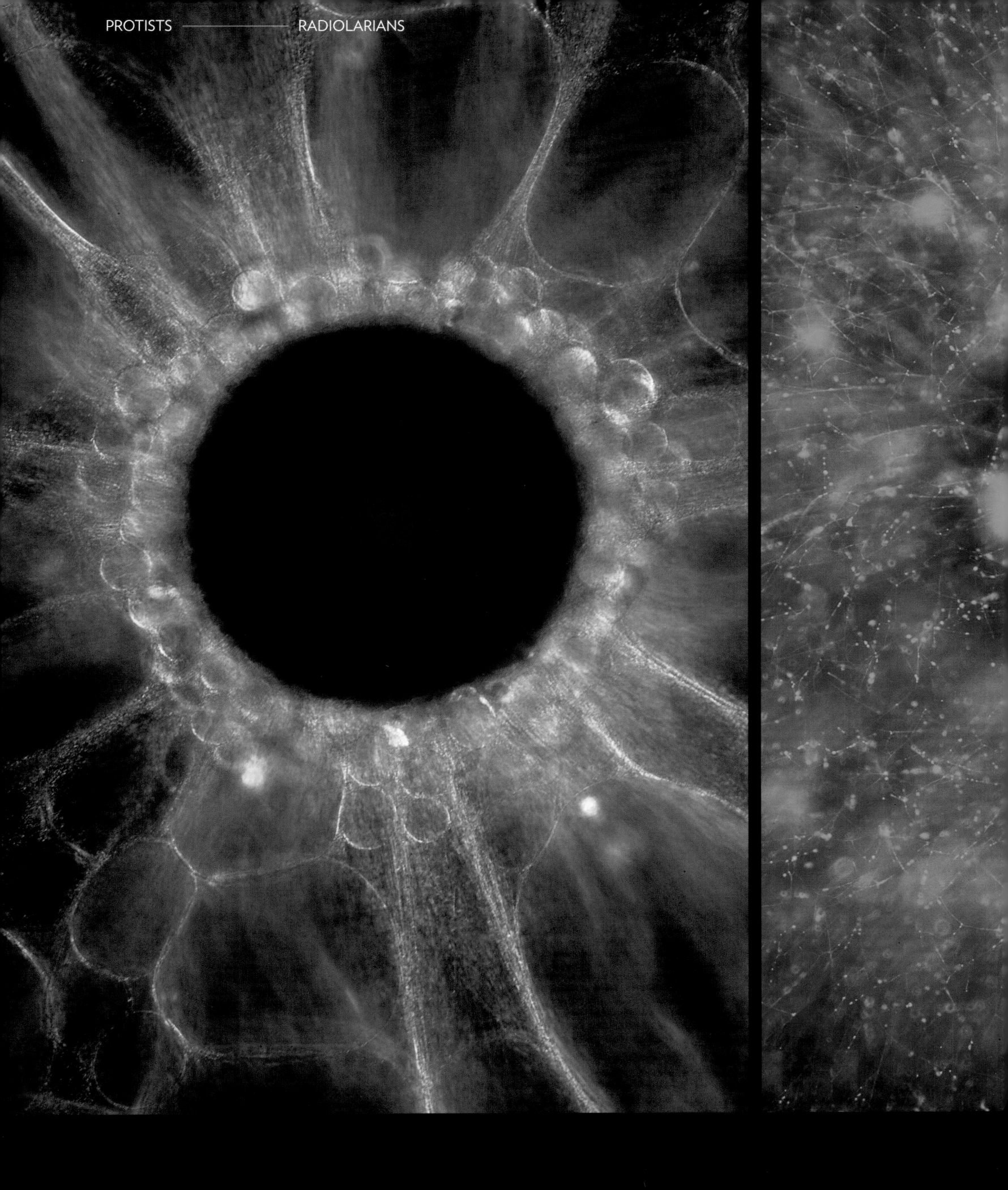

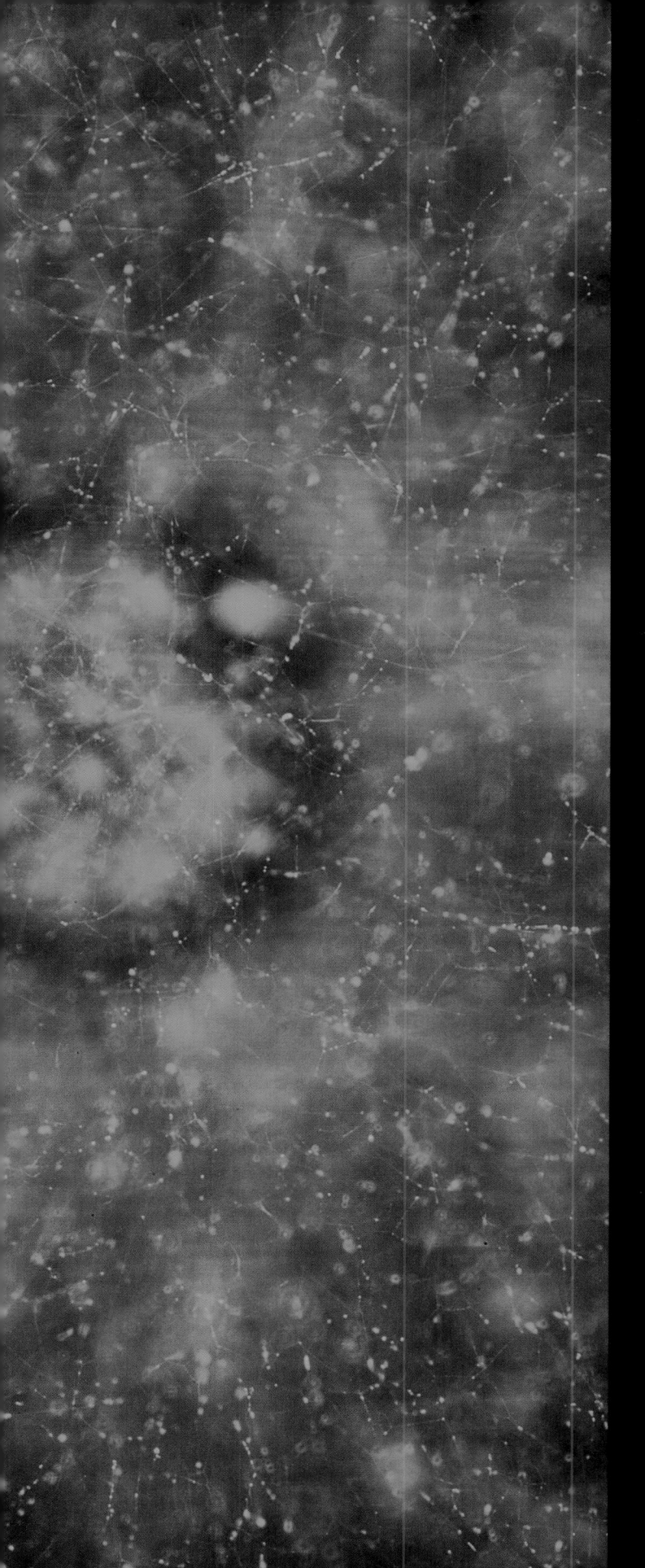

In the Heart of a Polycystine Radiolarian

The central capsule of a radiolarian (left) includes one or more nuclei and a mass of cytoplasm called endoplasm. The cytoplasm of this large cell (measuring a few millimeters) contains many mitochondria and an endoplasmic reticulum, a tubular network of intracellular membranes fluorescing in green (center image). In some radiolarians such as this *Thallassolampe* sp. the endoplasm contains many symbiotic microalgae filled with chlorophyll pigments that fluoresce red (right).

PLANKTON COLLECTED IN AUTUMN, IN THE BAY OF VILLEFRANCHE-SUR-MER, USING A 120-MICRON MESH NET.

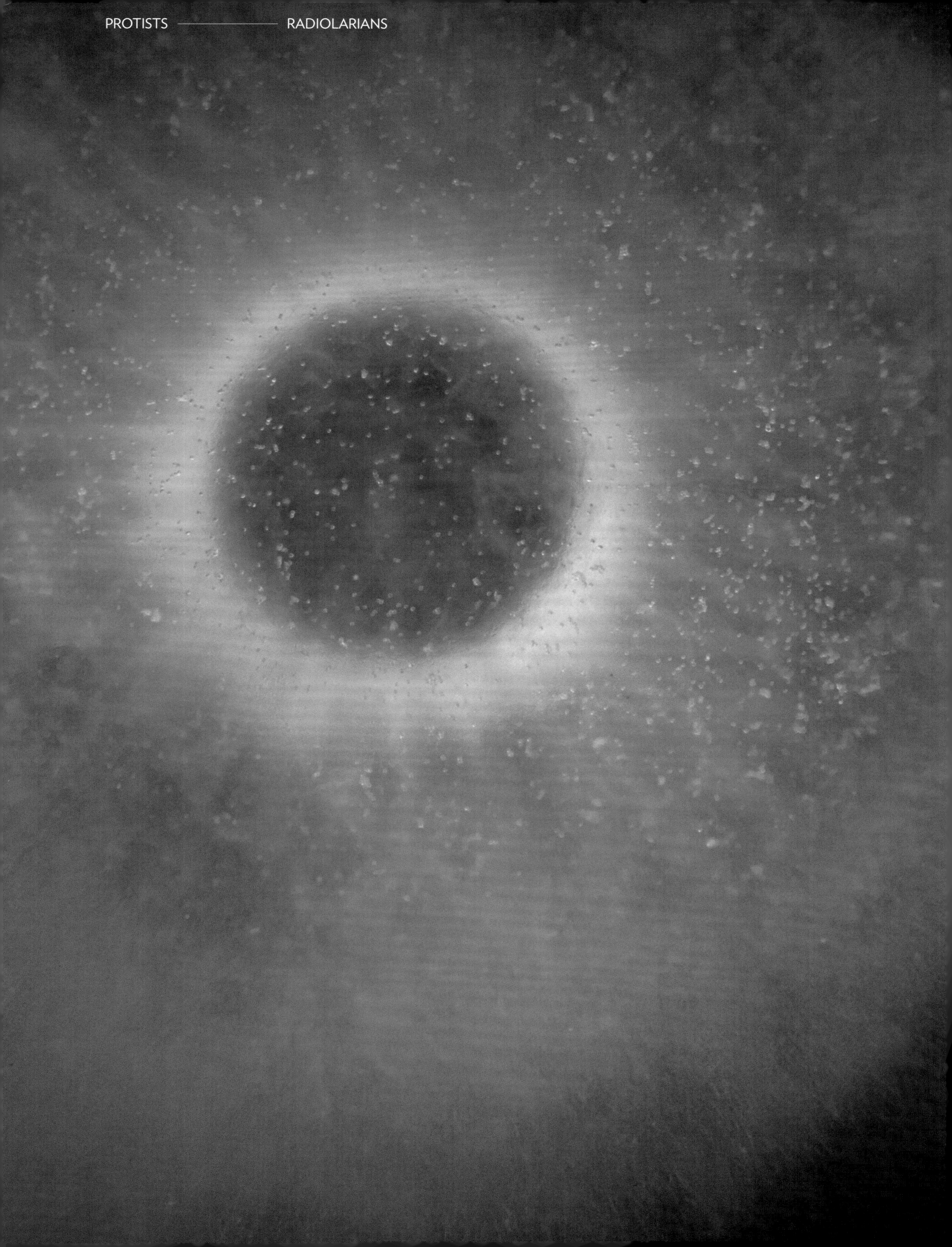

Two Kinds of Radiolarians

Opposite page: A large *Thalassicolanucleata* radiolarian with a central capsule enclosing the nucleus.

This page: A colonial collodarian radiolarian comprised of many individual cells, each with a central capsule, sharing a common jelly. In both images symbiotic microalgae are visible as small ochre-colored dots.

PLANKTON COLLECTED IN THE BAY OF VILLEFRANCHE-SUR-MER DURING THE FALL WITH A 120-MICRON MESH NET. PHOTO: CHRISTIAN SARDET AND NOÉ SARDET, PARAFILMS, MONTREAL.

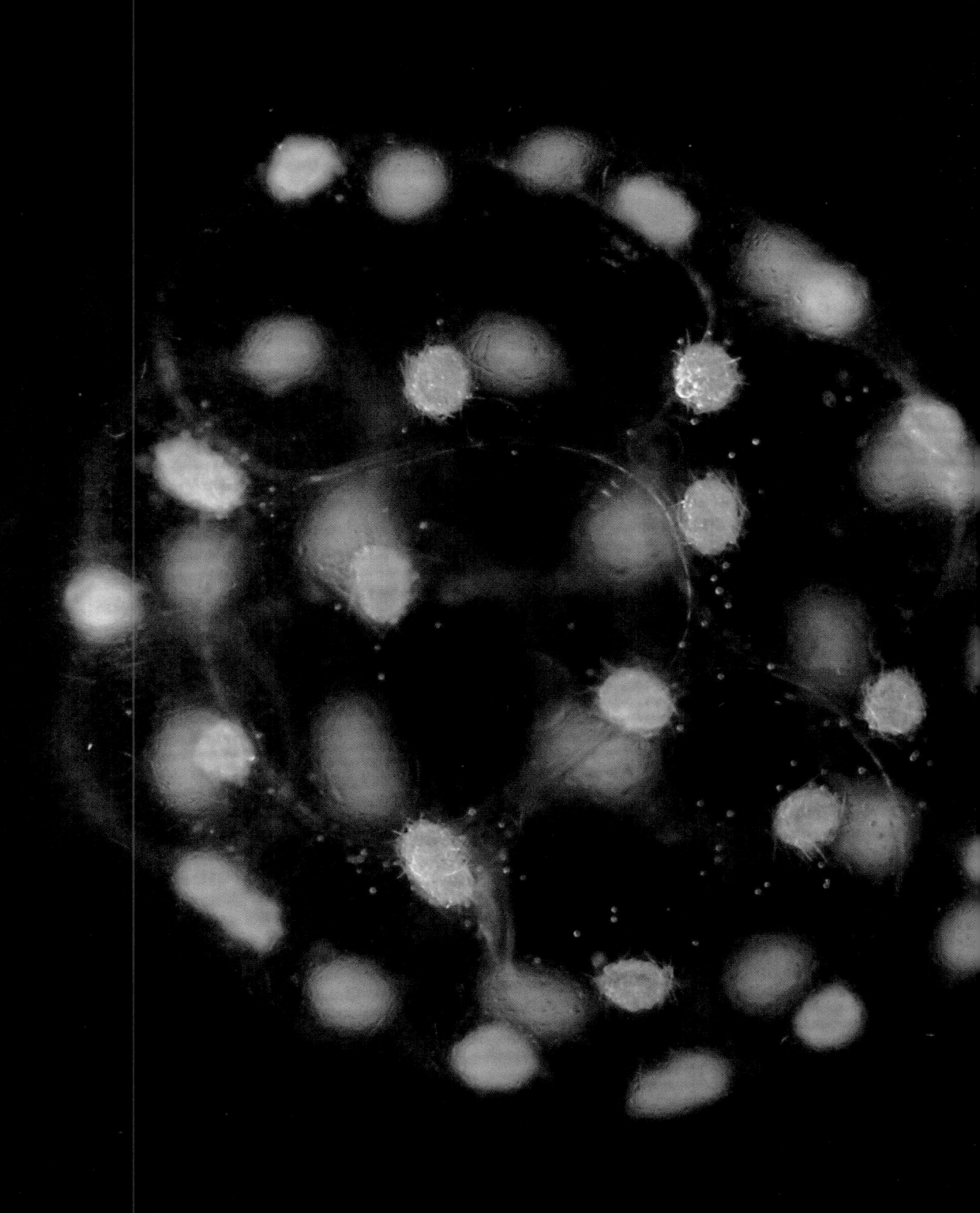

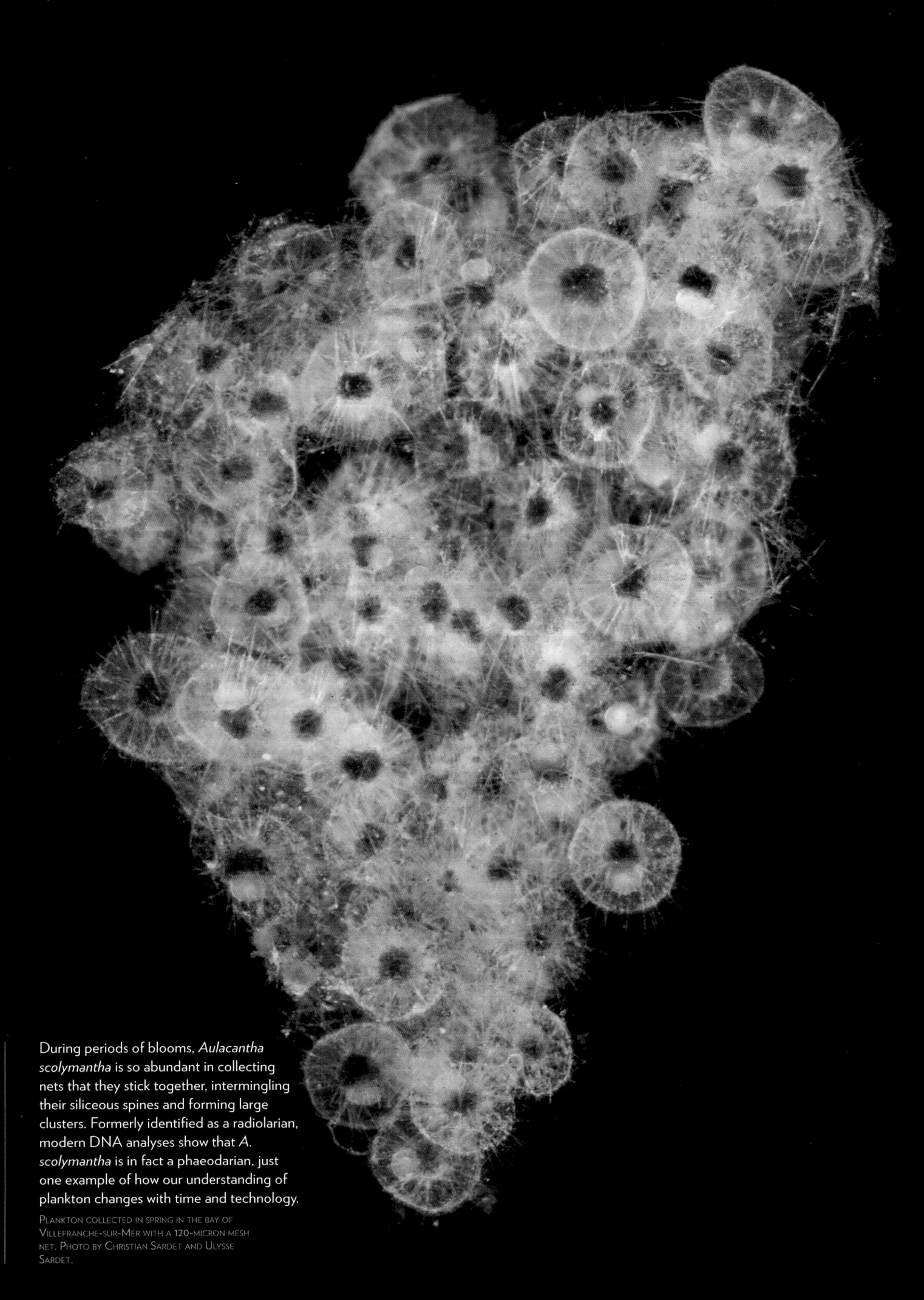

During periods of blooms, *Aulacantha scolymantha* is so abundant in collecting nets that they stick together, intermingling their siliceous spines and forming large clusters. Formerly identified as a radiolarian, modern DNA analyses show that *A. scolymantha* is in fact a phaeodarian, just one example of how our understanding of plankton changes with time and technology.

Plankton collected in spring in the bay of Villefranche-sur-Mer with a 120-micron mesh net. Photo by Christian Sardet and Ulysse Sardet.

Large white vesicles can be
distinguished around the ochre-
colored central capsule of this large
collodarian radiolarian, *Thalassolampe
margarodes.* These vesicles, along
with the jelly and skeleton, regulate
buoyancy and also constitute nutrient
reserves. The small yellow particles in
the jelly are symbiotic microalgae.

PLANKTON COLLECTED IN THE BAY OF VILLEFRANCHE-
SUR-MER IN AUTUMN WITH A 120-MICRON MESH NET.
PHOTO BY CHRISTIAN SARDET AND NOÉ SARDET.
PARAFILMS, MONTREAL.

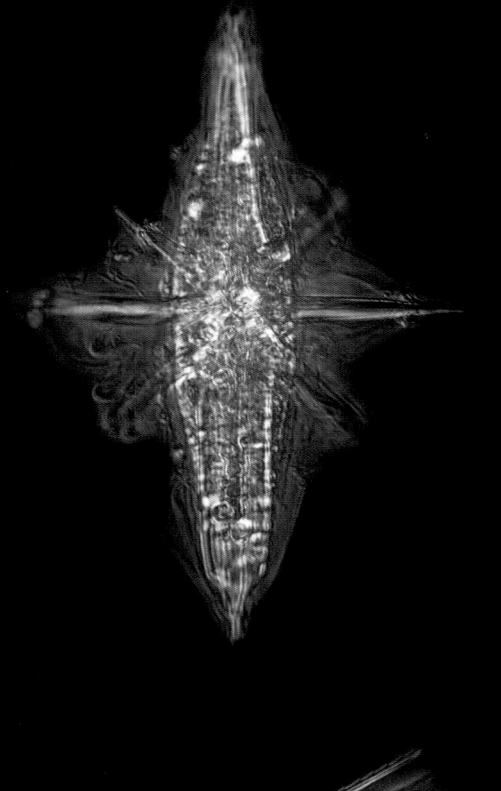

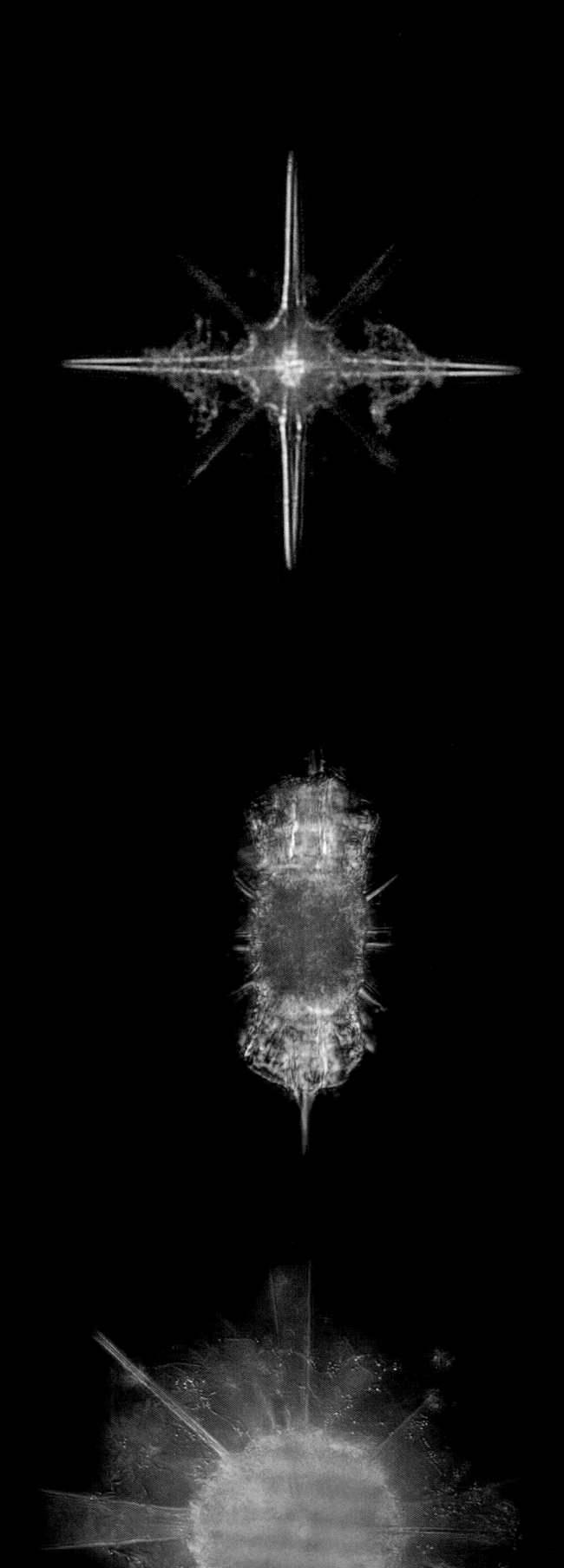

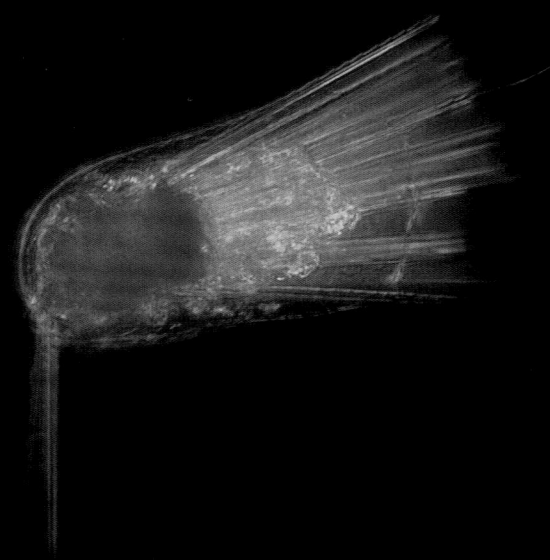

ACANTHARIAN RADIOLARIANS

These five acantharians have skeletons characteristic of the species, with either 10 or 20 spines (called spicules) made of strontium sulfate. Each spicule is a crystal, serving as an attachment point for contractile filaments called myonemes. The myonemes rapidly contract or expand the cytoplasm around the central capsule.

Above, clockwise from top left: *Amphibelone* sp., *Acanthostaurus purpurascens*, *Diploconus* sp., *Heteracon* sp. (pre-encystment stage).

Right: Unknown species of order Chaunacanthida.

PLANKTON COLLECTED IN SPRING IN THE BAY OF VILLEFRANCHE-SUR-MER.

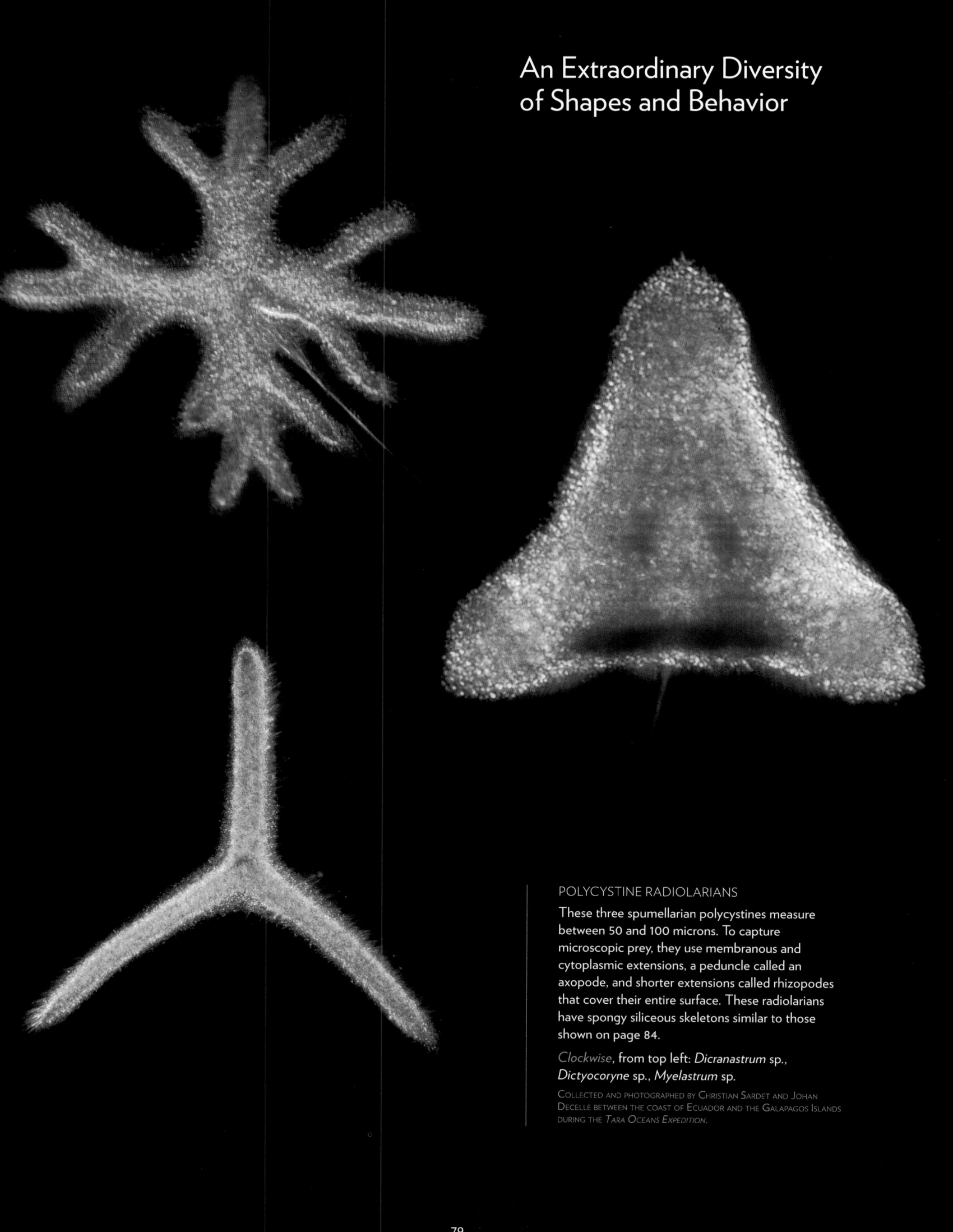

An Extraordinary Diversity of Shapes and Behavior

POLYCYSTINE RADIOLARIANS

These three spumellarian polycystines measure between 50 and 100 microns. To capture microscopic prey, they use membranous and cytoplasmic extensions, a peduncle called an axopode, and shorter extensions called rhizopodes that cover their entire surface. These radiolarians have spongy siliceous skeletons similar to those shown on page 84.

Clockwise, from top left: *Dicranastrum* sp., *Dictyocoryne* sp., *Myelastrum* sp.

Collected and photographed by Christian Sardet and Johan Decelle between the coast of Ecuador and the Galapagos Islands during the *Tara Oceans Expedition*.

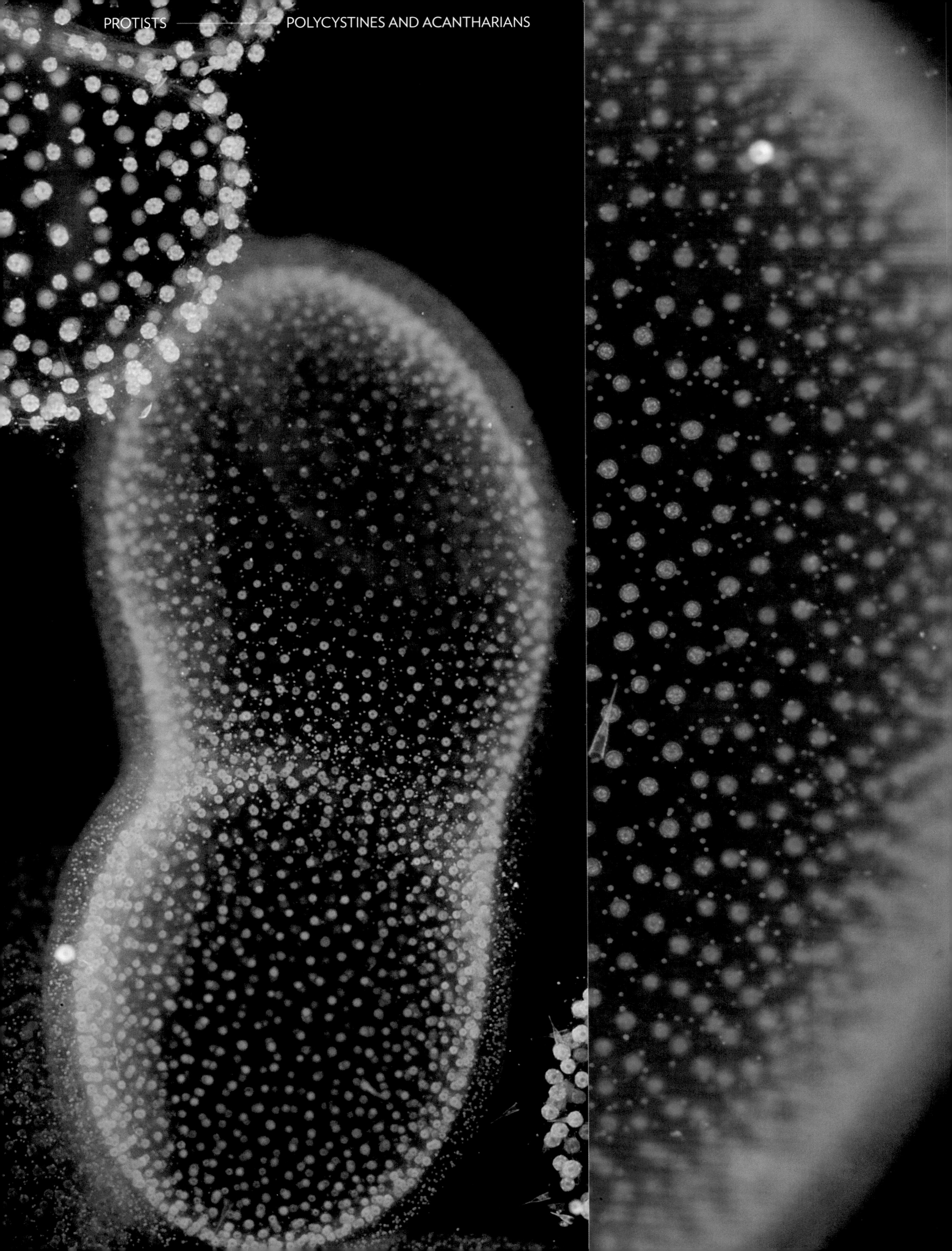

In the Jelly of a Radiolarian Colony

Collozoum sp. are radiolarians that form round or sausage-shaped colonies within a gelatinous mass varying in size from a few centimeters to 1 meter. Individual cells appear as whitish spheres within the jelly. The small ochre-colored particles are symbiotic microalgae.

Bottom right: Close-up of a young acantharian, *Lithoptera fenestrata*. In the top image, the *Lithoptera* appears stuck in the jelly of the colonial radiolarian *Collozoum* sp.

PLANKTON COLLECTED DURING A WINTER BLOOM OF THESE RADIOLARIANS IN THE BAY OF VILLEFRANCHE-SUR-MER.

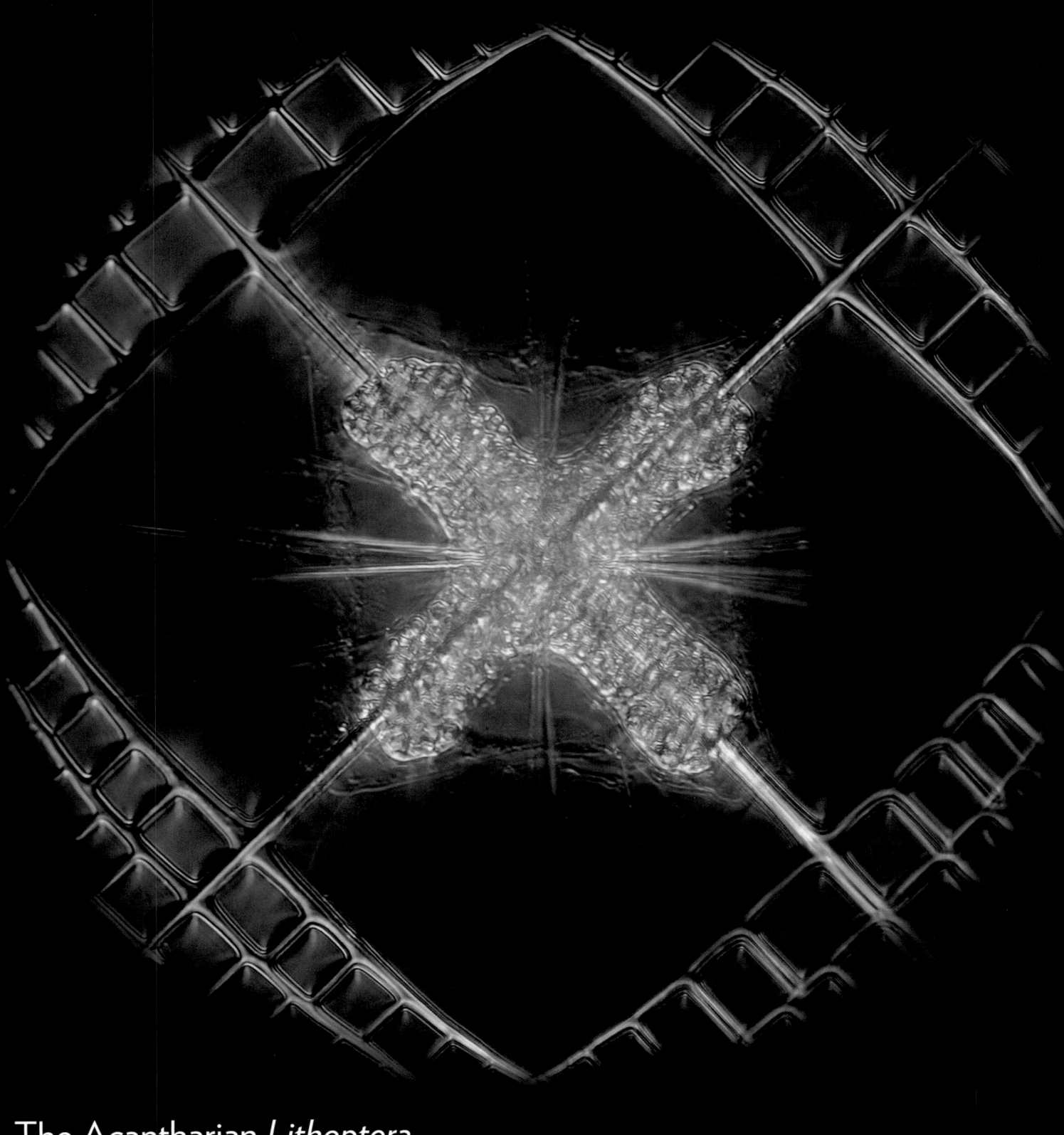

The Acantharian *Lithoptera*

Described for the first time at the end of the nineteenth century, this *Lithoptera fenestrata* was collected during winter in the bay of Villefranche-sur-Mer. The unicellular organism lives from one to several months, drifting in the ocean. Its skeleton of strontium sulfate grows in size with age.

Above: Cytoplasmic extensions are visible here. They explore the environment in search of prey. The four yellow masses are groups of symbiotic haptophyceae algae of the genus *Phaeocystis* living inside the cytoplasm.

Right: The cultured microalgae *Phaeocystis* sp. with its two flagella.

SCANNING ELECTRON MICROGRAPH BY JOHAN DECELLE AND FABRICE NOT, SBR-CNRS, UPMC, ROSCOFF.

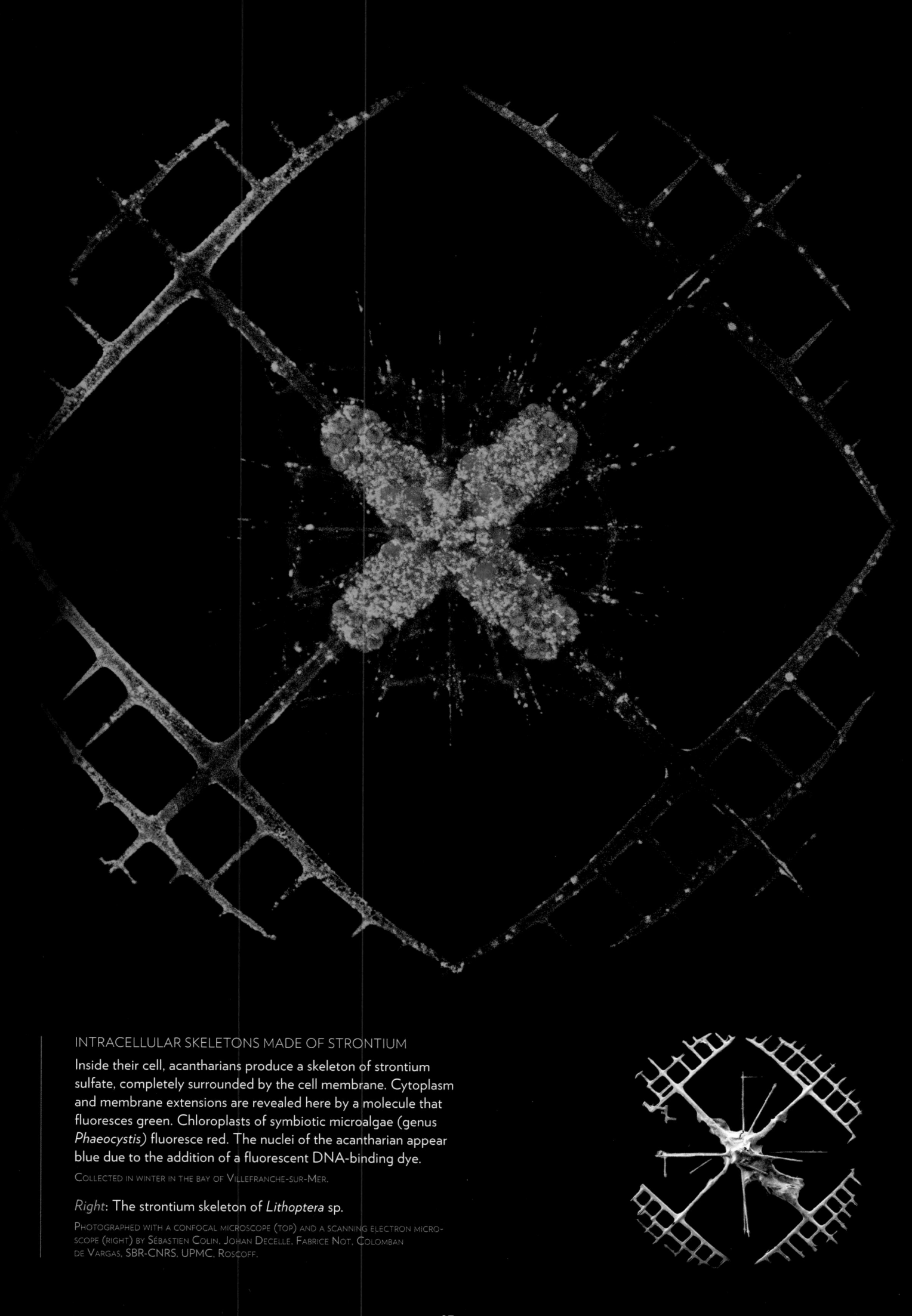

INTRACELLULAR SKELETONS MADE OF STRONTIUM

Inside their cell, acantharians produce a skeleton of strontium sulfate, completely surrounded by the cell membrane. Cytoplasm and membrane extensions are revealed here by a molecule that fluoresces green. Chloroplasts of symbiotic microalgae (genus *Phaeocystis*) fluoresce red. The nuclei of the acantharian appear blue due to the addition of a fluorescent DNA-binding dye.

Collected in winter in the bay of Villefranche-sur-Mer.

Right: The strontium skeleton of *Lithoptera* sp.

Photographed with a confocal microscope (top) and a scanning electron microscope (right) by Sébastien Colin, Johan Decelle, Fabrice Not, Colomban de Vargas, SBR-CNRS, UPMC, Roscoff.

Extraordinary Siliceous Skeletons

The siliceous skeletons of polycystine radiolarians are reticulated and display axial [nassellarians (1, 4)] or spherical [spumellarians (2, 3, 5)] symmetry. Paleontologists identify fossil radiolarians to characterize geological and hydrocarbon deposits.

THIS PAGE IS A TRIBUTE TO JEAN AND MONIQUE CACHON, WHO STUDIED RADIOLARIANS IN VILLEFRANCHE-SUR-MER BETWEEN 1960 AND 1990. THEY MADE THESE PHOTOS USING A SCANNING ELECTRON MICROSCOPE. PHOTO 5 IS A CONTRIBUTION OF JOHAN DECELLE AND FABRICE NOT, SBR-CNRS, UPMC, ROSCOFF. (1) *PTEROCANIUM* SP. (2) *TETRAPYLE* SP. (3) *HEXALONCHE* SP. (4) *LITHARACHNIUM* SP. (5) *DIDYMOSPYRIS* SP.

Opposite page: Skeletons of nasselarian radiolarians (order *Stephoidea*).

DRAWINGS BY ERNST HAECKEL IN *KUNSTFOREN DER NATUR* (1904).

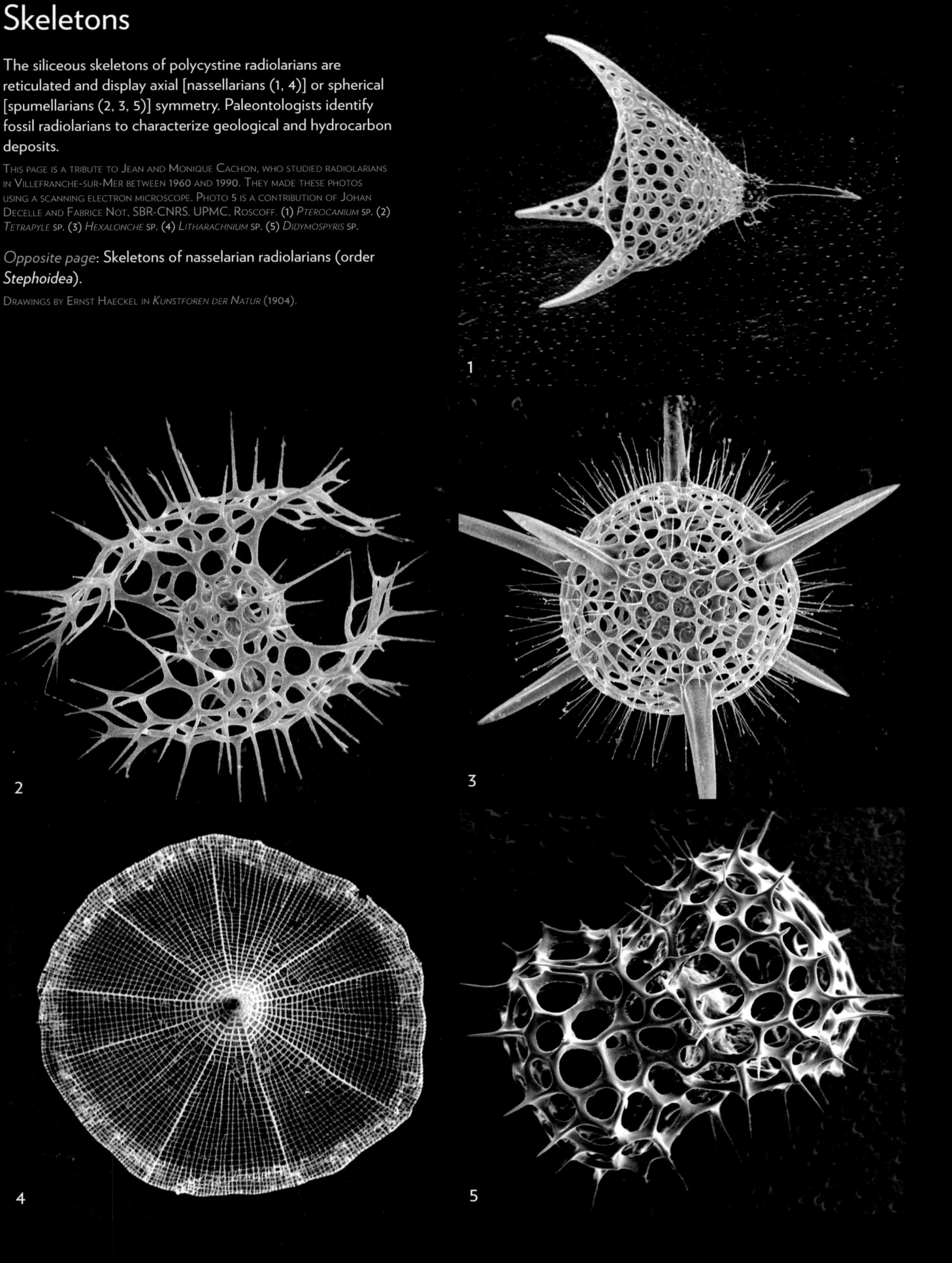

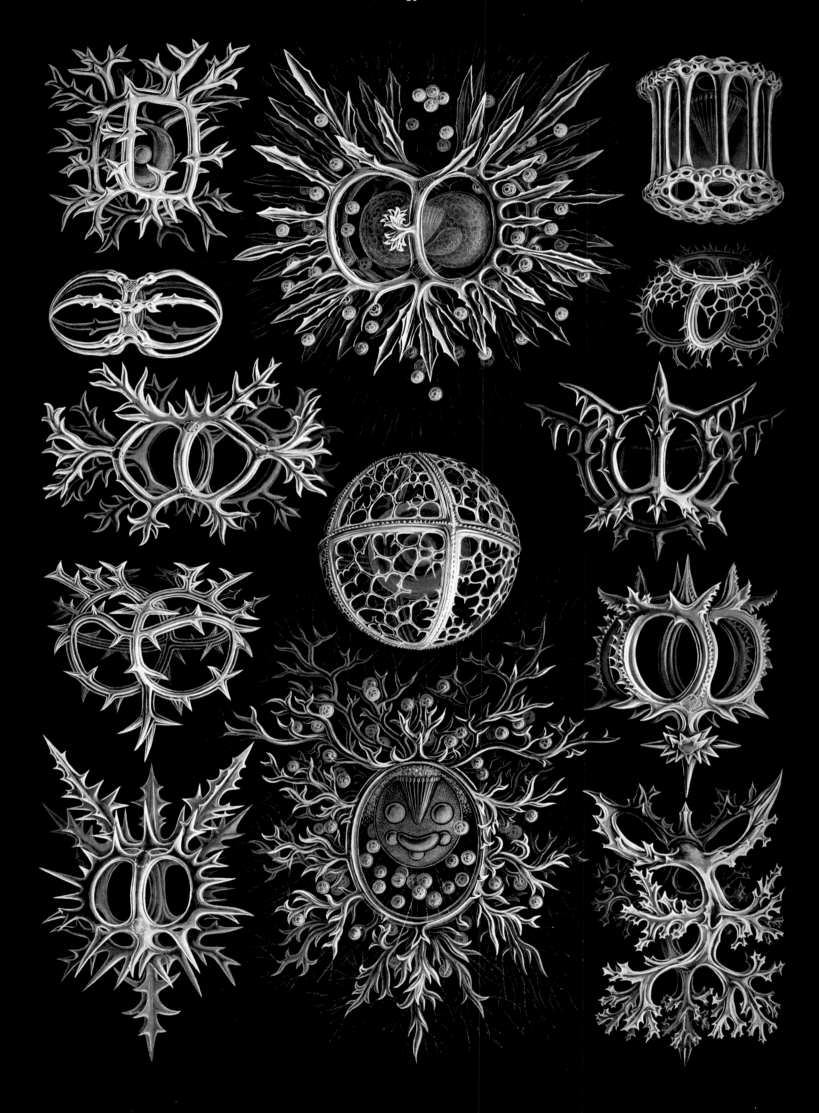

CILIATES, TINTINNIDS, AND CHOANOFLAGELLATES
MOTILITY AND MULTICELLULARITY

Among unicellular organisms, cilia and flagella are omnipresent. Cilia also exist in many human and other animal cells, and everyone knows that a sperm cell is propelled by its tail, a flagellum. All cilia and flagella share a multitubular internal structure, remarkably conserved throughout evolution. They receive and transmit signals from the environment and create the currents necessary for cells to move and feed. When conditions are favorable for their growth, several million ciliated and flagellated protists can be found in a single liter of seawater.

Some 10,000 species of ciliates inhabit the ocean and wetlands. They measure between 10 and 100 microns, with tiny hairs arranged in a corolla around a mouth, called the cytostome. On the cell surface, rows of cilia beat with coordinated movements. Ciliates and flagellates play an essential role in the food chain, grazing on the smaller protists and bacteria. In return, they are prey for larger protists and zooplankton.

Among the myriad species of ciliates in the ocean, tintinnids are the most recognizable, and some would say beautiful, due to their distinctive tunics called loricae, a name borrowed from the armor of Roman soldiers. Built with an armature of protein, loricae resemble trumpets, amphoras, or vases, and some are lavishly decorated with particles.

The ciliated protist attaches to the bottom of its tunic. It can stretch itself to extend a crown of cilia outside the lorica in order to move around. Ciliary motion also creates currents that gather prey near its mouth. If there is a disturbance, the ciliate retracts quickly inside its lorica. Tintinnids have several nuclei and can exchange genes by conjugation—a temporary union of compatible individuals. They divide by splitting in two.

Choanoflagellates are protists characterized by the presence of a flagellum surrounded by a collar, or corolla, of contractile muscle fibers. Flagellar movement creates a current, bringing bacteria to the mouth where the corolla snares the prey. About a hundred marine species of choanoflagellate have been identified, and some secrete an elegant shelter or lorica. Choanoflagellates strongly resemble the choanocyte feeding cells of the ancient animals, sponges. From an evolutionary point of view, this similarity suggests that choanoflagellates could be the sister group of the animal kingdom.

Choanoflagellates may reveal clues to the evolution of the multicellular lineages that characterize animals and plants. For example, small individual cells of the species *Salpingoeca rosetta* create colonies by sticking together at the side opposite their flagellum to form a rosette. This ability to create colonies is controlled by molecules secreted by bacteria. Could it be that bacteria in the ocean presided over the emergence of the first animals?

CILIATED TINTINNID INSIDE ITS LORICA

This tintinnid, *Rhabdonella spiralis*, was discovered in 1881 by Hermann Fol, one of the founders of the marine station at Villefranche-sur-Mer.

Left: The animal occupies the bottom of its lorica.

Right: The animal has moved up to the opening and deployed its ciliated corolla.

PHOTOS BY JOHN DOLAN, CNRS, OBSERVATOIRE OCÉANOLOGIQUE DE VILLEFRANCHE-SUR-MER.

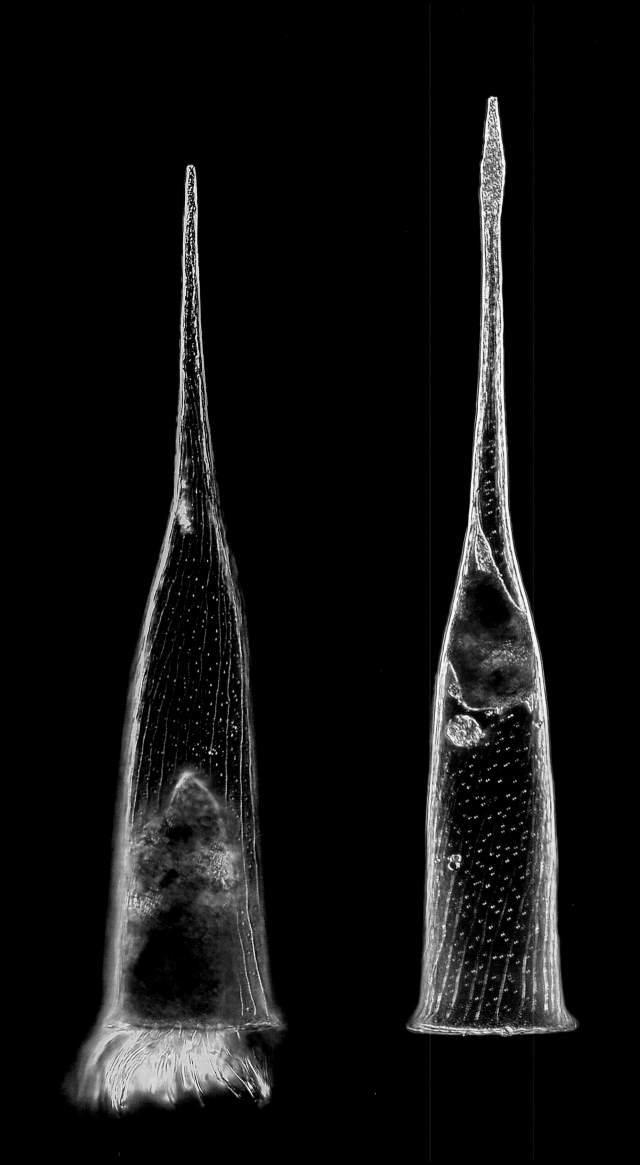

DECORATED LORICA

The lorica of the tintinnid *Codonellopsis orthoceras* measures 100 microns.

The lower part of the vase-shaped lorica is covered by hundreds of calcareous scales (coccoliths) originating from at least five species of coccolithophores. The upper part is the organic structure secreted by the tintinnid.

COLLECTED IN THE BAY OF VILLEFRANCHE-SUR-MER. SCANNING ELECTRON MICROGRAPH BY IMÈNE MACHOUK & CHARLES BACHY, CNRS PHOTOTHÈQUE.

Tintinnids Build and Decorate Their Lorica

Different species of tintinnids construct lorica as shelters. Each species has a characteristic form of lorica. Some look like amphoras, opaque and decorated: (1) *Codonellopsis schabi*, (2) *Dictyocysta lepida*, (3) *Stenosomella ventricosa*. Others resemble transparent tubes: (4) *Xystonella lohmanni*, (5) *Salpingella acuminata*, and (6) *Rhizodomus tagatzi*. The lorica likely protects the organism from predation.

PHOTOS BY JOHN DOLAN, CNRS, OBSERVATOIRE OCÉANOLOGIQUE DE VILLEFRANCHE-SUR-MER.

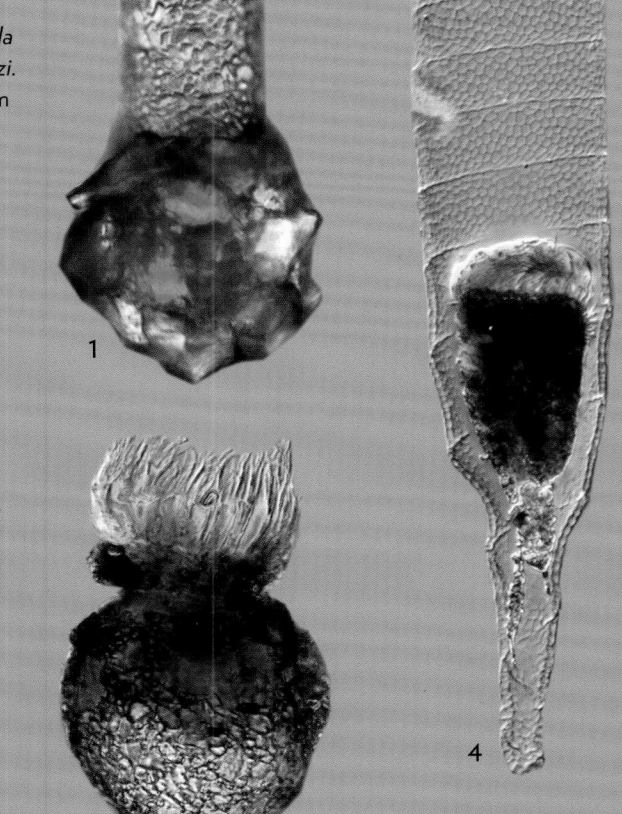

1

2

3

4

5

6

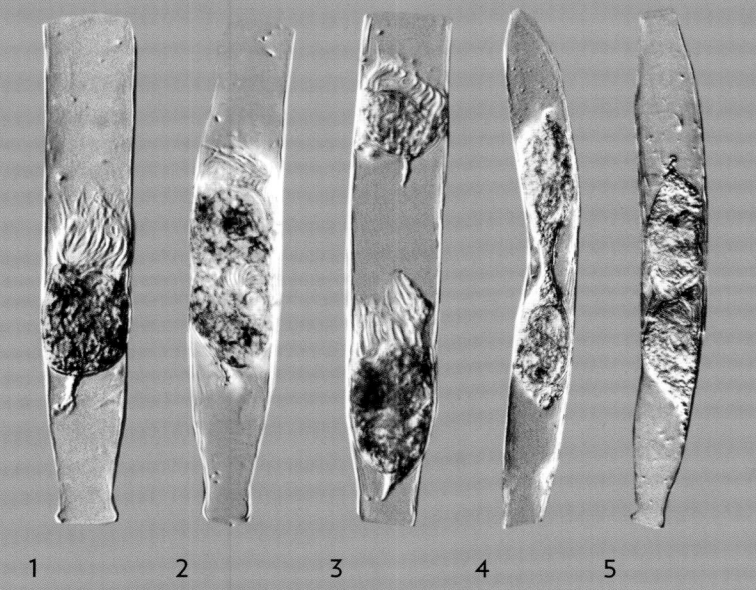

1   2   3   4   5

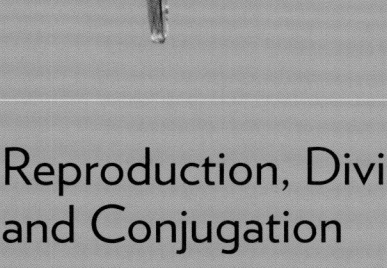

Reproduction, Division and Conjugation

The life cycle of this tintinnid, *Eutintinnus inquilinus*, can be seen through its transparent lorica:

1. The tintinnid with its corolla of cilia inside its lorica.

2. It begins to reproduce, splitting into two individuals.

3. It has divided. The top cell will emerge to build a lorica.

4. An early stage of conjugation between two individuals.

5. The two individuals exchange genetic material through a cytoplasmic bridge.

PHOTOS BY JOHN DOLAN, CNRS, OBSERVATOIRE OCÉANOLOGIQUE DE VILLEFRANCHE-SUR-MER.

89

Choanoflagellates with or without Lorica

Choanoflagellates are small protists, at most a few microns in size. There are three main families comprising about 150 species, primarily marine. The choanoflagellates of the Salpingoecidae and Anthocidae families manufacture simple shells or loricae covering the outside of the cell. In the Anthocidae family, loricae are reinforced by ribs made of silica, as in this *Plathypleura infundibuliformis*. This siliceous structure (about 10 microns in size) filters particles attracted by the flagellum's motion. The organism itself remains sheltered at the bottom of its lorica.

COLLECTED IN THE GULF STREAM, OFF THE FLORIDA COAST. ELECTRON MICROGRAPH BY PER FLOOD, BATHYBIOLOGICA.

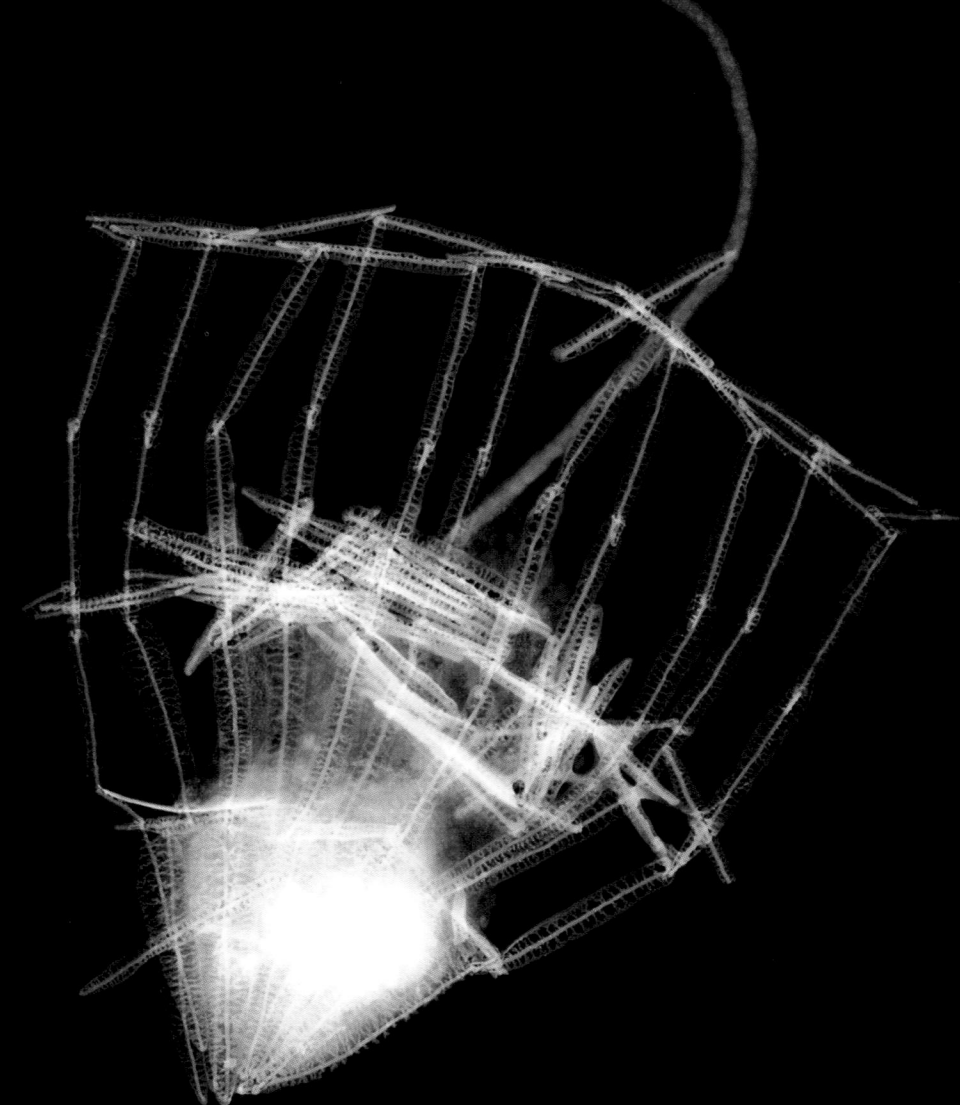

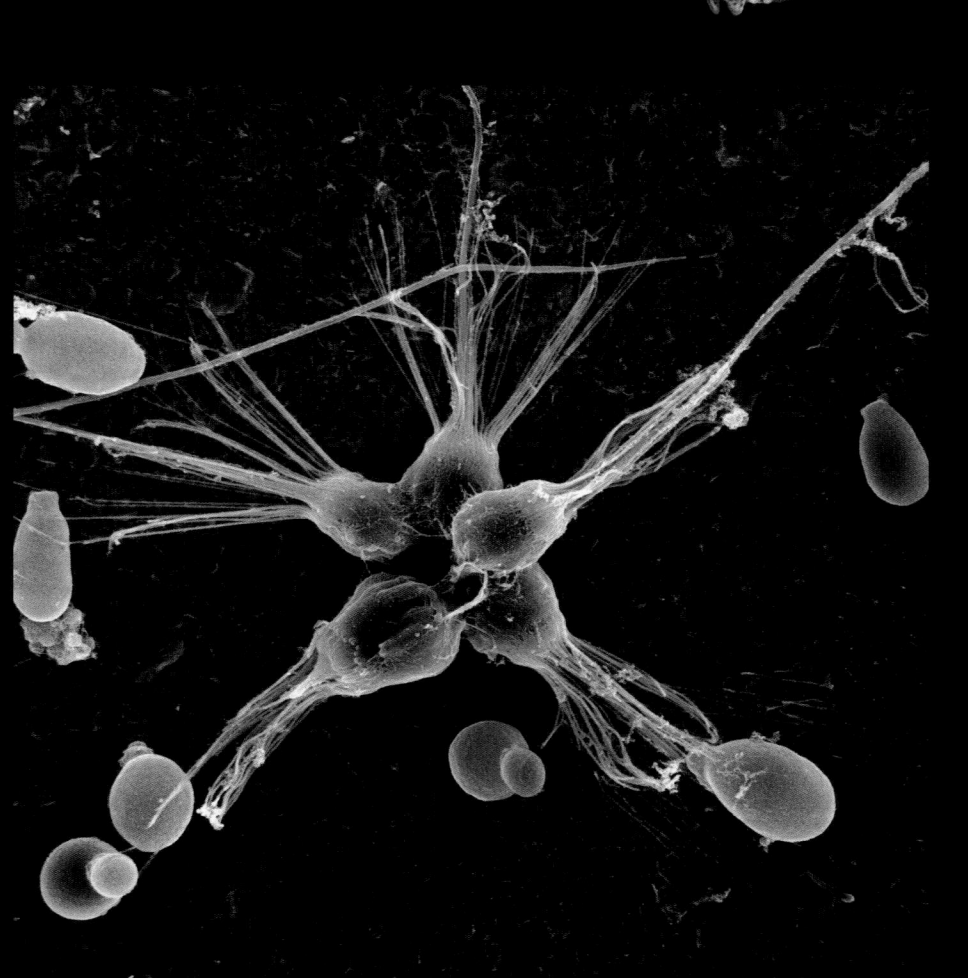

An Origin of Multicellularity

Cells of choanoflagellates are characterized by a corolla of microvilli surrounding a long flagellum. Microvilli are dynamic due to the presence of actin proteins resembling those in our muscles. The corolla/microvilli structure serves for movement and feeding. Individuals are solitary, but sometimes attach themselves to a substratum (opposite page), or gather in colonies. Their colonial habit may have led to the multicellular state that defines plants and animals.

Left: After division, the cells of the species *Salpingoeca rosetta* remain joined at the oval end, opposite the flagellum. The small cells colorized in yellow are budding yeast, measuring 2 microns.

SCANNING ELECTRON MICROGRAPH BY MARK DAYEL, MARK@DAYEL.COM.

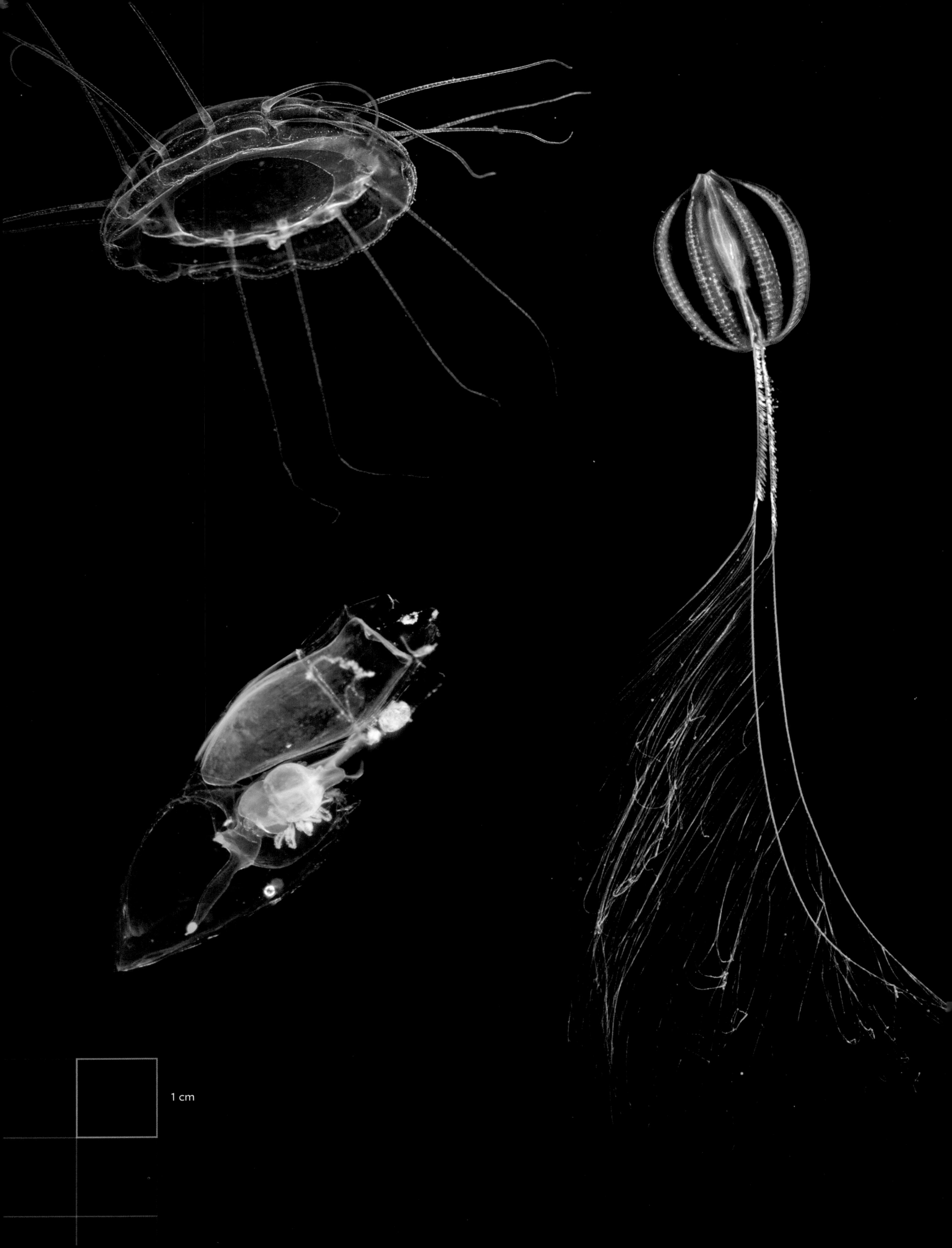

1 cm

CTENOPHORES AND CNIDARIANS

Ancestral Forms

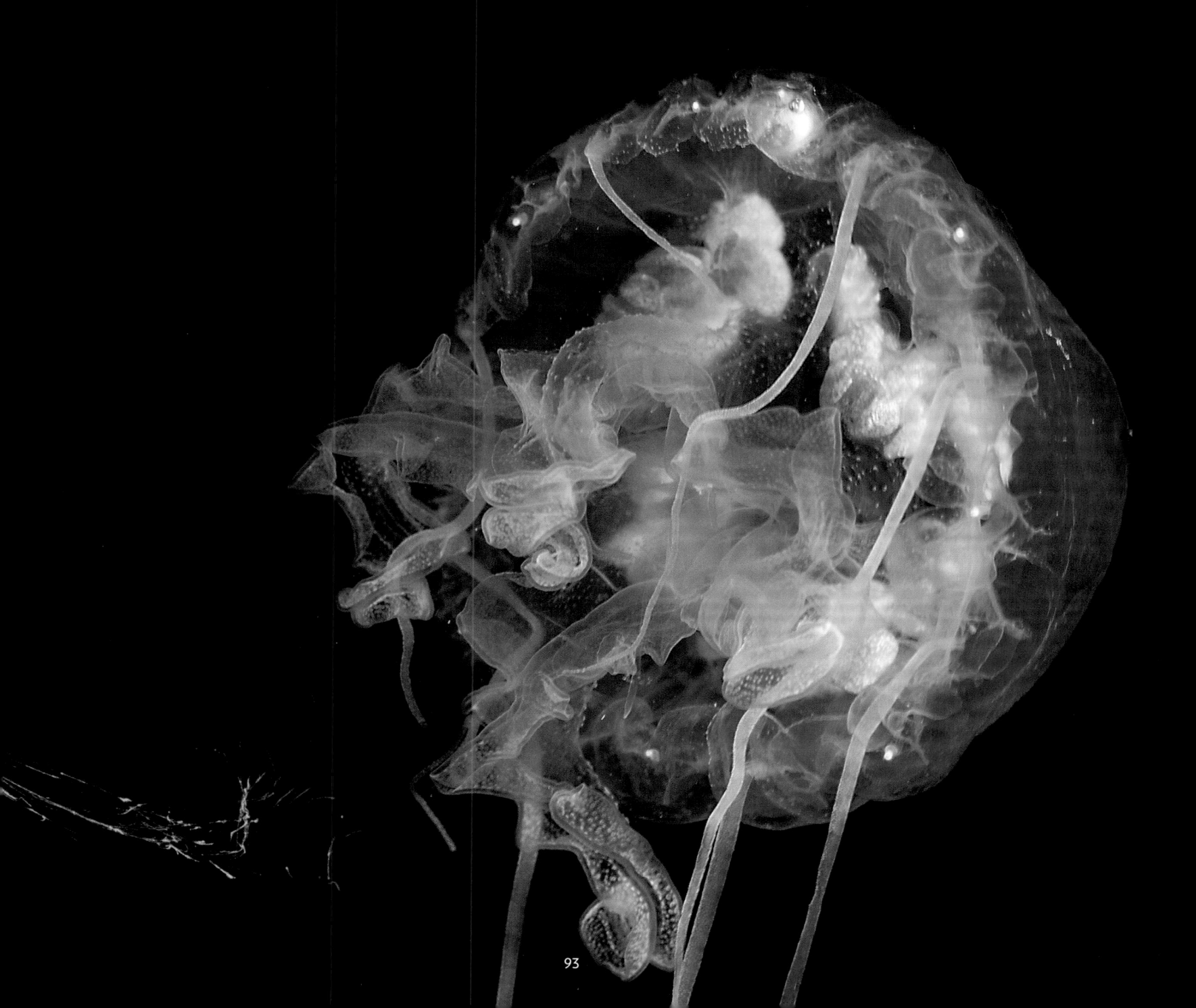

CTENOPHORES
CARNIVOROUS COMB JELLIES

Almost all of the 200 known species of ctenophores are planktonic, found in the open ocean. Many of these roam the depths of the abyss. They are often mistaken for jellyfish since most bear tentacles on gelatinous, transparent bodies. But ctenophores have a distinguishing feature—eight rows of "comb plates" consisting of thousands of tightly arrayed cilia. When the cilia move, they diffract light, like prisms, in iridescent waves of rainbow colors. The word *ctenophore* comes from the Greek *ctenos* ("comb") and *phoros* ("carrier"). The cilia in the comb plates propel the animal in quick, acrobatic movements. Champion of this sport is the elegant and slender *Cestum veneris*, nicknamed "Venus girdle."

A ctenophore's sense of balance and orientation comes from a small dome-shaped organ called a statocyst, located on the opposite side of its body from its mouth. The statocyst uses small grains of calcium carbonate to detect gravity, then communicates the information to a simple network of sensory neurons. This elementary nervous system controls a range of behaviors, including swimming, predation, and reproduction. The ctenophore's statocyst is analogous to our otolith, the sensory organ in the inner ear of humans that controls our balance.

Unlike jellyfish that have stinging cells on their tentacles, ctenophores use glue. Their tentacles are covered with sticky cells called colloblasts that ensnare their prey. *Pleurobrachia*, a comb jelly commonly called "sea gooseberry" or "cat's eyes," constantly extends and contracts two long-branched tentacles to catch food. *Leucothea*, a larger and slower comb jelly, relies on troll fishing. It maneuvers two large lobes to

Photo by Christoph Gerig

concentrate prey near its sticky tentacles. Some comb jellies, such as *Beroe*, are naked with no tentacles. They hunt rather than snare their prey. Equipped with sharp teeth made of packed cilia, *Beroe* attack by stretching their mouths wide open, biting, and engulfing ctenophores much bigger than themselves.

When it comes to reproduction, ctenophores are hermaphrodites—female and male at the same time. Each individual produces and stores both eggs and sperm in eight separate channels situated beneath the eight rows of comb plates. Every day the animal releases large numbers of male and female gametes that disperse into the sea. At fertilization, several sperm may penetrate the egg, but only one male nucleus fuses with the female nucleus. Comb jellies develop rapidly: it takes only a few days for the fertilized egg to become a *Pleurobrachia* larva already equipped with sticky tentacles. As for the young *Beroe*, it immediately begins biting into other ctenophores, its favorite prey.

Plankton Chronicles website
Ctenophores

VENUS GIRDLE
Contortions of *Cestum veneris*.
Photographed by Christoph Gerigk while diving near the Galapagos Islands during the *Tara Oceans Expedition*.

Ocyropsis maculata is a ctenophore with two
fleshy lobes bearing dark brown spots.

PHOTOGRAPHED OFF THE ATLANTIC COAST OF THE UNITED STATES
BY CASEY DUNN, BROWN UNIVERSITY, USA.

Juvenile *Beroe ovata.* This naked
ctenophore is often cannibalistic,
hungrily devouring other ctenophores.

COLLECTED IN THE BAY OF VILLEFRANCHE-SUR-MER AND
PHOTOGRAPHED BY CLAUDE CARRÉ, UPMC.

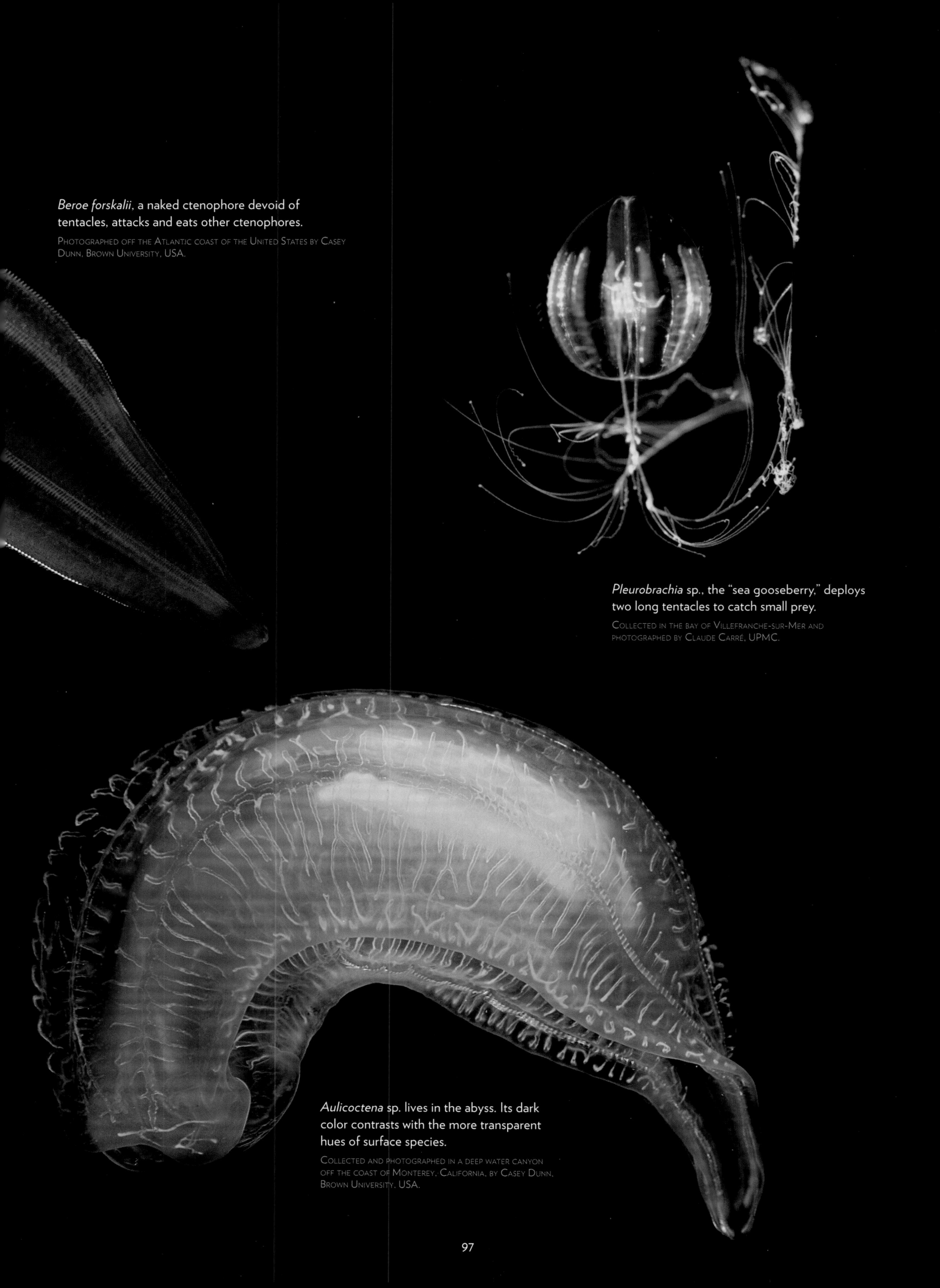

Beroe forskalii, a naked ctenophore devoid of tentacles, attacks and eats other ctenophores.

PHOTOGRAPHED OFF THE ATLANTIC COAST OF THE UNITED STATES BY CASEY DUNN, BROWN UNIVERSITY, USA.

Pleurobrachia sp., the "sea gooseberry," deploys two long tentacles to catch small prey.

COLLECTED IN THE BAY OF VILLEFRANCHE-SUR-MER AND PHOTOGRAPHED BY CLAUDE CARRÉ, UPMC.

Aulicoctena sp. lives in the abyss. Its dark color contrasts with the more transparent hues of surface species.

COLLECTED AND PHOTOGRAPHED IN A DEEP WATER CANYON OFF THE COAST OF MONTEREY, CALIFORNIA, BY CASEY DUNN, BROWN UNIVERSITY, USA.

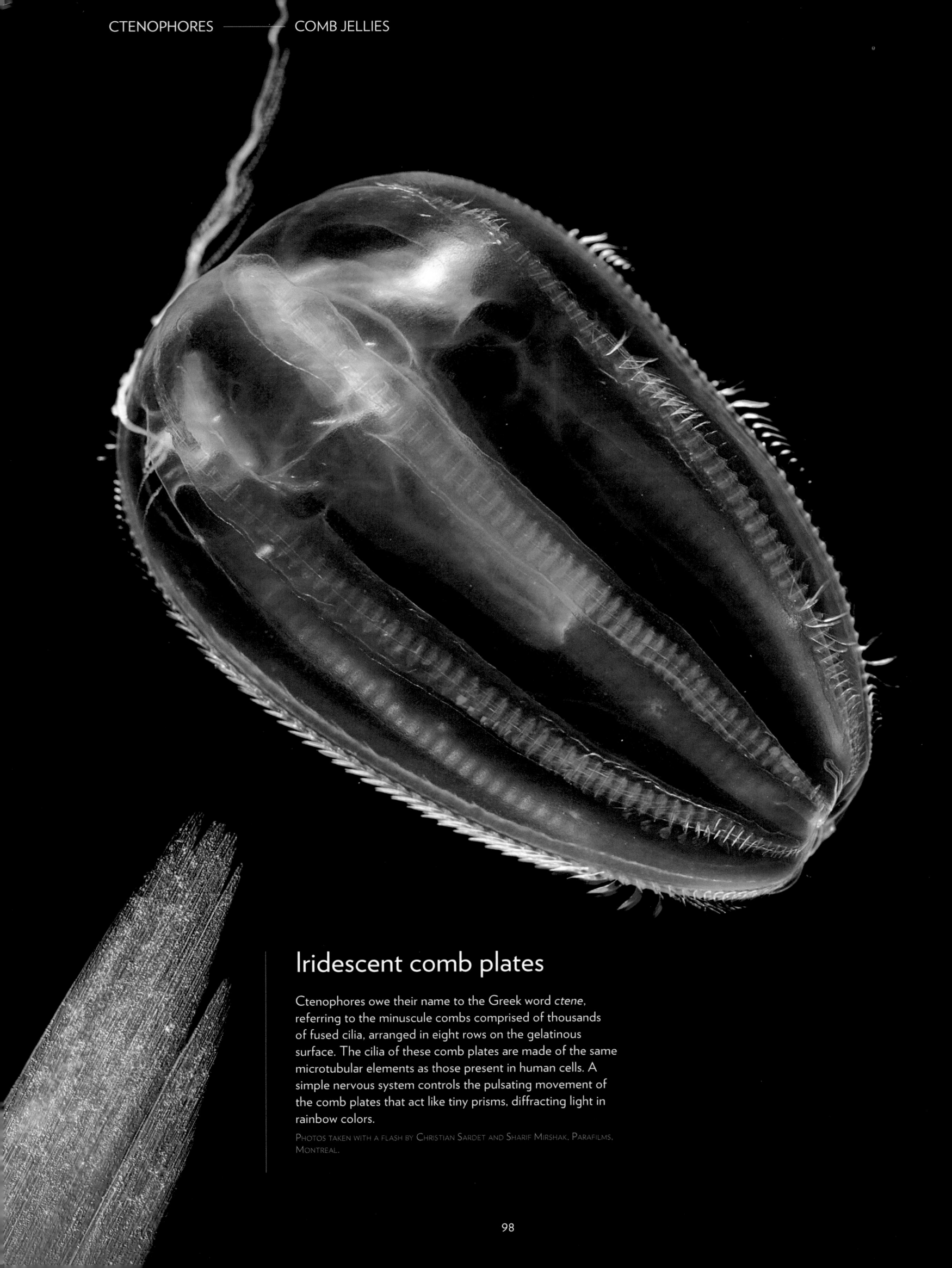

Iridescent comb plates

Ctenophores owe their name to the Greek word *ctene*, referring to the minuscule combs comprised of thousands of fused cilia, arranged in eight rows on the gelatinous surface. The cilia of these comb plates are made of the same microtubular elements as those present in human cells. A simple nervous system controls the pulsating movement of the comb plates that act like tiny prisms, diffracting light in rainbow colors.

PHOTOS TAKEN WITH A FLASH BY CHRISTIAN SARDET AND SHARIF MIRSHAK, PARAFILMS, MONTREAL.

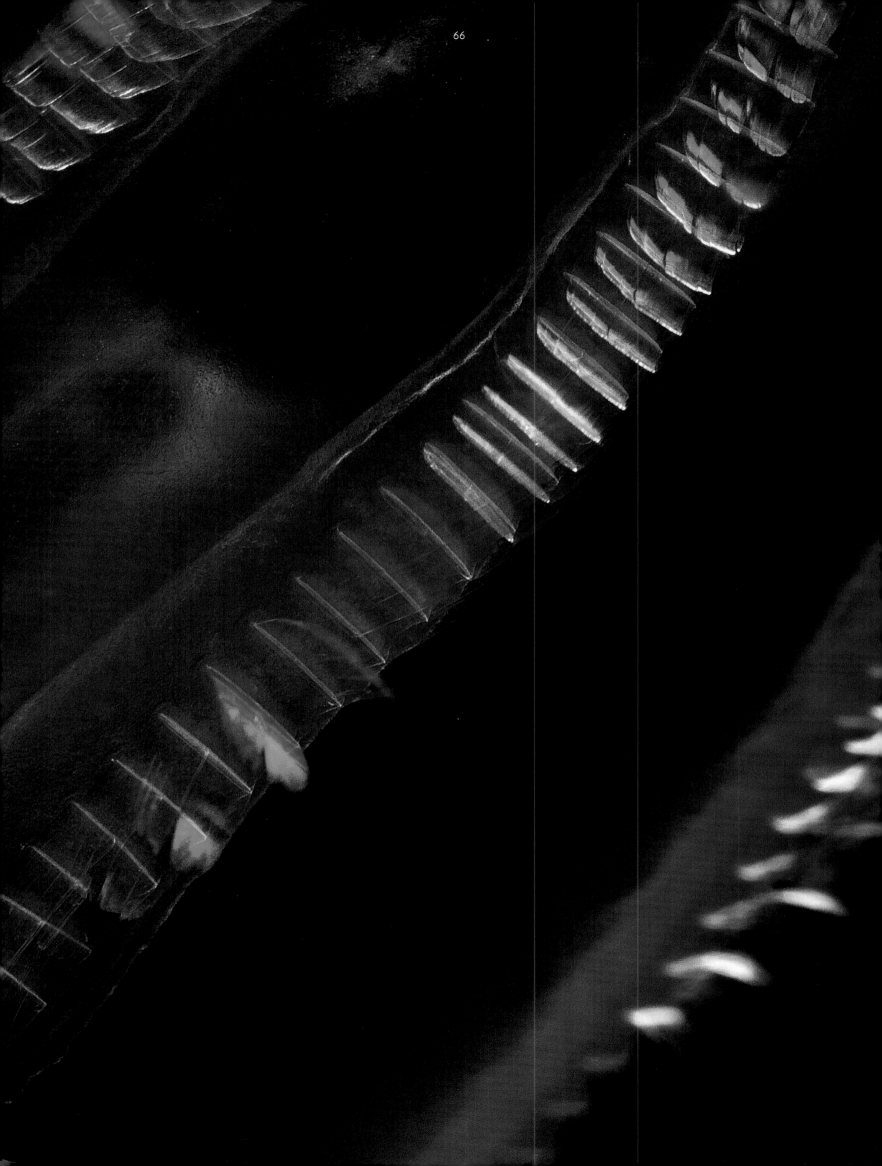

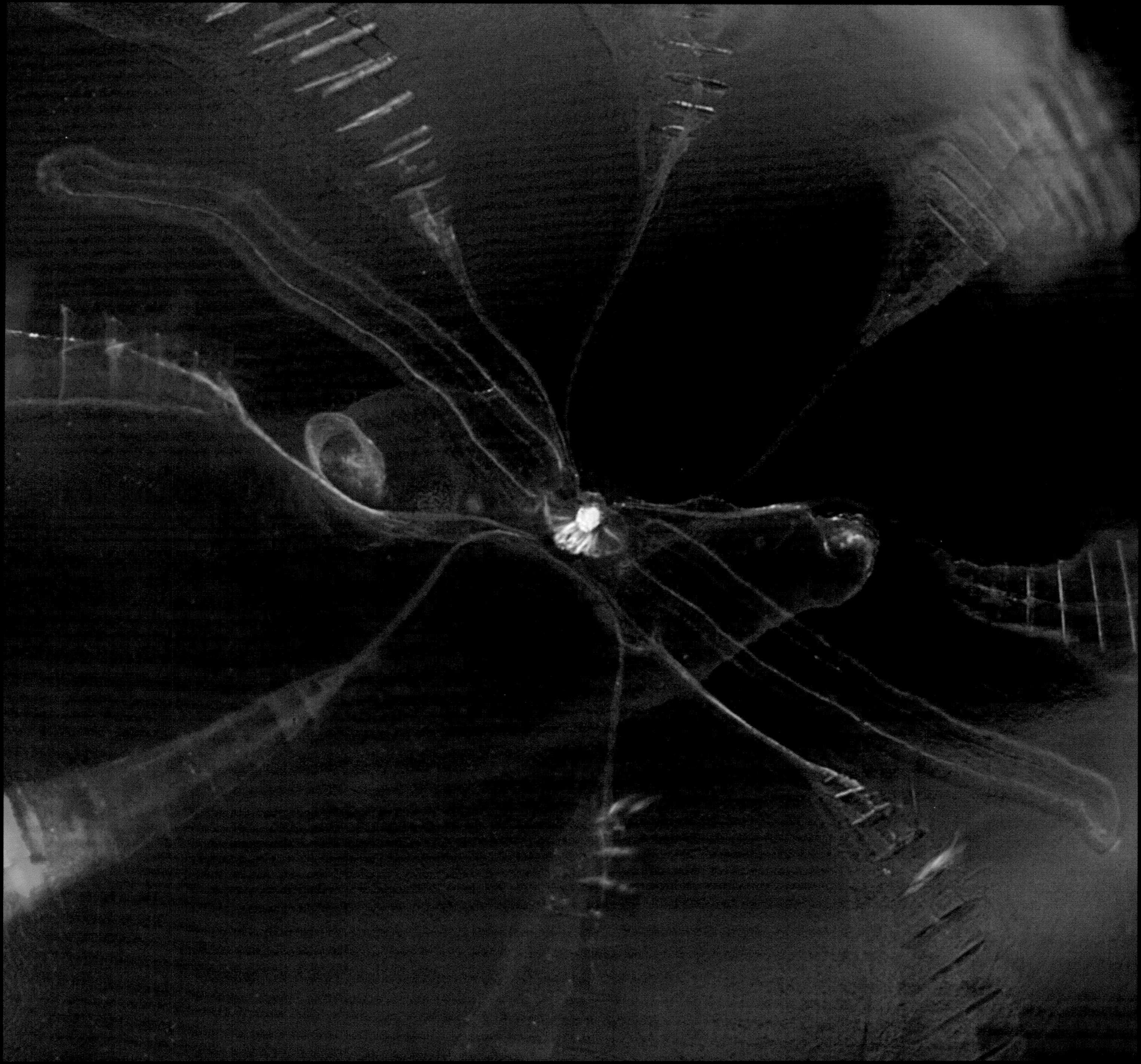

Photos by Christian Sardet and Sharif Mirshak. Parafilms. Montreal.

A Sense of Balance

Ctenophores detect their orientation and maintain balance using an organ called a statocyst. This sensory structure contains a statolith, a dome-shaped sac resting on four bundles of cilia called balancers. The statolith is filled with small grains of calcium carbonate (pictured at right). The statocyst detects movements of the calcium carbonate grains and alerts a simple nerve network that controls the motion and coordination of the comb plates.

Photos by Christian Sardet and Sharif Mirshak. Parafilms. Montreal.

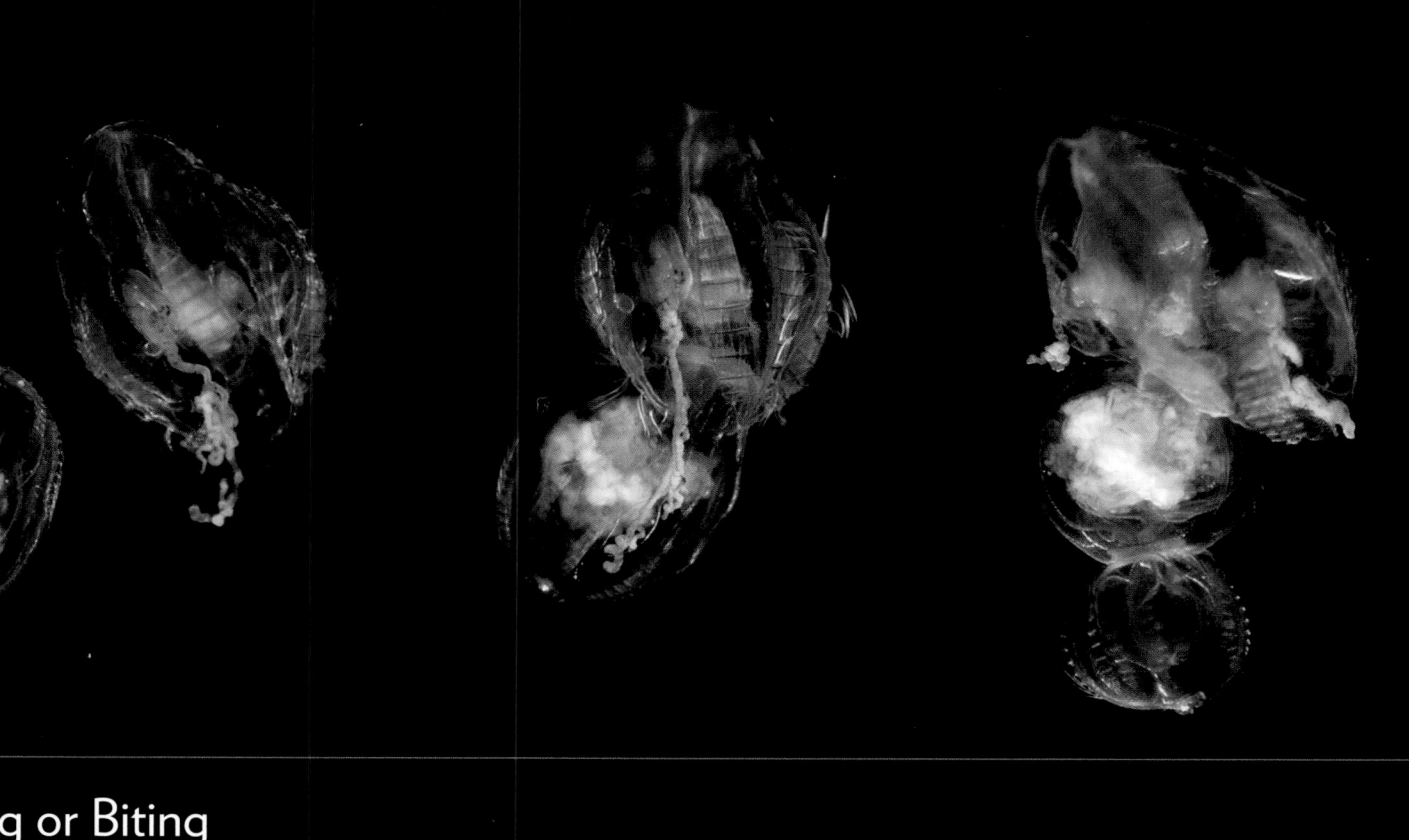

Fishing or Biting

Above: A naked ctenophore juvenile attacks a young tentaculate ctenophore bigger than itself.
FILMED AND PHOTOGRAPHED IN SHIMODA, JAPAN.

Right: A close-up of tentacles covered with colloblasts, the sticky cells with which ctenophores catch small prey.

Below: At the beginning of our filming of this scene, *Beroe ovata*, the smaller, naked comb jelly, and *Leucothea multicornis*, the larger tentaculate, were peacefully lying side by side. Suddenly, the smaller *Beroe* took a bite of the *Leucothea* and swallowed a big chunk, including the comb plates.

FILMED AND PHOTOGRAPHED IN VILLEFRANCHE-SUR-MER BY CHRISTIAN SARDET AND NOÉ SARDET, PARAFILMS, MONTREAL.

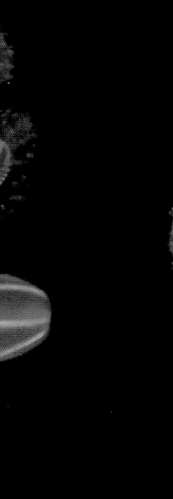

JELLYFISH

EQUIPPED TO SURVIVE

Jellyfish are the plankton that everybody knows, and most often for their bad reputation. Certain species, for example *Pelagia*, the "mauve stinger," inflict nasty burns. Others, such as *Chironex*, the famous "box jellyfish," are extremely dangerous: their stinging cells inject potent toxins that kill more people every year than sharks do. Called cnidocysts, the stinging cells are a common feature of the phylum Cnidaria, which also includes siphonophores, anemones, and corals. Jellyfish are the largest planktonic predators, along with siphonophores and ctenophores. The carnivorous jellies compete with fish and marine mammals for their common food—zooplankton.

Medusa, another name for jellyfish, can be enormous, like the *Nomura* jelly from the China Sea. Accidentally captured in nets, *Nomura* are so large and numerous their combined weight can cause fishing trawlers to overload and capsize. But the majority of the 3,500 identified jellyfish species are barely visible with the naked eye, and many are truly microscopic. This makes them practical for use in research. The leptomedusa *Clytia hemispherica* is the size of a small coin and can easily be raised in the laboratory. Biologists study their extraordinary capacities for survival and adaptation, including sexual reproduction, budding, regeneration, and even immortality in the case of the small jellyfish *Turritopsis*.

Clytia, like the vast majority of jellyfish, bud from colonies of polyps attached to algae, rocks, or shells. Polyps look like underwater flowers, with stalks topped by gelatinous bags or tentacles. The polyps with tentacles feed the colony, catching and bringing prey to their mouths. The others are reproductive polyps that produce generation after generation of baby jellyfish. After budding, tiny *Clytia* leave the colony and drift away in search of their first prey. Every day at dawn, male and female *Clytia* release sperm or eggs from little sacs under their gelatinous umbrella. After fertilization in the open sea, the eggs develop into embryos, then become ovoid-shaped larvae called planulae. Covered with a thin layer of cilia, the planulae swim and drift until they settle down to form a new colony of polyps. This new colony will eat and grow, and in turn, bud new tiny jellyfish, called ephyrae.

But not all jellyfish pass through a polyp stage. Some such as *Pelagia* and *Liriope* develop directly from embryos. Female and male *Pelagia* release eggs or sperm in huge quantities. The fertilized eggs divide into embryos that in a single day develop into planulae larvae. As each planula grows, it develops a mouth and eight lobes. Gradually, tentacles emerge, sensory organs develop, and four arms grow around the mouth. A new stinger, pink or purple in color, is born.

Plankton Chronicles website
Clytia: A laboratory favorite

Plankton Chronicles website
Pelagia: Fearsome jellyfish

LARGEST MEDITERRANEAN MEDUSA
The umbrella of *Rhizostoma pulmo* is huge, up to a meter in diameter. This scyphozoan jellyfish gets its name from the Greek words *rhiza* ("arm"), *stoma* ("mouth"), and *pulmo* ("lung"), meaning "buccal arms resembling lungs."
Photographed while diving in the bay of Villefranche-sur-Mer by Sharif Mirshak, Parafilms, Montreal.

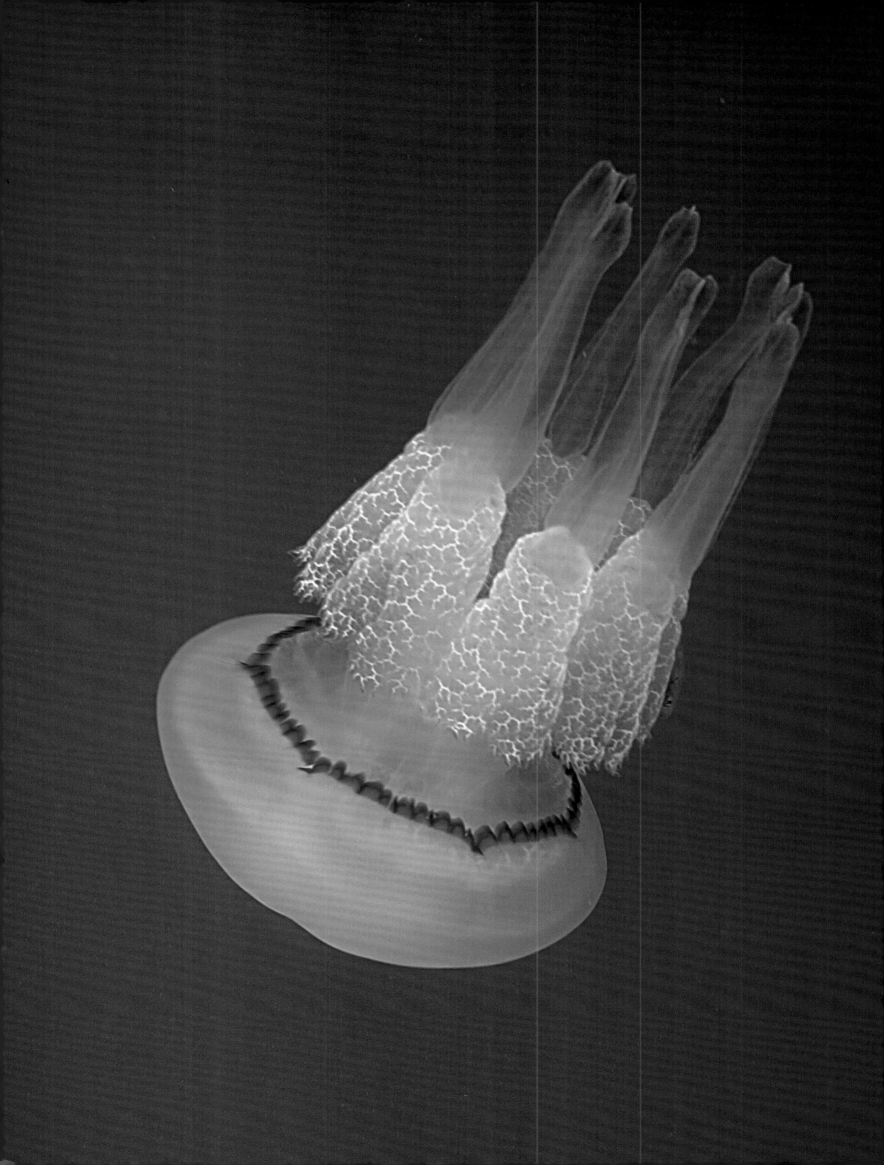

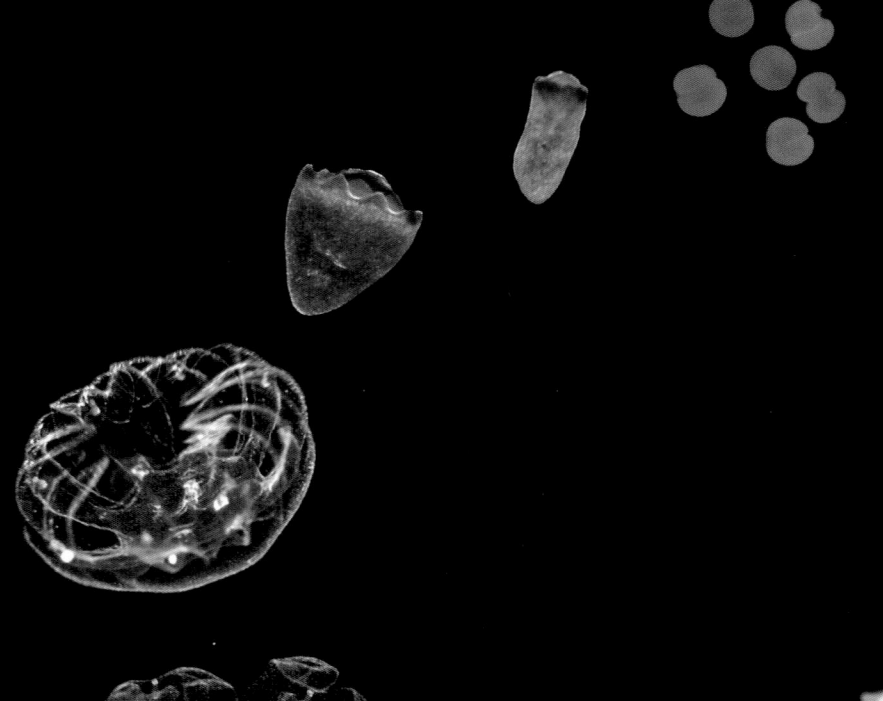
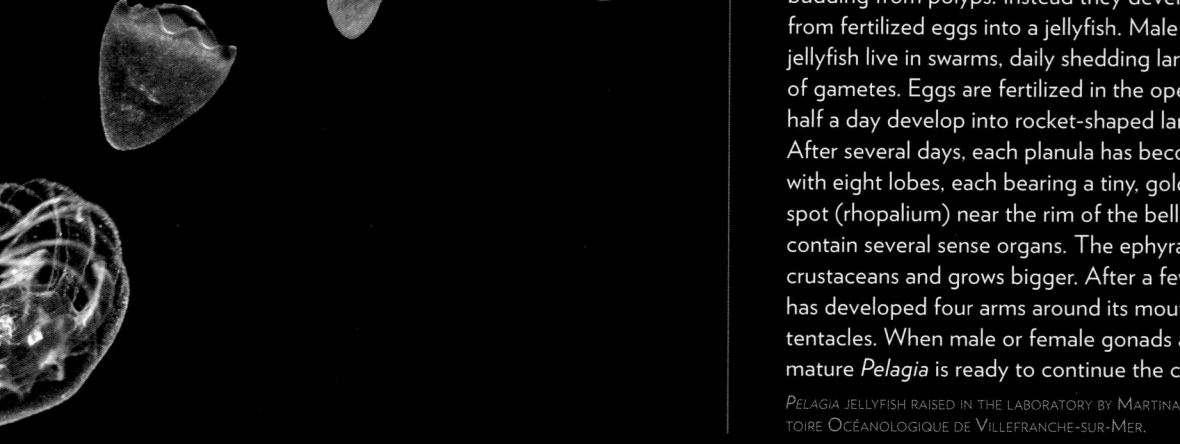

Pelagia, from Egg to Medusa

Unlike most jellyfish, *Pelagia* do not reproduce by budding from polyps. Instead they develop directly from fertilized eggs into a jellyfish. Male and female jellyfish live in swarms, daily shedding large quantities of gametes. Eggs are fertilized in the open sea and in half a day develop into rocket-shaped larvae (planulae). After several days, each planula has become an ephyra with eight lobes, each bearing a tiny, golden-colored spot (rhopalium) near the rim of the bell. The rhopalia contain several sense organs. The ephyra catches small crustaceans and grows bigger. After a few weeks, it has developed four arms around its mouth, and eight tentacles. When male or female gonads appear, a mature *Pelagia* is ready to continue the cycle.

PELAGIA JELLYFISH RAISED IN THE LABORATORY BY MARTINA FERRARIS, OBSERVATOIRE OCÉANOLOGIQUE DE VILLEFRANCHE-SUR-MER.

MEDUSA MUSCLE

The ephyra larva has a belt of muscles at the base of its eight lobes. Here we can see actin, the main protein component of all muscle fibers, revealed by a red fluorescent marker molecule.

JELLYFISH PREPARED AND PHOTOGRAPHED IN VILLEFRANCHE-SUR-MER BY REBECCA HELM, BROWN UNIVERSITY, USA.

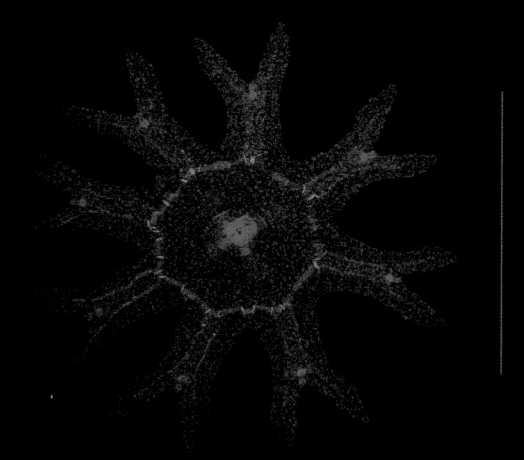

FEMALE MEDUSA

Four pink-colored female gonads surrounded by four oral arms and eight tentacles lie beneath the umbrella of *Pelagia noctiluca*. Eight small golden grains, the rhopalial lappets, situated at the periphery of the umbrella contain several sensory organs: an olfactory dimple, a visual organ (the ocellus), and a statocyst that senses gravity.

Clytia:
from Egg to
Polyp to Jellyfish

The small hydromedusa *Clytia hemisphaerica* goes through a polyp stage and buds baby *Clytia*, as shown in the sequence at the bottom of page 107.

Right: a female and male *Clytia*. Each umbrella has four gonads that release eggs or sperm (see close-up below each umbrella). Fertilization of eggs takes place in open waters. The fertilized egg divides rapidly and in a single day develops into an ovoid ciliated planula larva. The planula attaches itself to substrata. In a few more days it becomes a new colony of polyps.

JELLYFISH RAISED IN THE LABORATORY BY TSUYOSHI MOMOSE AND EVELYN HOULISTON, BIODEV LABORATORY, OBSERVATOIRE OCÉANOLOGIQUE DE VILLEFRANCHE-SUR-MER.

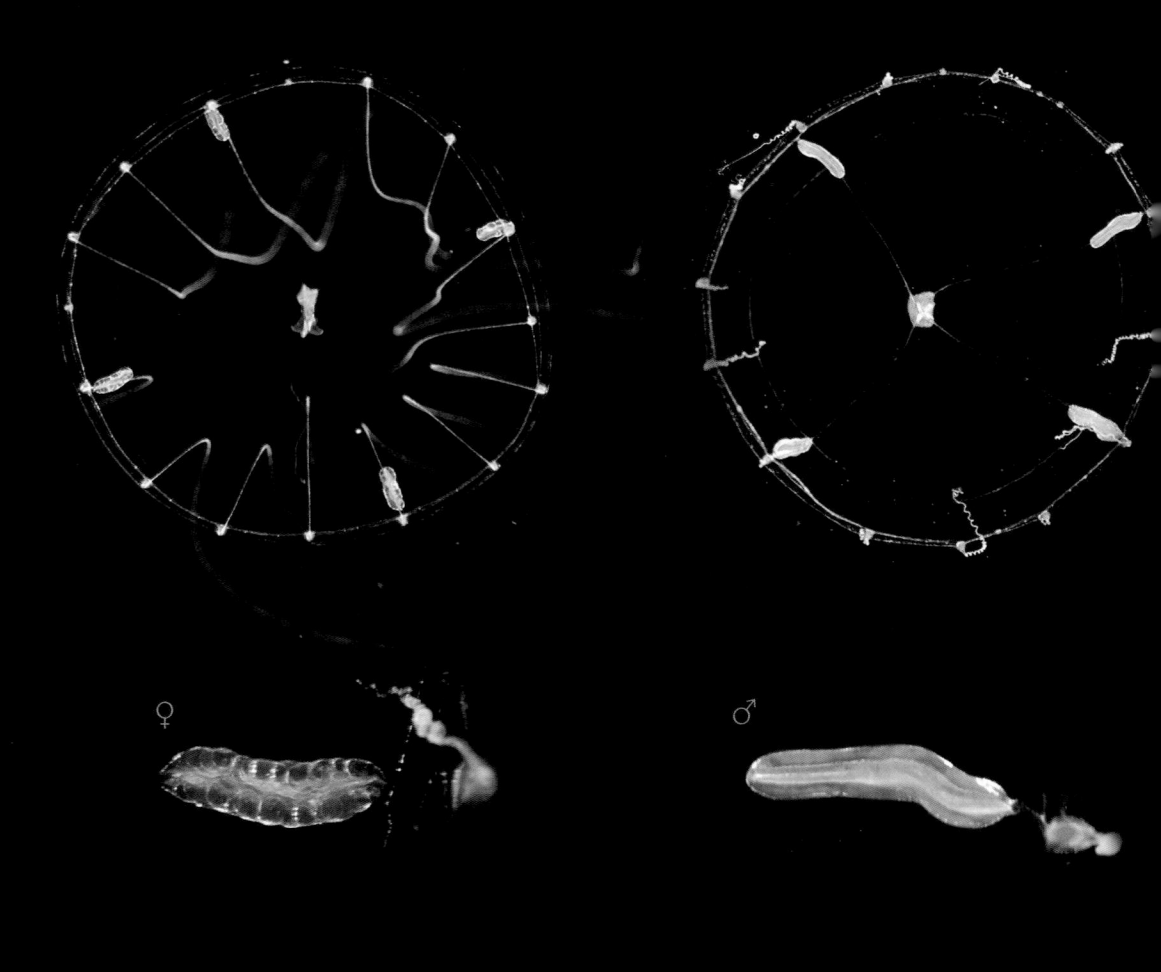

♀ ♂

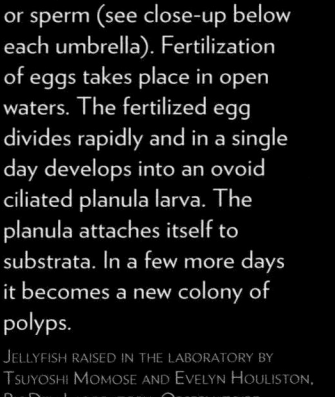

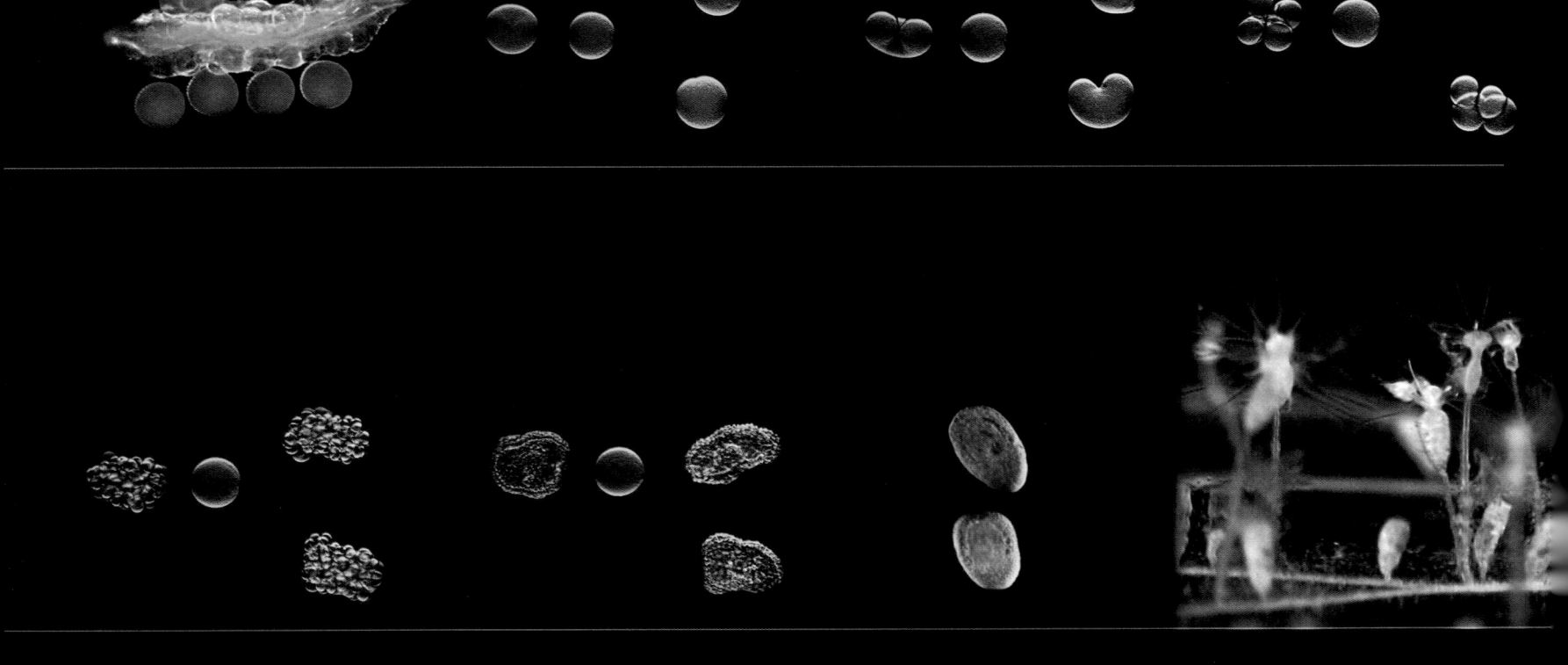

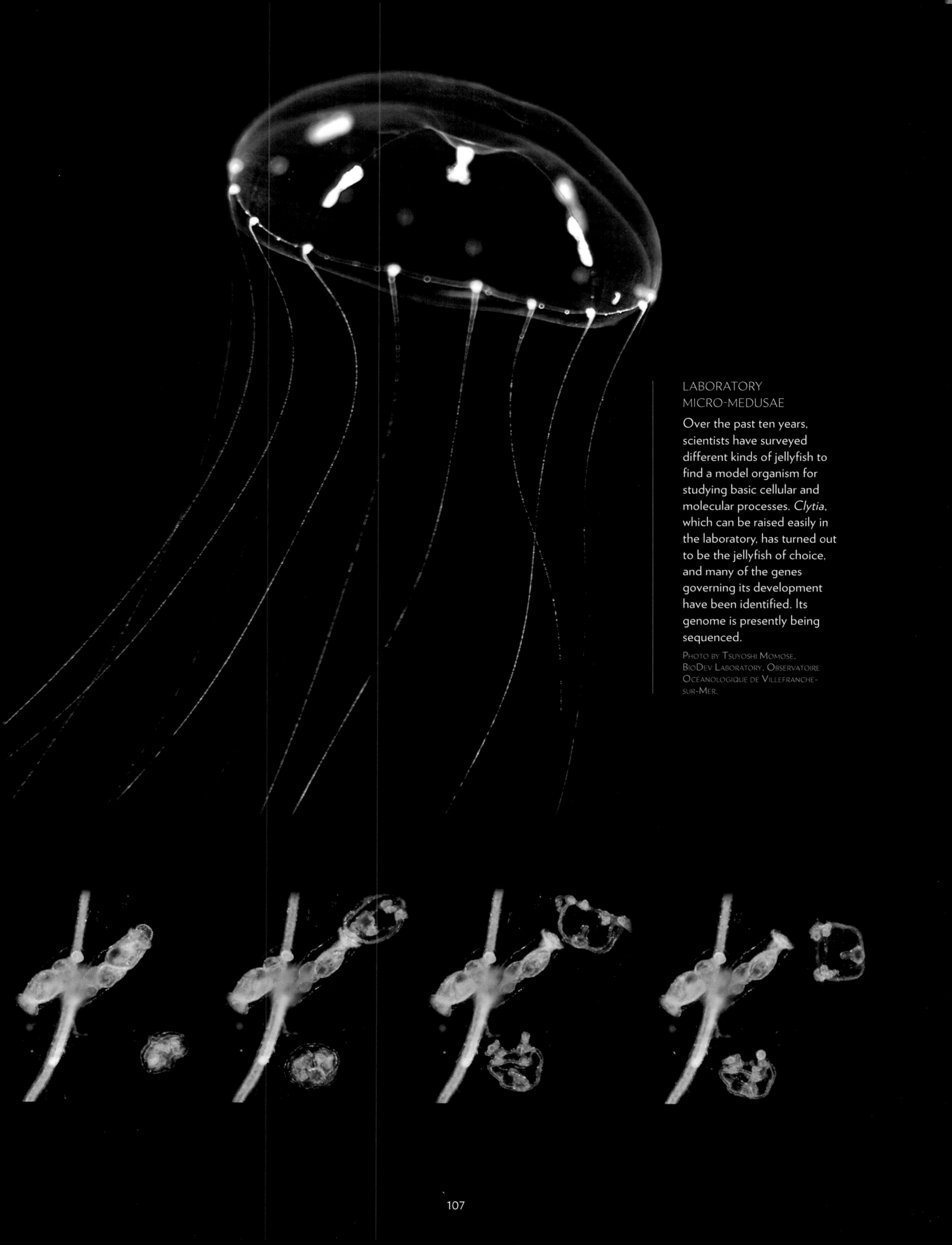

LABORATORY MICRO-MEDUSAE

Over the past ten years, scientists have surveyed different kinds of jellyfish to find a model organism for studying basic cellular and molecular processes. *Clytia*, which can be raised easily in the laboratory, has turned out to be the jellyfish of choice, and many of the genes governing its development have been identified. Its genome is presently being sequenced.

Photo by Tsuyoshi Momose, BioDev Laboratory, Observatoire Océanologique de Villefranche-sur-Mer.

A Medusa's Feast

Unlike *Pelagia*, whose mouth and stomach are under the umbrella, in *Liriope tetraphylla* the fluorescent-green mouth is located at the end of a long, flexible proboscis. For two hours one morning at the Marine Station of Nagoya University in Toba, I filmed and photographed these jellyfish capturing and ingesting a fish hatchling. Captured by the tentacle of a first *Liriope*, the small fish was soon grabbed by another *Liriope*. The victorious medusa gradually expanded its fluorescent mouth, wrapping it around the prey until it emerged on the other side. Once the juices of the small fish had been sucked through the proboscis, the *Liriope* expelled a residue.

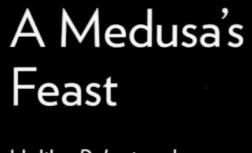

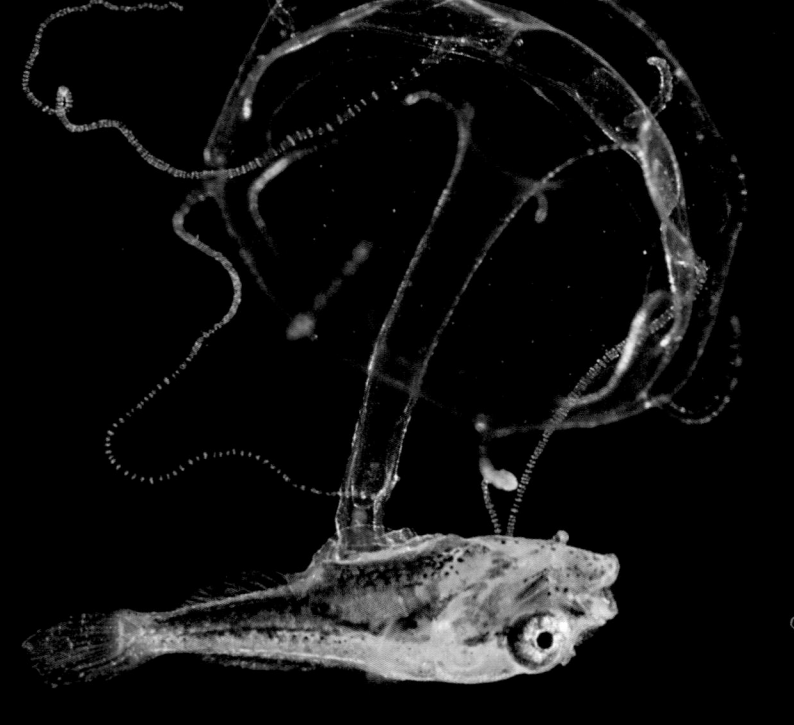

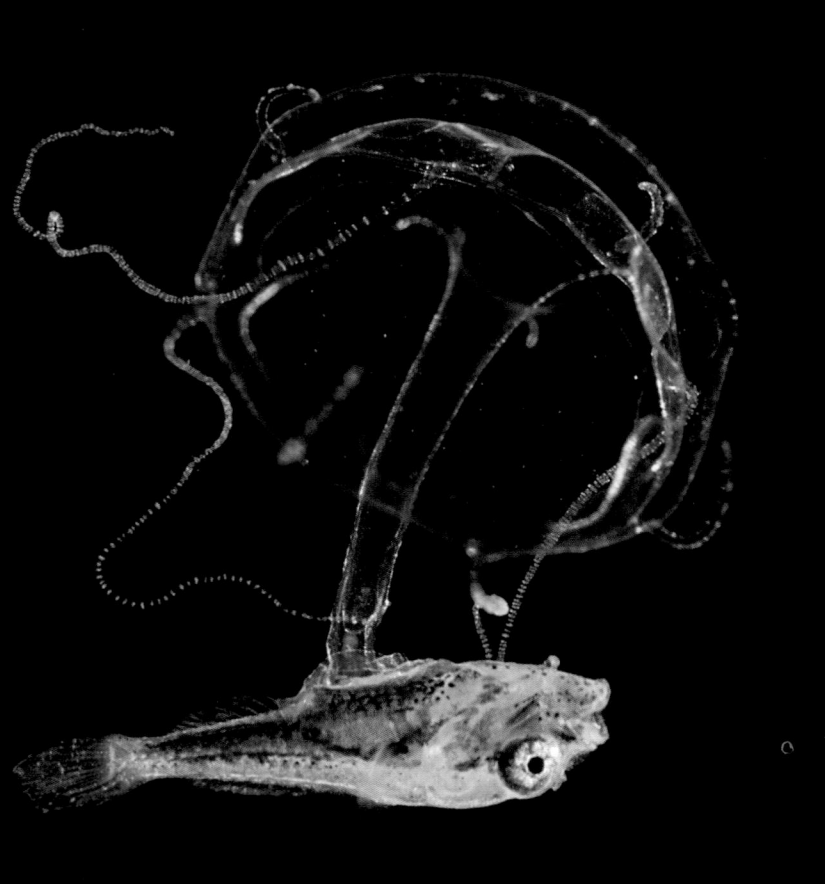

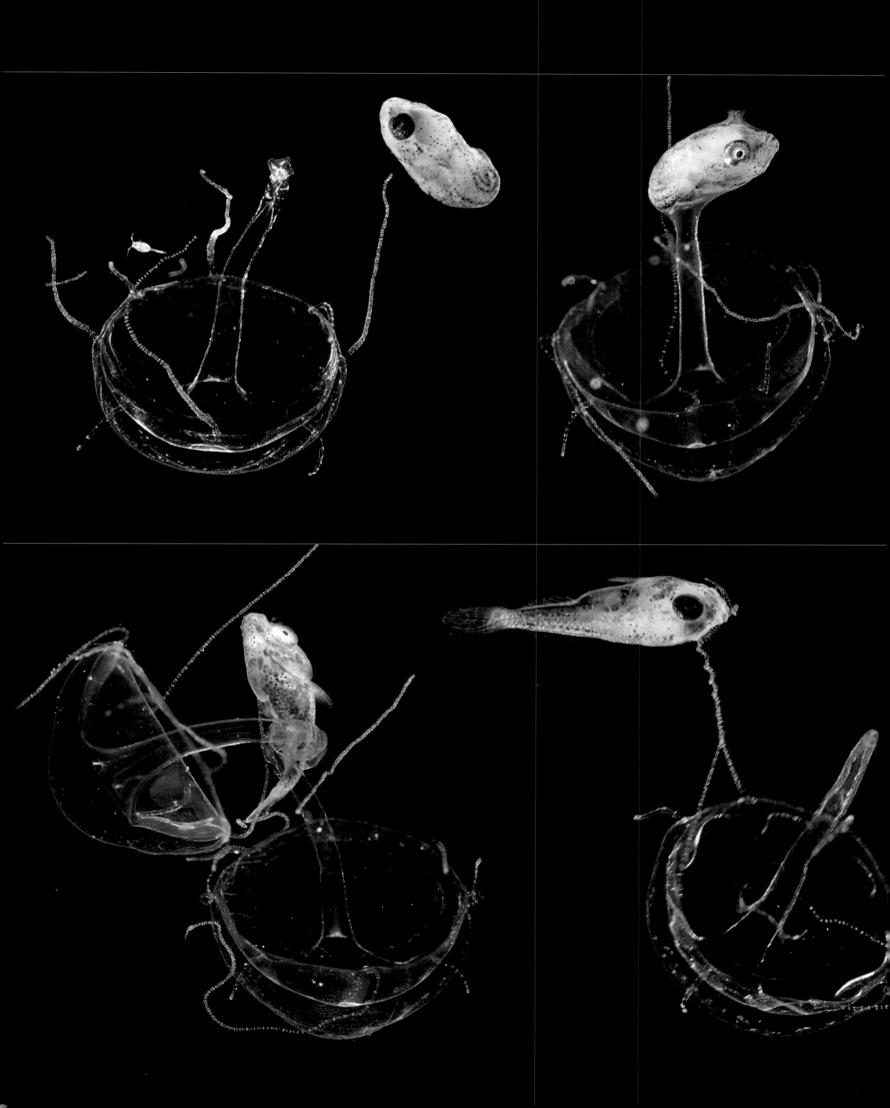

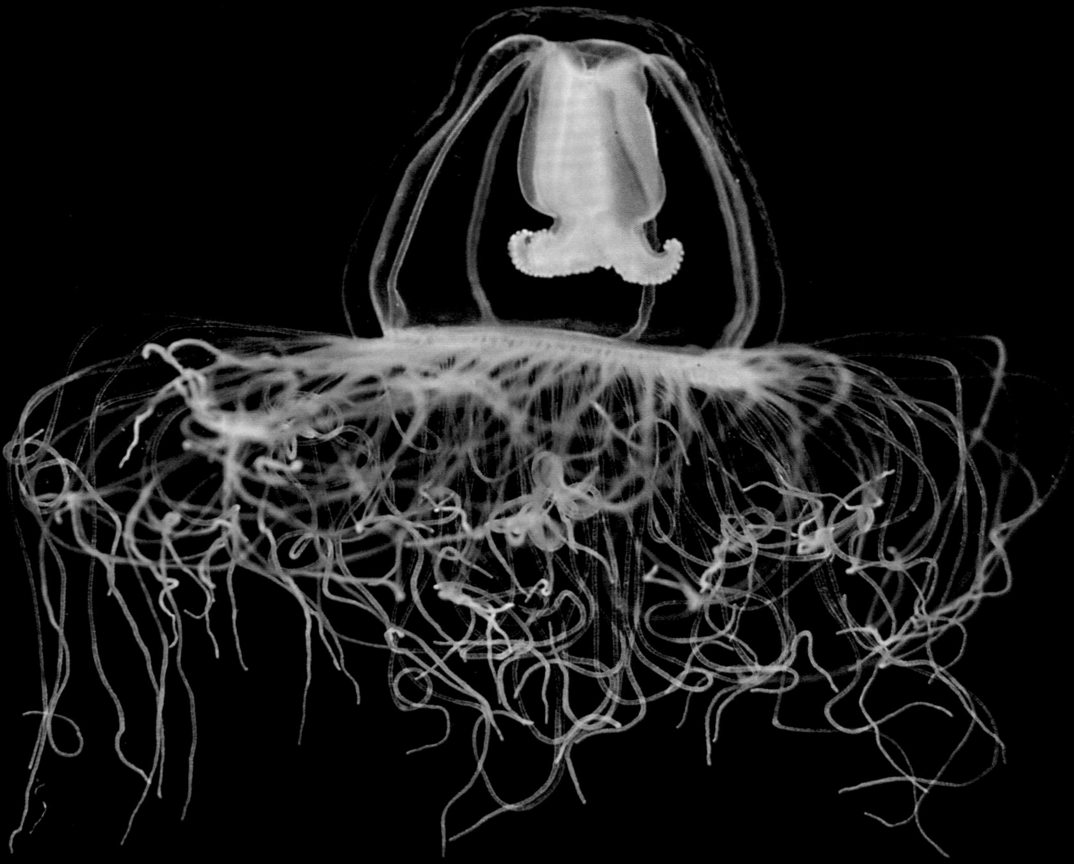

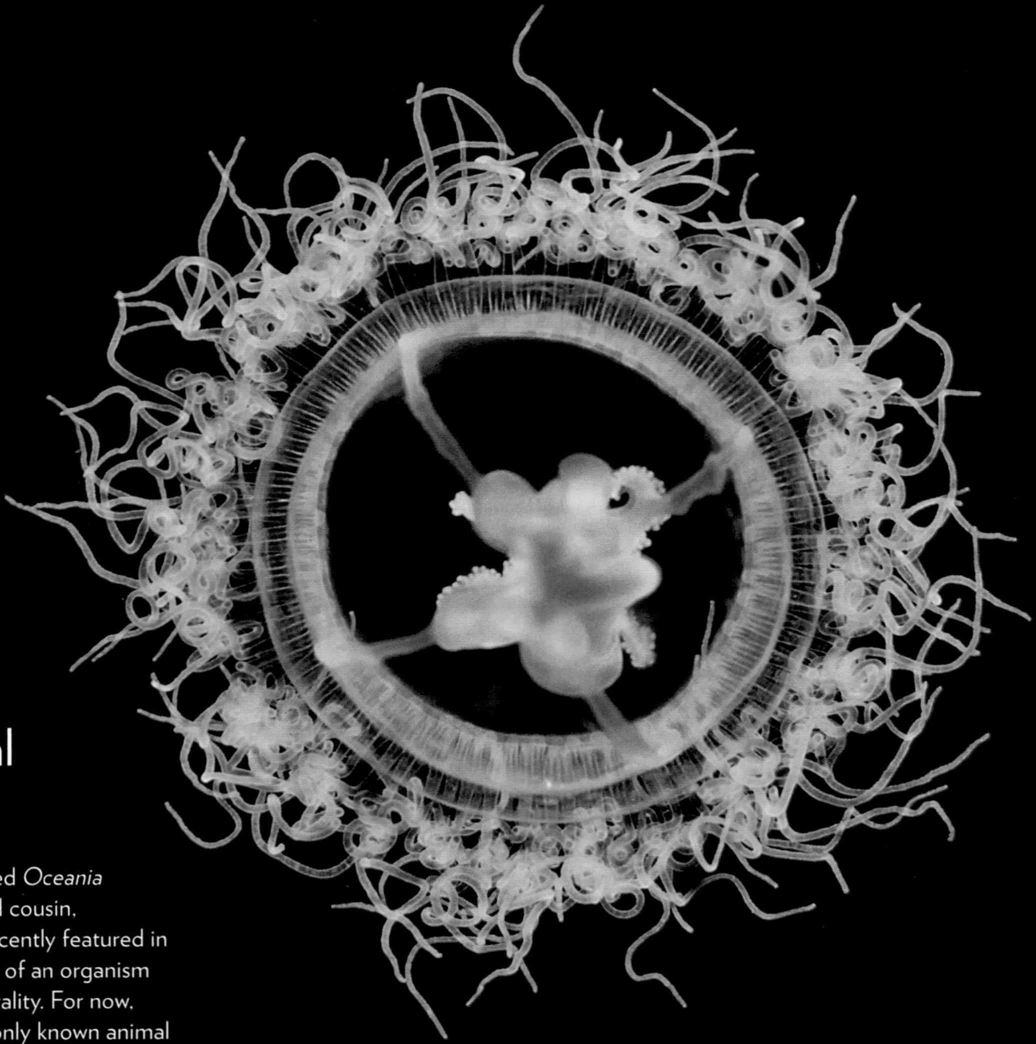

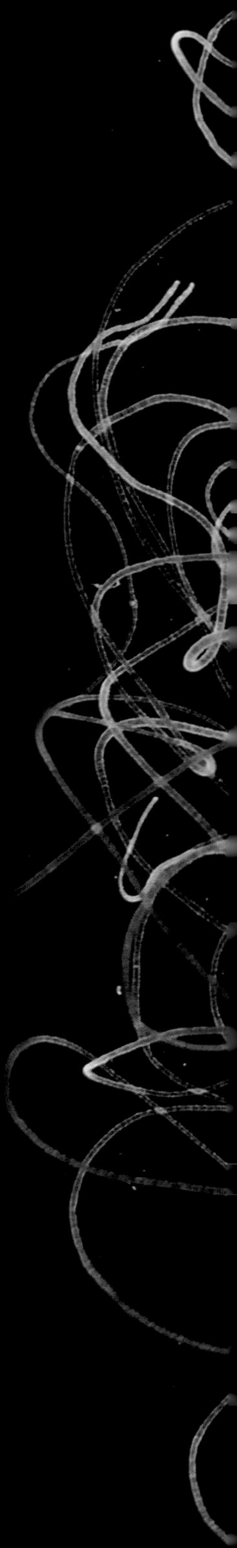

An Immortal Jellyfish?

This small jellyfish is called *Oceania armata*. Its closely related cousin, *Turritopsis dohrnii*, was recently featured in the media as an example of an organism that may achieve immortality. For now, *Turritopsis dohrnii* is the only known animal that can develop in reverse, transforming from a medusa back into a polyp.

Collected while diving by David Luquet, Observatoire Océanologique de Villefranche-sur-Mer.

SIPHONOPHORES
THE LONGEST ANIMALS IN THE WORLD

Whales are the largest animals on the planet but the longest are the siphonophores. *Praya*, a calycophoran siphonophore, and *Apolemia*, a physonect siphonophore, can both exceed 30 meters in length. Floating on the surface, the infamous *Physalia*, nicknamed "Portuguese man-of-war," trails toxic filaments tens of meters long. Rocket-shaped calycophoran siphonophores such as *Chelophyes* are much smaller, but also extend their filaments over long distances.

The 175 identified species of siphonophores are almost all planktonic and carnivorous. Many inhabit the abyss. They belong to the phylum Cnidaria, along with jellyfish, anemones, and corals. Siphonophores are colonial organisms comprised of specialized individuals called zooids. Zooids look very different from each other: some resemble polyps, others medusae. All the zooids have the same genome and develop from the same embryo, yet they perform distinct functions within the colony. This division of labor differentiates siphonophores from other colonial animals.

Siphonophores constantly bud new zooids to extend the colony or replace parts that have been shed or eaten by predators. All zooids of a colony are linked together by a stolon, a kind of umbilical cord. Arranged in repetitive sequences along the stolon, different types of zooids ensure nutrition (gastrozooids), flotation (pneumatophores), movement (nectophores), and reproduction of the colony (gonozooids). Gastrozooids feed the colony by extending long fishing filaments studded with stinging cells that harpoon and immobilize crustaceans, mollusks, larvae, and even fish. As they

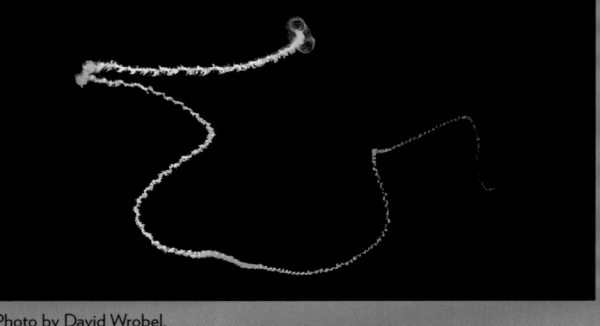

Photo by David Wrobel.

retract, the filaments bring prey into the mouths and digestive organs, supplying nutrition to the entire colony via the stolon. The gas-filled pneumatophore acts as a float, and contracting bells, the nectophores, propel the colony. Smaller calycophoran siphonophores have no float. They fuse by contracting their nectophores, releasing packets of zooids called cormidia that develop into medusa-like eudoxia bearing male and female gametes.

The male and female reproductive gonozooids resemble small jellyfish. Released in open water, they expel packages of eggs or sperm. How do these gametes meet in the vast ocean to ensure reproductive success? In fact, the eggs diffuse special molecules that attract sperm of the same species. The sperm gather around a particular site on the egg in order to fertilize it. The fertilized egg then becomes an embryo that soon buds its first zooids, perpetuating the colony and the species.

A PHYSONECT SIPHONOPHORE

This siphonophore, *Nanomia cara*, was collected in the Gulf of Maine at a depth of 600 meters by the submersible *Johnson Sea Link II*. This species measures from 20 centimeters to 2 meters, and deploys long fishing filaments. A favorite prey, the copepod *Calanus finmarchicus*, can be seen at bottom right.

PHOTOGRAPHED BY PER FLOOD, BATHYBIOLOGICA A/S.

Plankton Chronicles website
Siphonophores

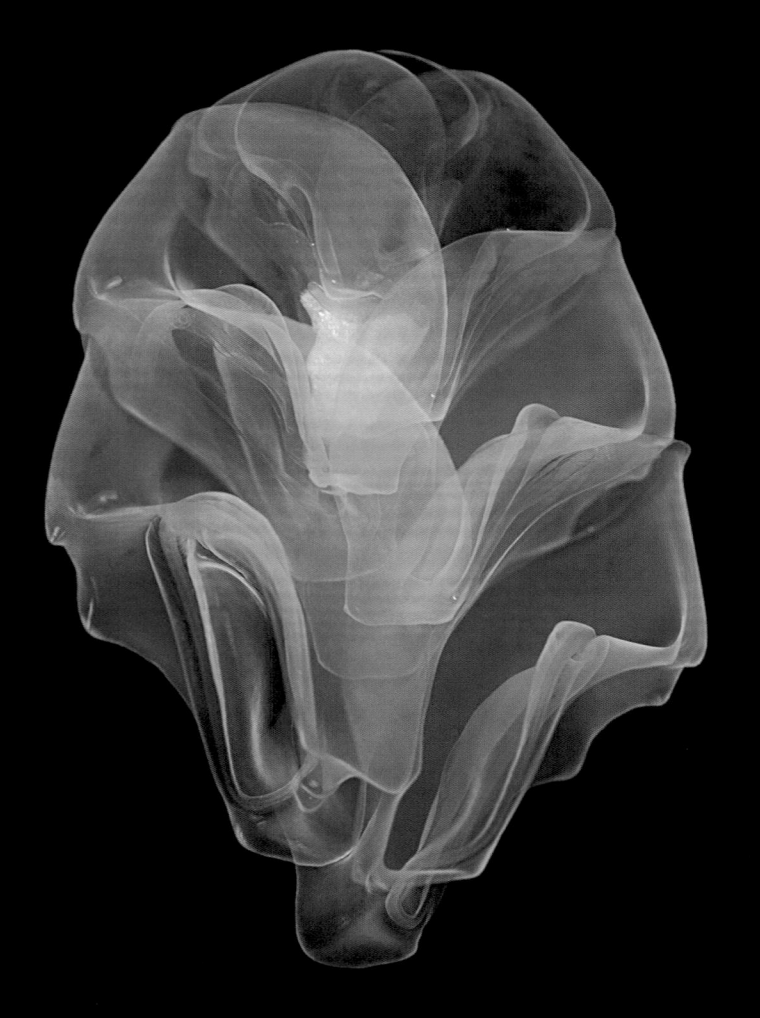

Lensia conoidea (below) is a small
siphonophore with two rocket-shaped
contractile bells that propel the colony.
Beneath the bells lies a siphosome
consisting of reproductive and feeding
zooids with fishing filaments that are
contracted here (also see pages 116–117).
Collected in the bay of Villefranche-sur-Mer.

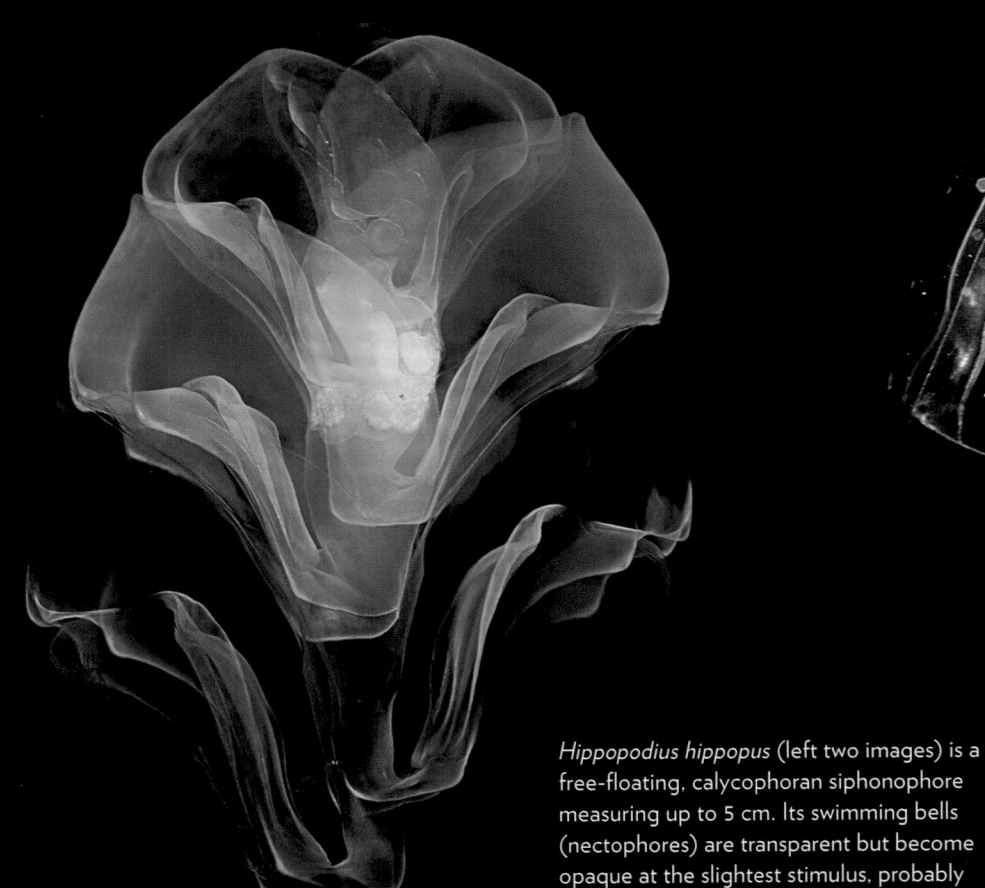

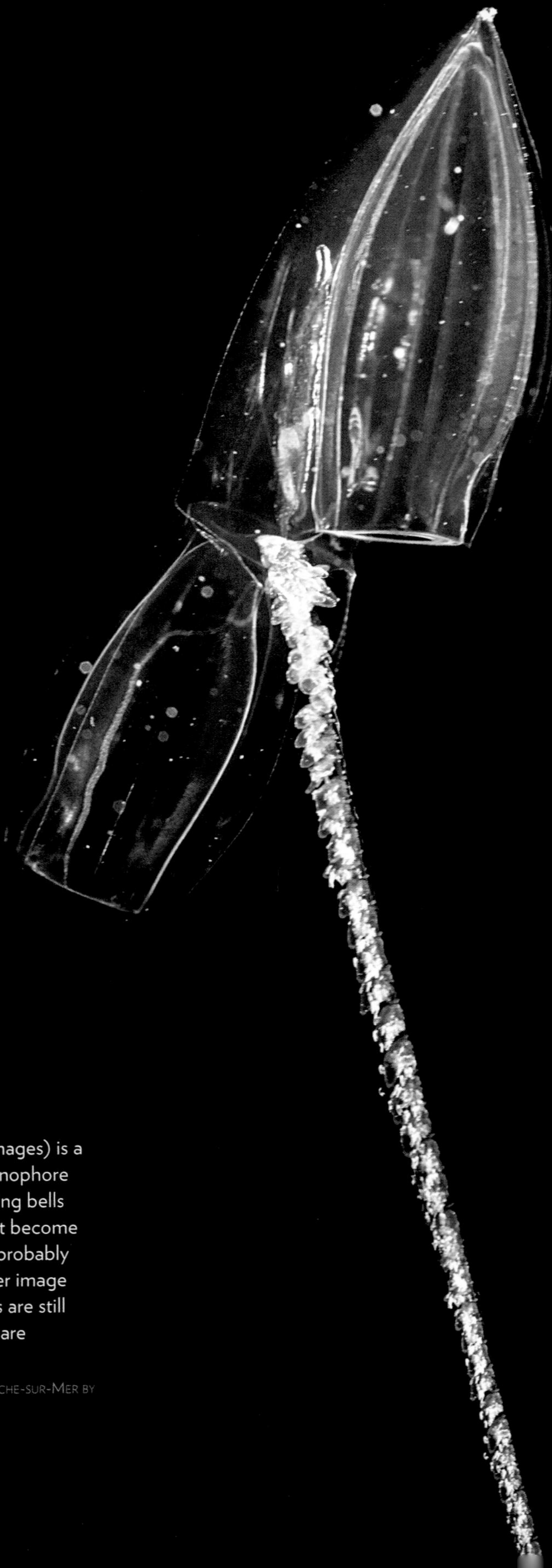

Hippopodius hippopus (left two images) is a
free-floating, calycophoran siphonophore
measuring up to 5 cm. Its swimming bells
(nectophores) are transparent but become
opaque at the slightest stimulus, probably
a defense mechanism. In the lower image
the siphonophore's posterior bells are still
transparent but the anterior bells are
becoming opaque.
Collected and photographed in Villefranche-sur-Mer by
Stefan Siebert, Brown University, USA.

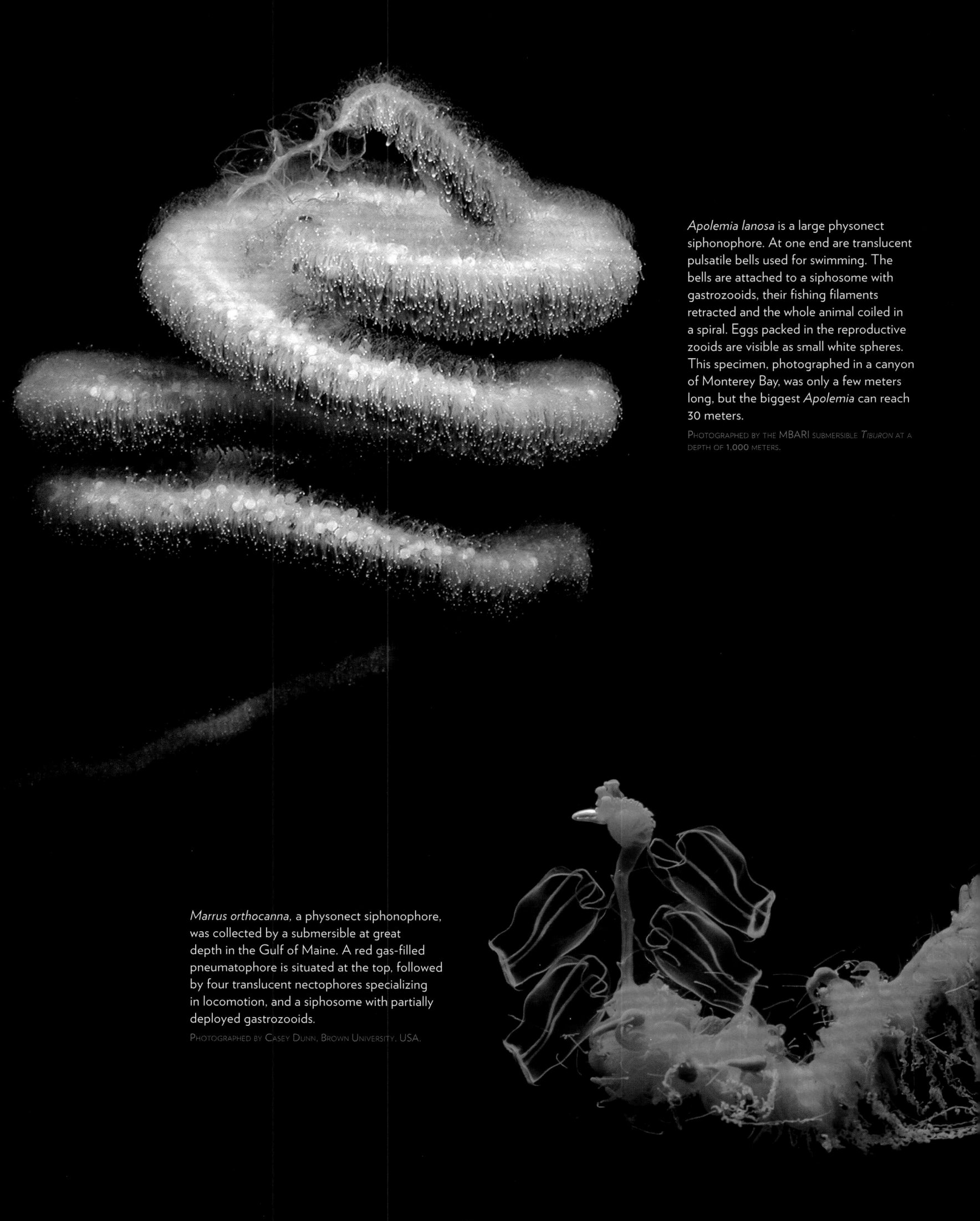

Apolemia lanosa is a large physonect siphonophore. At one end are translucent pulsatile bells used for swimming. The bells are attached to a siphosome with gastrozooids, their fishing filaments retracted and the whole animal coiled in a spiral. Eggs packed in the reproductive zooids are visible as small white spheres. This specimen, photographed in a canyon of Monterey Bay, was only a few meters long, but the biggest *Apolemia* can reach 30 meters.

PHOTOGRAPHED BY THE MBARI SUBMERSIBLE *TIBURON* AT A DEPTH OF 1,000 METERS.

Marrus orthocanna, a physonect siphonophore, was collected by a submersible at great depth in the Gulf of Maine. A red gas-filled pneumatophore is situated at the top, followed by four translucent nectophores specializing in locomotion, and a siphosome with partially deployed gastrozooids.

PHOTOGRAPHED BY CASEY DUNN, BROWN UNIVERSITY, USA.

Calycophoran Siphonophore

Chelophyes appendiculata (1 to 2 cm long) darts around rapidly, propelled by the contraction of its two rocket-shaped swimming bells. Its pink-colored fishing filaments are contracted. *Chelophyes*, like *Lensia* (opposite page), deploy their fishing filaments studded with stinging cells to capture prey.

COLLECTED IN THE BAY OF VILLEFRANCHE-SUR-MER.

Fishing Siphonophore

Calycophoran siphonophores, such as this *Lensia conoidea* (see close-up on page 114), use long fishing filaments to catch small crustaceans, visible here as small whitish particles. They pull their prey toward the mouths of gastrozooids by contraction of the filaments.

<small>COLLECTED IN THE BAY OF VILLEFRANCHE-SUR-MER.</small>

Physophora, the "Hula Skirt" Siphonophore

Physophora hydrostatica is an elegant siphonophore, 10 to 20 cm long.

From top to bottom: the small gas-filled pneumatophore is followed by two rows of nectophore swimming bells, then a remarkable crown of large orange-pink dactylozooid polyps. The tips of the dactylozooids are loaded with stinging buttons used for defensive purposes. Under the crown of polyps lies a siphosome composed of gastrozooids and reproductive gonozooids organized into packets.

<small>PHOTOGRAPHED IN MONTEREY BAY, CALIFORNIA, BY DAVID WROBEL.</small>

Close-up below: On the fishing filaments, spiral structures called tentilla are composed of several types of stinging cells.

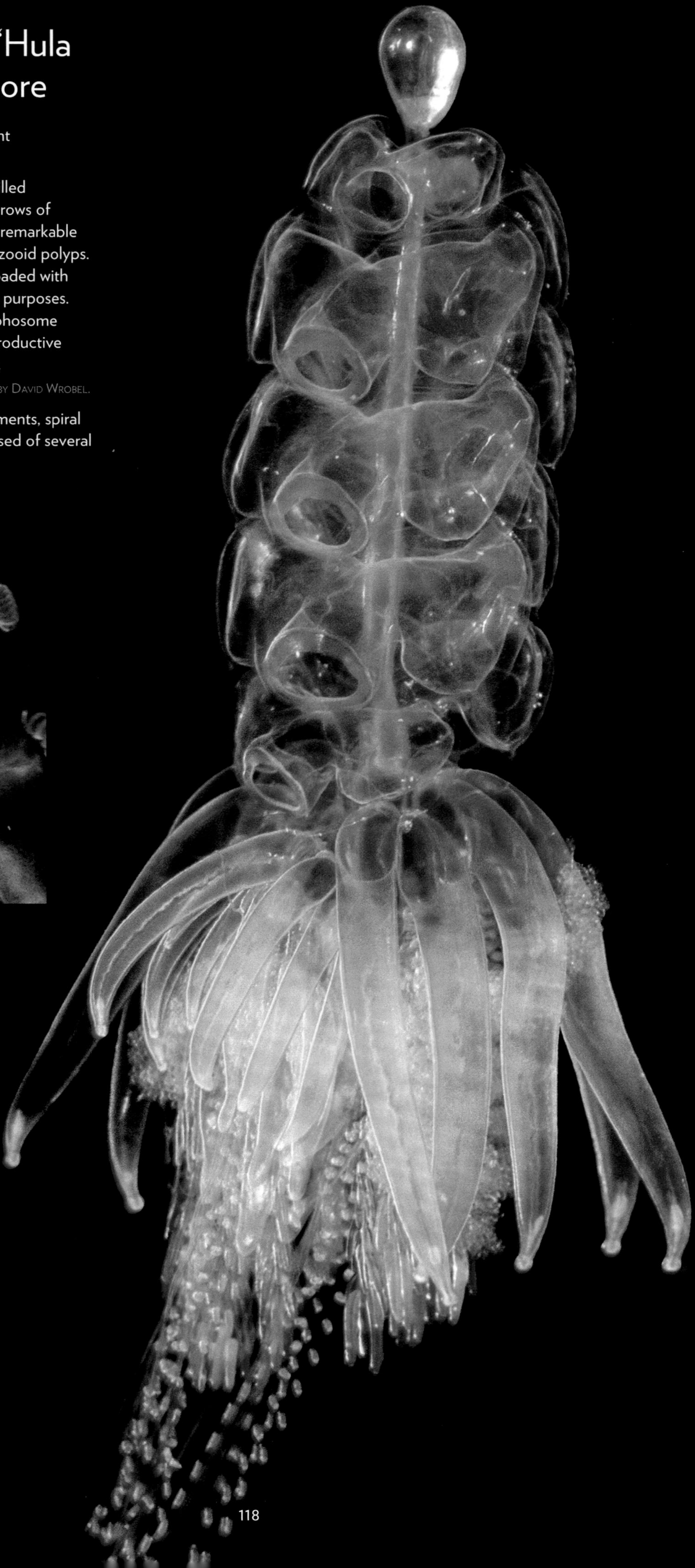

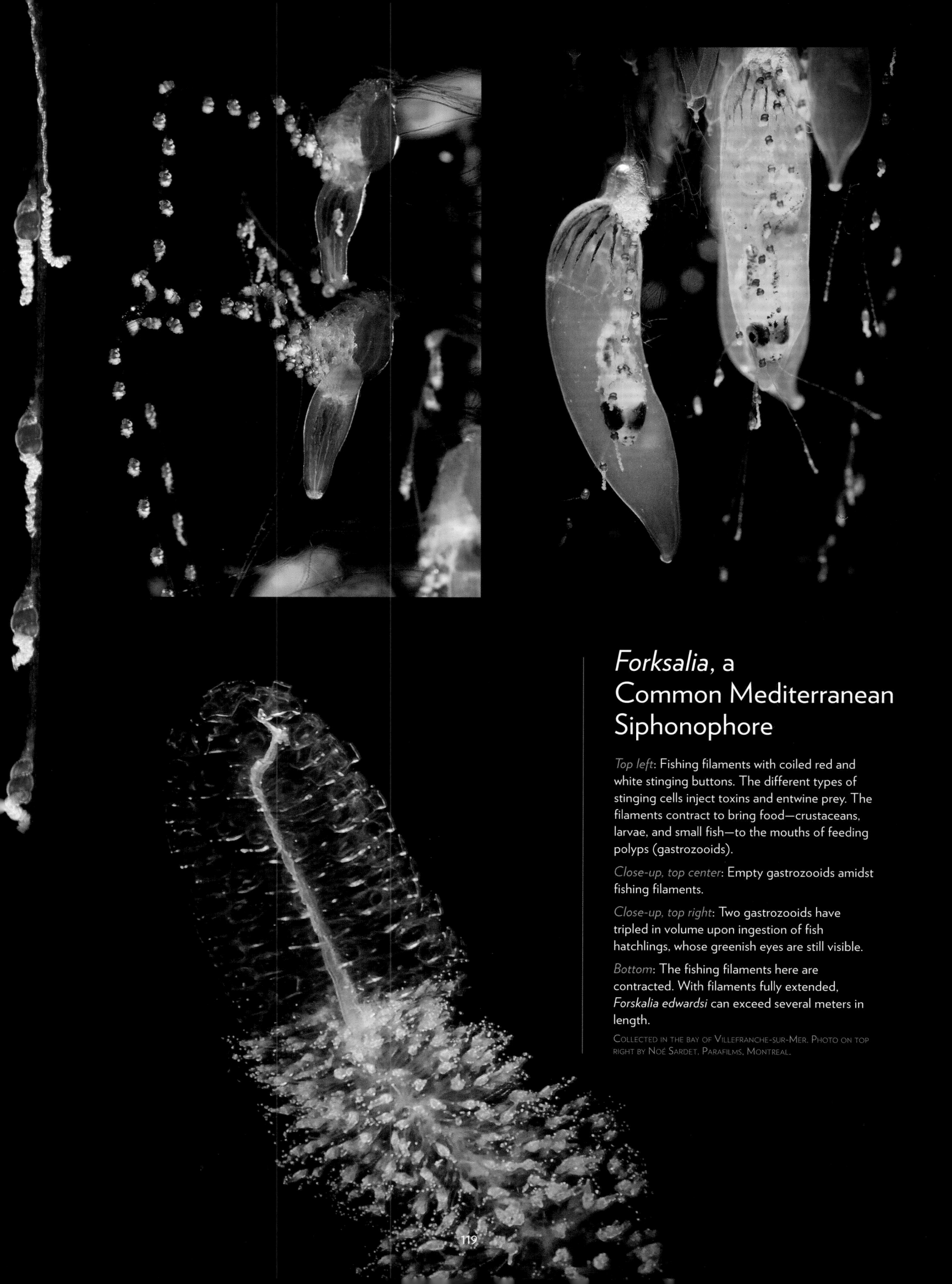

Forksalia, a Common Mediterranean Siphonophore

Top left: Fishing filaments with coiled red and white stinging buttons. The different types of stinging cells inject toxins and entwine prey. The filaments contract to bring food—crustaceans, larvae, and small fish—to the mouths of feeding polyps (gastrozooids).

Close-up, top center: Empty gastrozooids amidst fishing filaments.

Close-up, top right: Two gastrozooids have tripled in volume upon ingestion of fish hatchlings, whose greenish eyes are still visible.

Bottom: The fishing filaments here are contracted. With filaments fully extended, *Forskalia edwardsi* can exceed several meters in length.

COLLECTED IN THE BAY OF VILLEFRANCHE-SUR-MER. PHOTO ON TOP RIGHT BY NOÉ SARDET. PARAFILMS, MONTREAL.

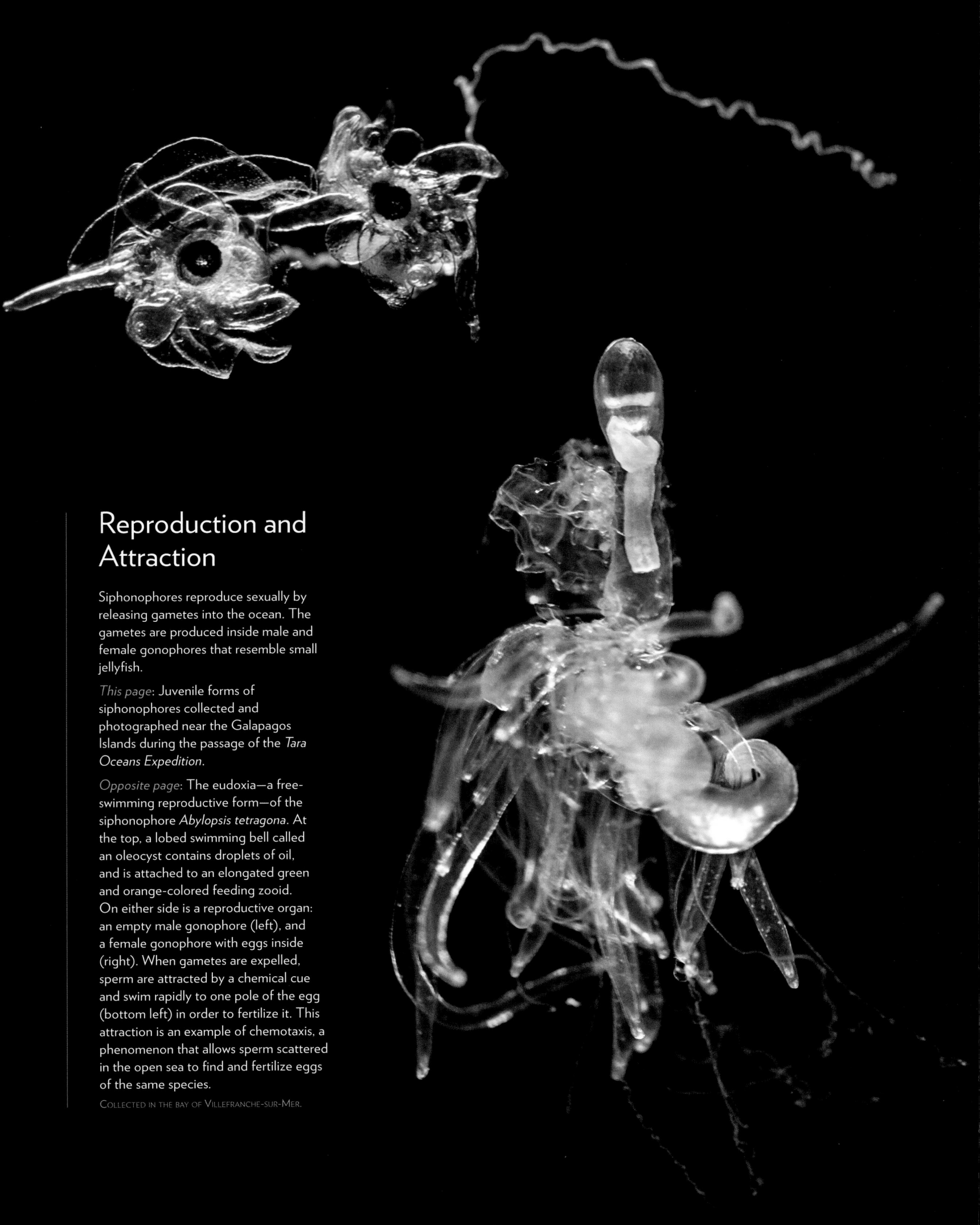

Reproduction and Attraction

Siphonophores reproduce sexually by releasing gametes into the ocean. The gametes are produced inside male and female gonophores that resemble small jellyfish.

This page: Juvenile forms of siphonophores collected and photographed near the Galapagos Islands during the passage of the *Tara Oceans Expedition*.

Opposite page: The eudoxia—a free-swimming reproductive form—of the siphonophore *Abylopsis tetragona*. At the top, a lobed swimming bell called an oleocyst contains droplets of oil, and is attached to an elongated green and orange-colored feeding zooid. On either side is a reproductive organ: an empty male gonophore (left), and a female gonophore with eggs inside (right). When gametes are expelled, sperm are attracted by a chemical cue and swim rapidly to one pole of the egg (bottom left) in order to fertilize it. This attraction is an example of chemotaxis, a phenomenon that allows sperm scattered in the open sea to find and fertilize eggs of the same species.

COLLECTED IN THE BAY OF VILLEFRANCHE-SUR-MER.

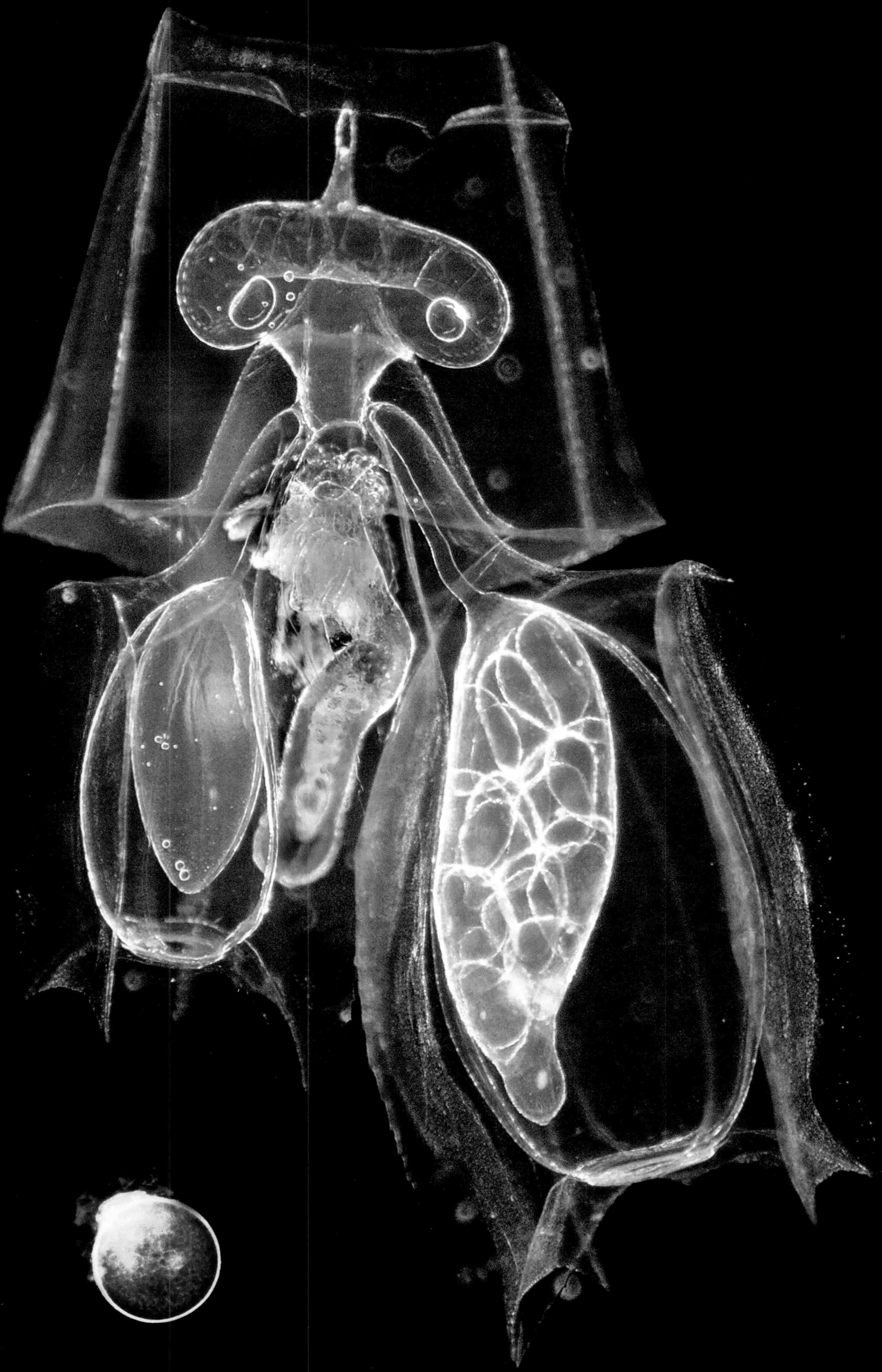

VELELLA, PORPITA AND PHYSALIA

PLANKTONIC SAILORS

Have you ever seen swarms of tiny blue sails drifting on the ocean's surface? They were probably *Velella* or *Porpita*. With their triangular sails and bright blue floats, *Velella* resemble miniature boats and earned the nickname "by-the-wind sailors." In a population of *Velella*, the sails are fixed obliquely to the right or left so the little boats are dispersed in different directions by the wind. In the Mediterranean, like their cnidarian cousins jellyfish and siphonophores, *Velella* often show up in late spring. Pushed by winds, they eventually reach the coast, forming a blue fringe along the beach.

Velella, *Porpita*, and the bigger *Physalia*, known as "Portuguese man-of-war," are colonial organisms consisting of many individual polyps situated under a gas-filled bladder or float. Under *Physalia*'s float are fishing filaments tens of meters long, studded with stinging cells. They can spear and paralyze fish and all sorts of zooplankton. Prey are brought to the mouth of the gastrozooids, engulfed, and digested to feed the whole colony.

Physalia can empty its gas-filled float to dive beneath the surface and escape from predators—turtles, or "blue dragons," the elegant nudibranchs of the genus *Glaucus*. These voracious mollusks appropriate the stinging cells of *Physalia* to use for their own armory. Another predator, *Tremoctopus*, the "blanket octopus," is naturally immune to *Physalia* toxins, so it can steal *Physalia*'s filaments to use for its own defense as well.

Underneath its float, *Velella* is equipped

Photo by Monique Picard.

with short tentacles and reproductive polyps surrounding a large feeding polyp. It eats fish embryos and larvae, small gelatinous zooplankton, and crustaceans. *Velella* are themselves prey to the "sunfish" *Mola mola*, or the purple snail *Janthina janthina* that huddles under a nest of bubbles on the ocean's surface.

Velella reproduce sexually. Their reproductive polyps bud microscopic male or female medusae, spotted with yellowish symbiotic algae. These medusae sink into the depths where they produce eggs and sperm. Although details about its fertilization and development are not well known, *Velella* larvae go through several stages before they float to the surface to sail with the winds.

VELELLA: A BIRD'S-EYE VIEW

This could be how a bird approaching the ocean surface sees a floating *Velella*. Measuring about 3 cm, *Velella velella* has a sail above its float and a crown of blue tentacles. In the center, seen through the float, is a tangle of hundreds of gonozooid polyps. On one of its tentacles, this *Velella* appears to have captured a foraminiferan protist.

COLLECTED DURING THE SPRING IN THE BAY OF VILLEFRANCHE-SUR-MER. PHOTO BY CHRISTIAN SARDET AND SHARIF MIRSHAK, PARAFILMS, MONTREAL.

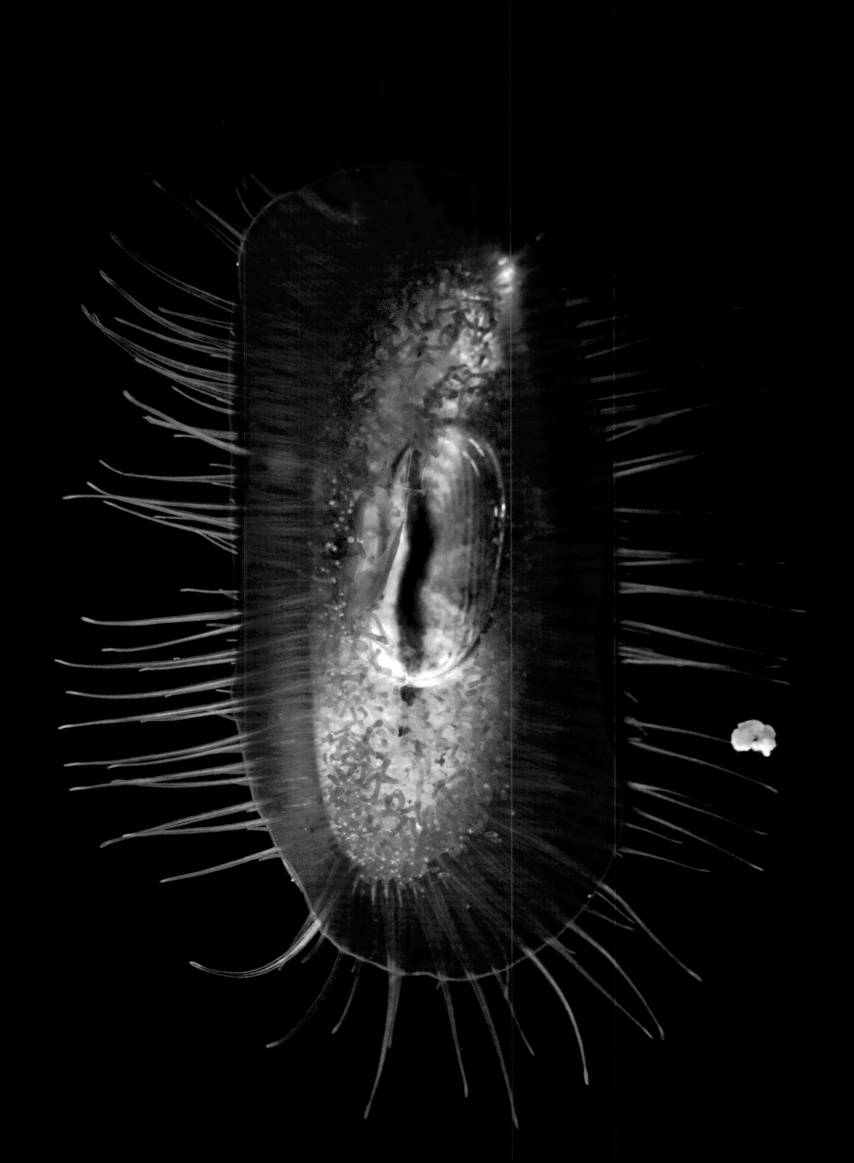

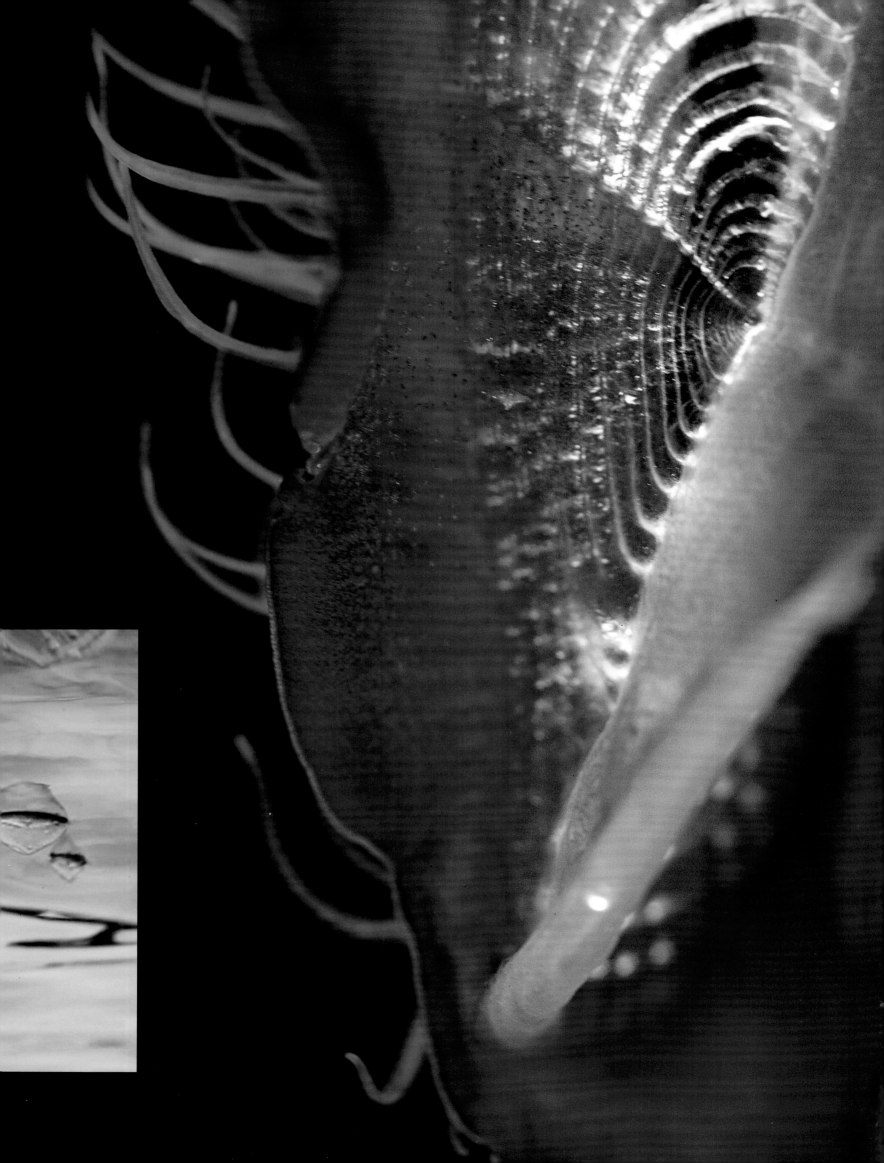

Velella,
Symphony in Blue

The intense blue color of *Velella* is due to pigmented proteins in the tissues of the float and tentacles (close-up, opposite page). The rigid triangular sail and air-filled tubes of the float are made of chitin, a polysaccharide chemically related to cellulose. Beached along the coastline, *Velella* dry out and twirl in the wind like pieces of paper.

Left: *Velella* observed in profile and from below. Small medusae shed by the reproductive polyps under the float ensure reproduction. Their yellow spots are pockets of symbiotic algae.

COLLECTED DURING SPRING IN THE BAY OF VILLE-FRANCHE-SUR-MER. PHOTOS BY CHRISTIAN SARDET, NOÉ SARDET, AND SHARIF MIRSHAK (THE DIVER). PARAFILMS, MONTREAL.

Life Cycle

In *Velella*, reproduction
is performed by tiny
medusae that bud from the
reproductive polyps (see
full-page photo opposite).
These medusae have four
radial canals, and many
yellow patches revealing
the presence of symbiotic
zooxanthellae.

To the right: a juvenile *Velella*.

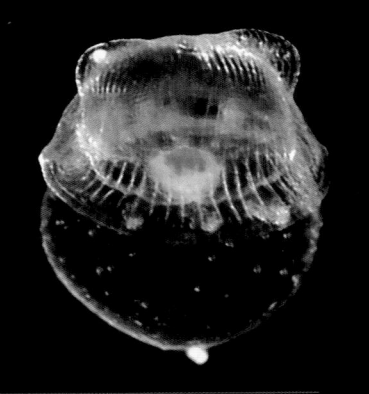

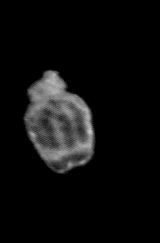

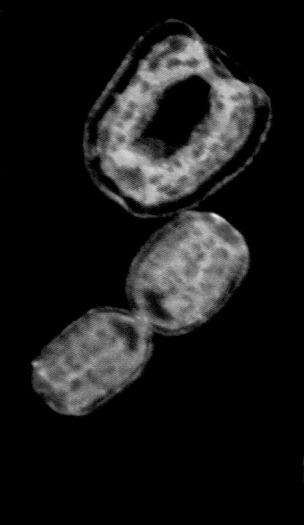

Yellow-colored zooxanthellae
provide nutrients to the
medusa. The various phases
of reproduction are not known
with certainty. It is believed
that medusae sink to the
depths where they produce
male or female gametes. After
fertilization, a larva called a
conaria develops (below). The
conaria larva has an orange-
colored gastric bag. Its future
float is starting to develop.

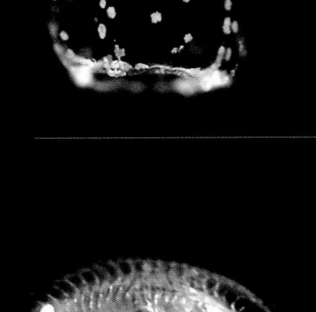

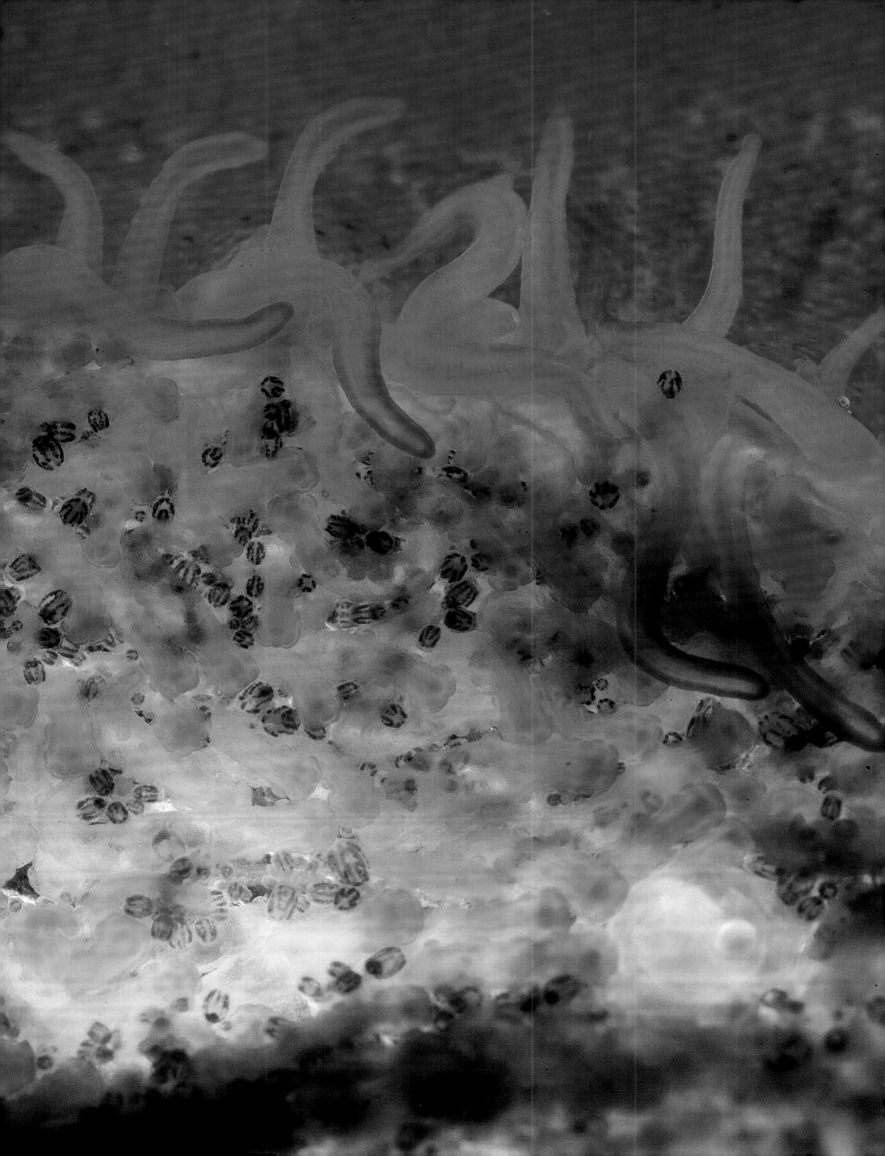

Physalia, the Toxic Portuguese Man-of-War

Physalia, along with *Velella* and *Porpita*, are part of the pleuston, a class of planktonic organisms living at the air/water interface. The float is filled with gasses (10–15 percent carbon monoxide) and sits atop the water. Beneath are four types of reproductive and feeding polyps as well as a multitude of filaments. Here they are contracted, but when deployed, the filaments spread over tens of meters. Stinging cells arranged along the filaments have powerful venoms that paralyze prey. The fishing filaments contract, pulling prey into the mouths of gastrozooid polyps beneath the float.

PHOTO, THIS PAGE: *PHYSALIA PHYSALIS*, BY CASEY DUNN, BROWN UNIVERSITY, USA.
PHOTO, OPPOSITE PAGE: *PHYSALIA UTRICULUS*, BY KEOKI STENDER, WWW.MARINELIFEPHOTOGRAPHY.COM.

Specific Predators

Only certain organisms can feed on *Velella*, *Porpita*, and *Physalia*.

Opposite page: *Glaucus*, a nudibranch mollusk dubbed “sea dragon” attacking a *Velella*.

Photo by Peter Parks, www.imagequestmarine.com

This page, left: The gastropod mollusk *Janthina janthina* stays at the surface under carpet of bubbles to devour *Porpita pacifica*

Photo by Keoki Stender, www.marinelifephotography.co

Below: A sunfish *Mola mola* swallows a *Vele*

Photo by Laurent Colombet.

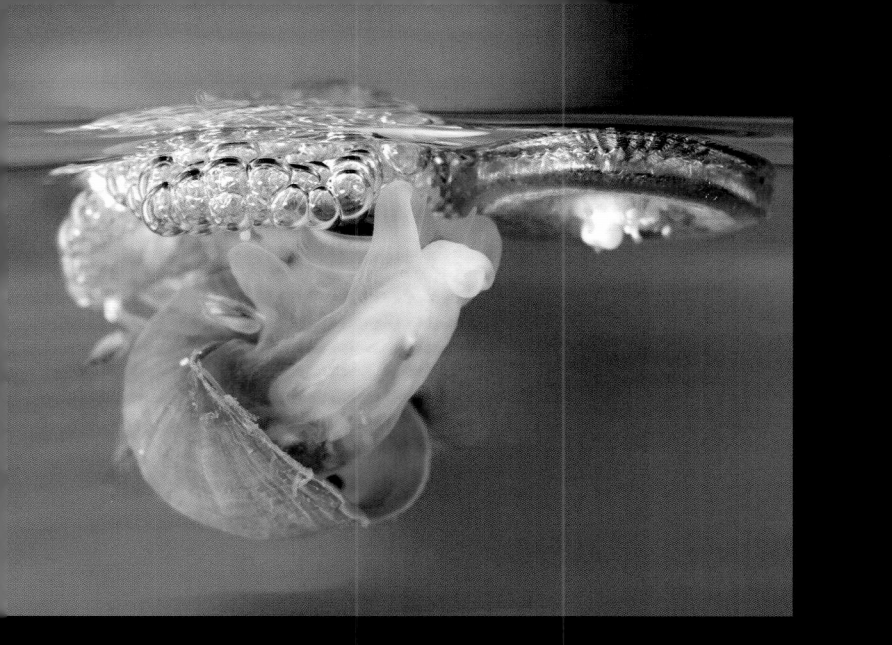

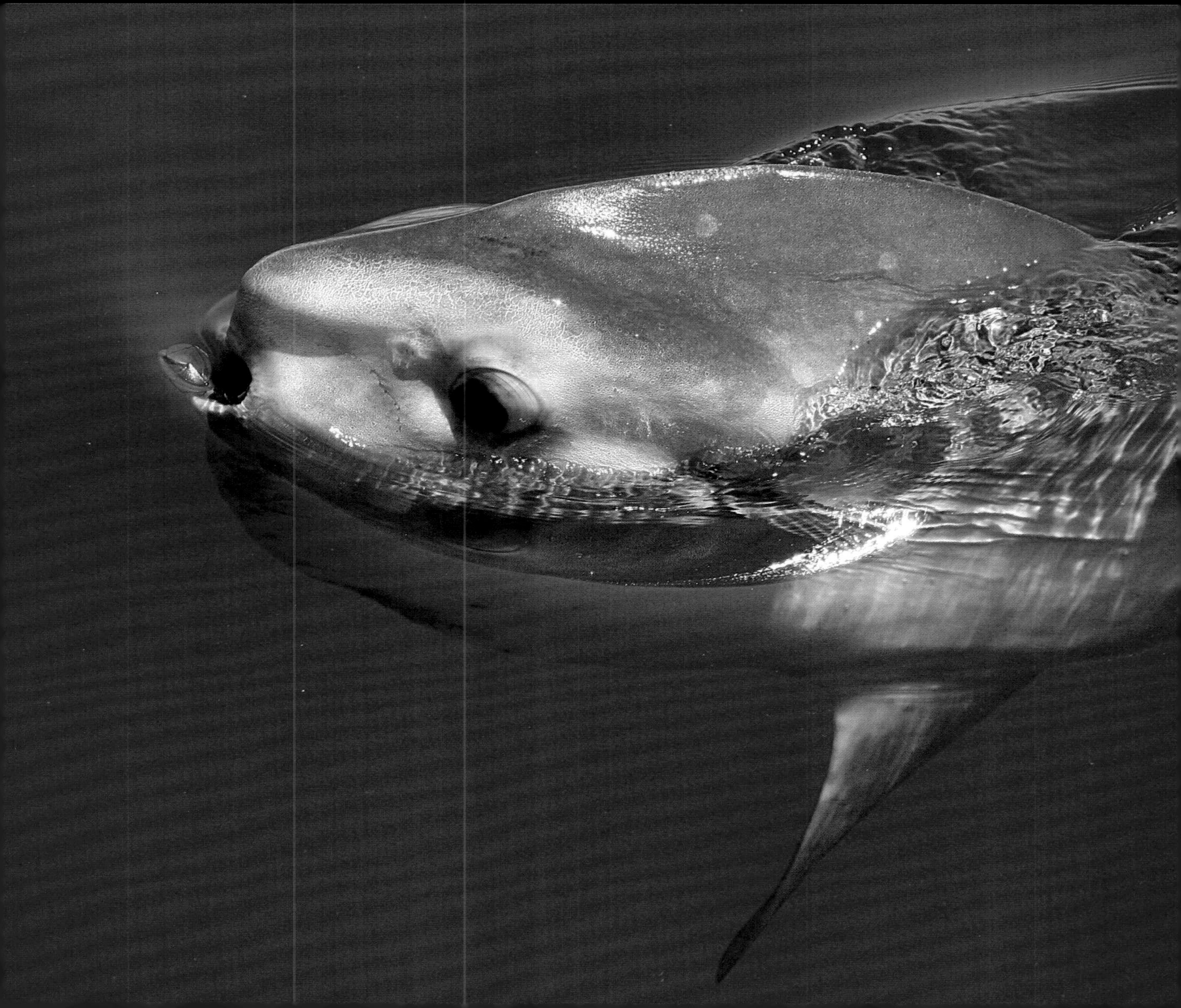

100 microns
(= 0,1 mm)

CRUSTACEANS AND MOLLUSKS

Champions of Diversity

CRUSTACEAN LARVAE

MOLTING AND METAMORPHOSIS

No matter where or how deep in the ocean we lower the zooplankton net, it invariably comes up with myriad tiny crustaceans. Some minuscule crabs and shrimp in the catch are easily recognizable, but their larvae have unexpected shapes. For example, barnacles, the small crustaceans studied extensively by Darwin, release large populations of nauplius larvae near rocky coasts. These larvae bear little resemblance to their sessile parents who permanently attach to a solid substrate.

The most numerous organisms caught in the plankton nets are usually tiny shrimp-like copepods. Adults and half a dozen stages of nauplius larvae coexist during the periods of reproduction. Other crustaceans called decapods, including shrimp, crabs, and squat lobsters, also go through multiple larval stages during their weeks or months drifting among the plankton.

The larval stages, called zoea larvae, have large compound eyes, articulated tails, many pairs of appendages, and chitinous shells with sharp spines that discourage predators. Some of their appendages are covered with feathery bristles used for capturing phytoplankton. Multiple molts and metamorphoses occur before the zoea larva begins to resemble a crab. The final larval stage, the megalopa, has a short abdominal tail equipped with swimming appendages. As soon as megalopae molt, the young crabs leave the plankton and start crawling on the seabed.

Some juvenile decapod crustaceans look like acrobats or miniature robots. The caprellids, or skeleton shrimp, are reminiscent of contortionists. The late-stage larvae of squat lobsters use jerky, repetitive motions to collect food particles clinging to their bristles and employ appendages to rip their prey to shreds. When mature, they leave the plankton to join the adults in rocky crevices. Among the most remarkable larvae are those of the mantis shrimp, a crustacean equipped with eyes that provide some of the keenest 3-D vision in the animal kingdom.

CRAB LARVAE: ZOEAE AND MEGALOPAE

Plankton nets deployed in the winter and spring in the bay of Villefranche-sur-Mer collect a great variety of crustacean and shrimp larvae just a few millimeters in size.

Opposite page: a zoea crab larva (top left), two megalopa crab larvae with short tails, and two zoea shrimp larvae with longer tails.

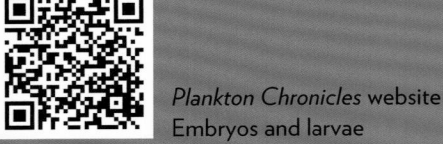

Plankton Chronicles website
Embryos and larvae

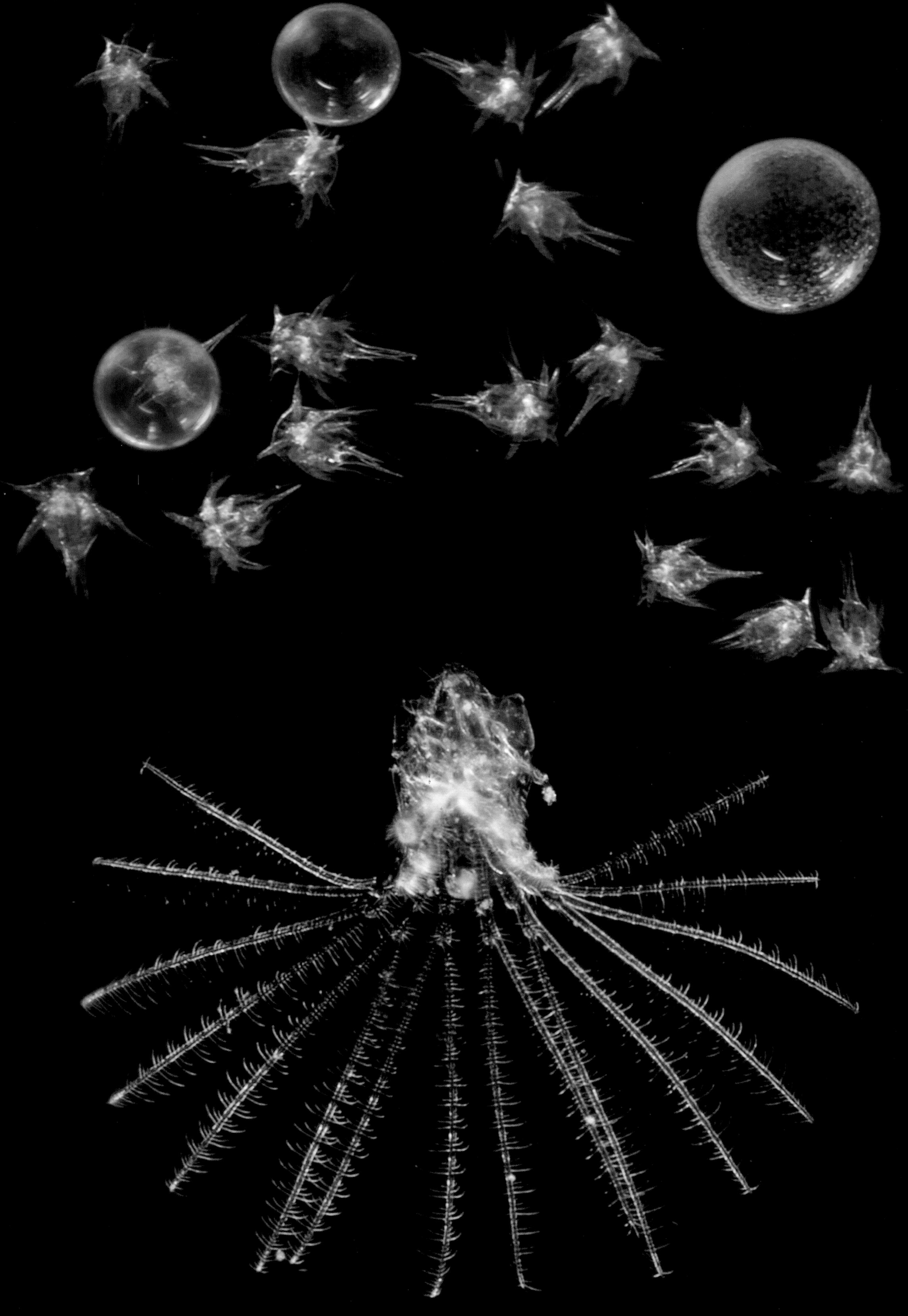

An Abundance of Nauplius Larvae

Opposite page: Near the coast, the most abundant larvae among the plankton are often the nauplii released by barnacles. Adult barnacles are the familiar cone-shaped cirriped crustaceans attached to rocks and docks in tidal zones. Many nauplii are shown here in the presence of an exuvium (the remains of barnacle cuticles after molting) and three spherical colonies of the green algae *Halosphaera* sp.

<small>PLANKTON COLLECTED IN ROSCOFF.</small>

This page, top right: A nauplius larva of a cirriped crustacean.

Below: Nauplius larvae from four different species of copepods at various stages of development. The first in the series has not yet hatched. The larva on the lower right is the calyptopis of a euphausiid shrimp, the next life history phase following the nauplius.

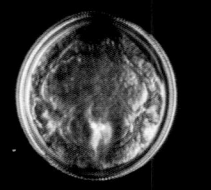

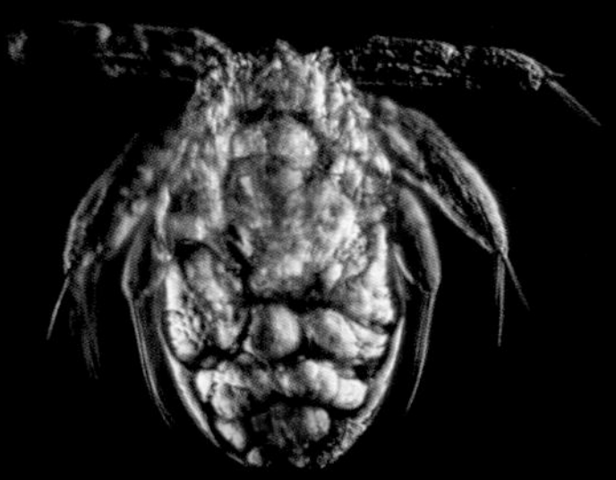

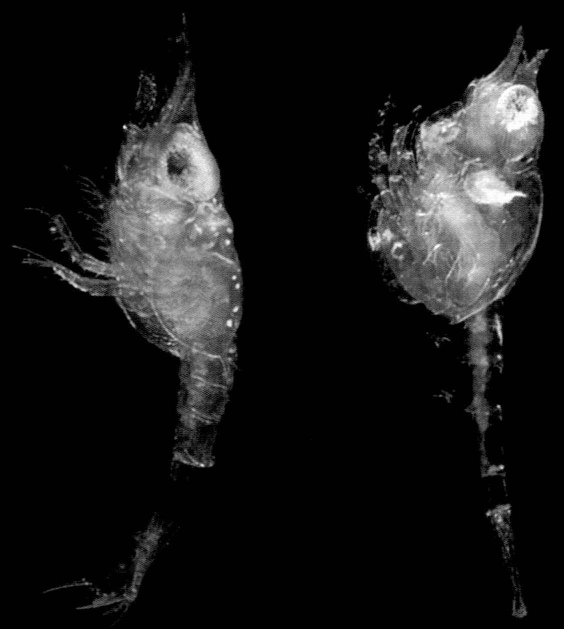

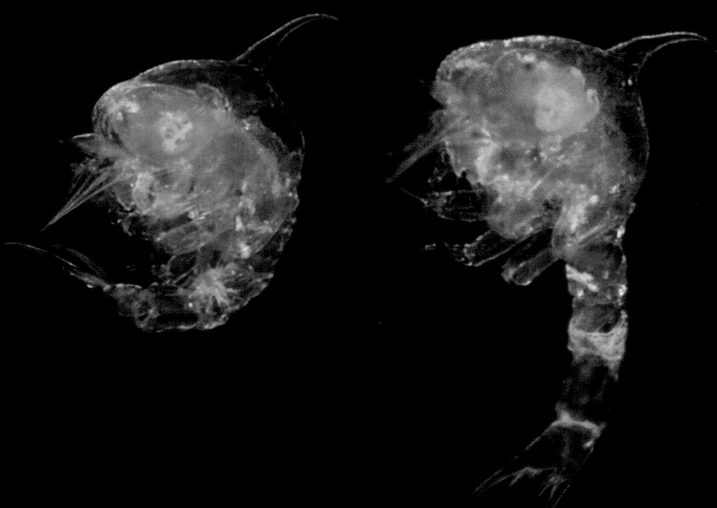

ZOEA LARVAE

Above: Two zoea shrimp larvae.

Above right: Two greenish zoea larvae of fiddler crabs abundant in marshlands. An adult fiddler crab appears on page 134.

Plankton collected in summer in the marshes of South Carolina.

Right: Two orange-colored zoea larvae of brachyuran crustaceans (crabs or sea spiders).

Collected in the bay of Villefranche-sur-Mer.

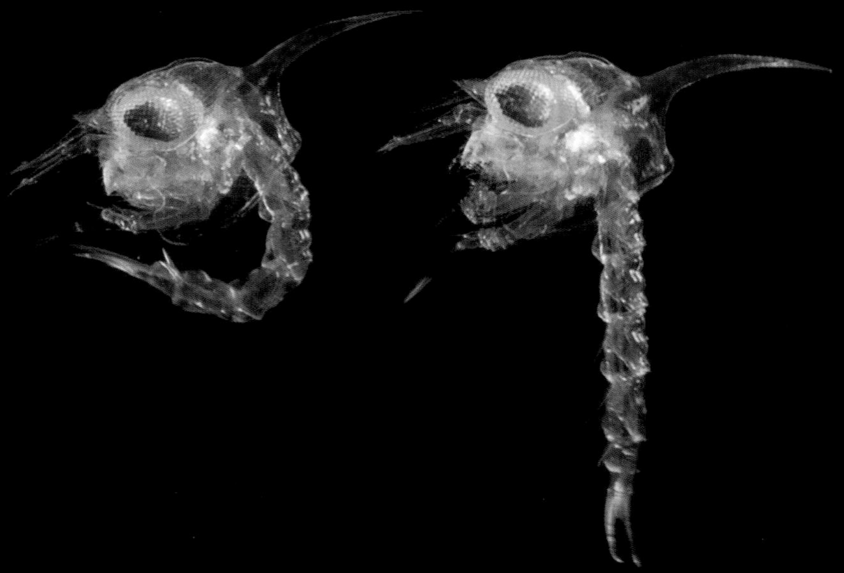

MEGALOPA LARVAE

Megalopa is the final larval stage of a decapod crustacean after several weeks of planktonic life.

Opposite page: A close-up of the compound eyes of a megalopa larva. Dark, branch-like chromatophore cells are seen through the cuticle. These pigment cells spread out or contract to change the appearance of the megalopa, probably a camouflage technique.

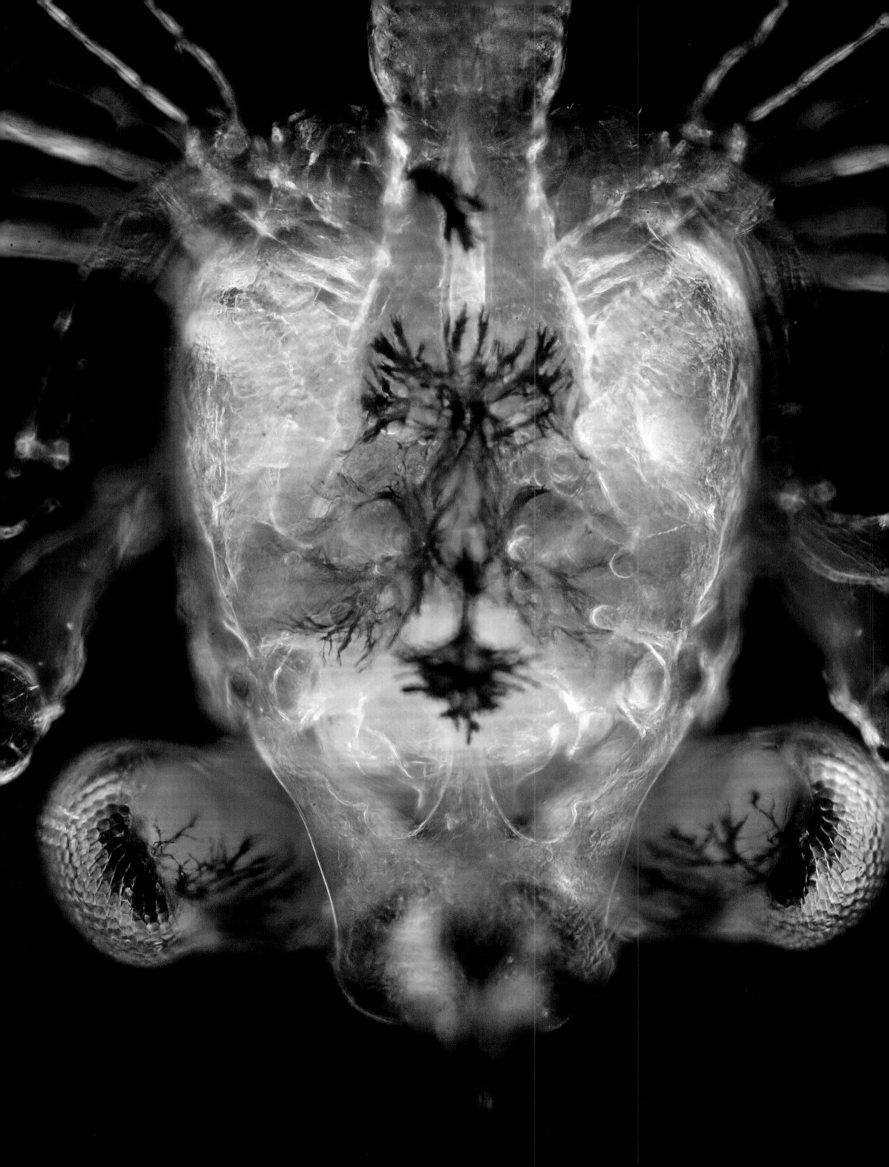

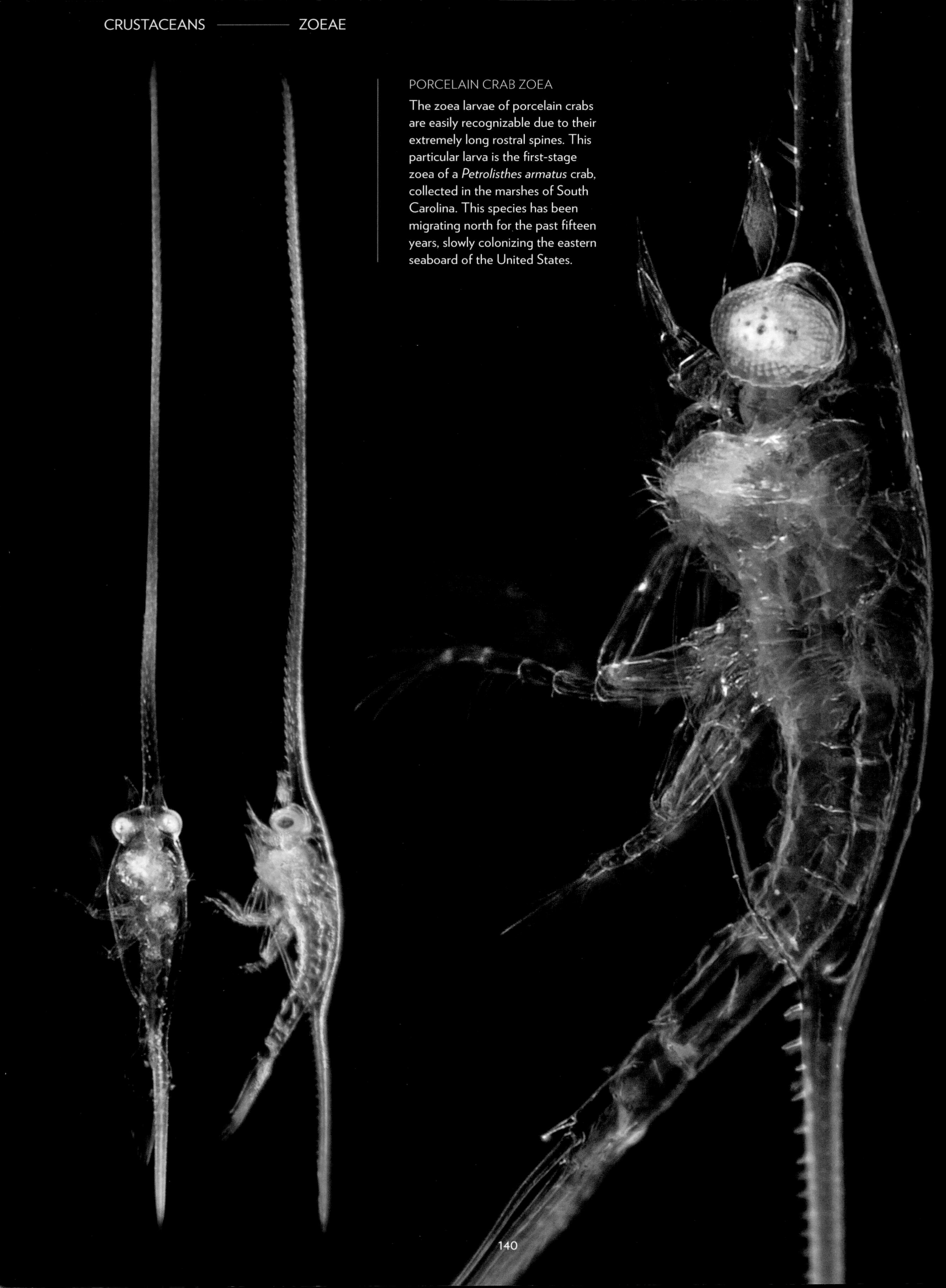

PORCELAIN CRAB ZOEA

The zoea larvae of porcelain crabs are easily recognizable due to their extremely long rostral spines. This particular larva is the first-stage zoea of a *Petrolisthes armatus* crab, collected in the marshes of South Carolina. This species has been migrating north for the past fifteen years, slowly colonizing the eastern seaboard of the United States.

140

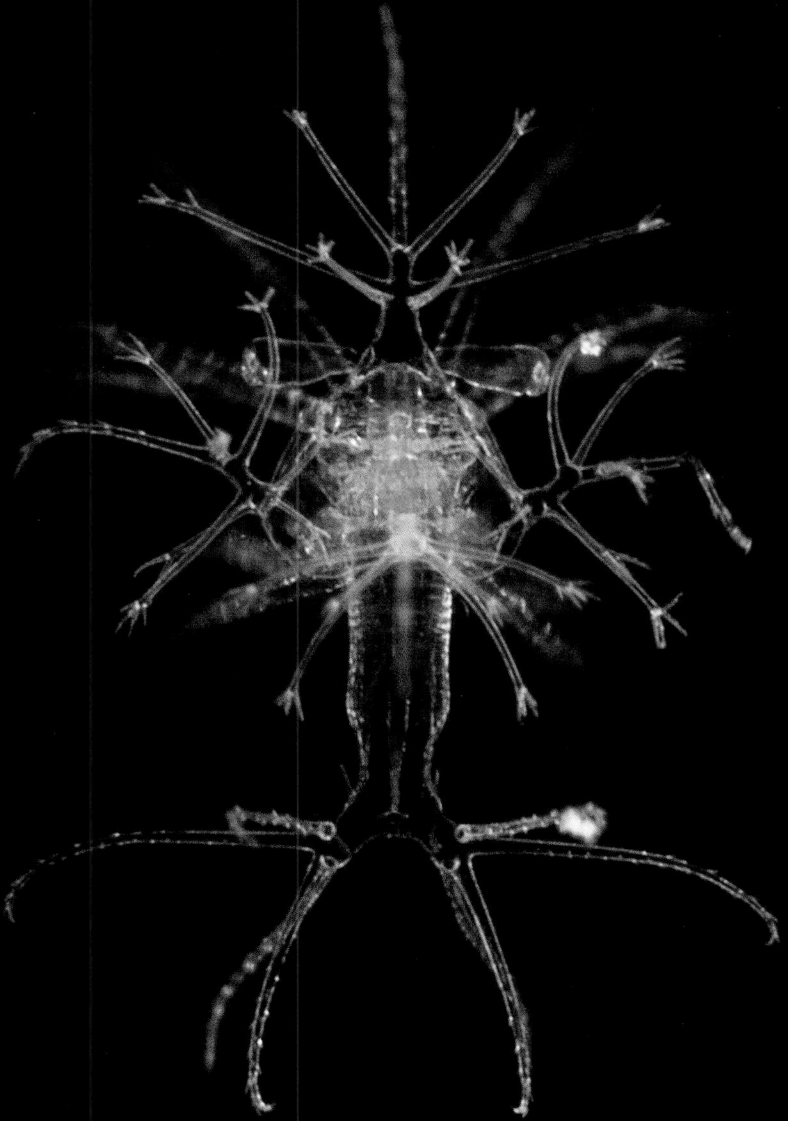

DECAPOD LARVAE

Top: Protozoea larva (stage 3) of a pelagic shrimp *Sergestes* sp. We see the stomach (central red spot), two eyes (small red spots), a rostrum, and a multitude of appendages and branching spines.

Collected in the bay of Villefranche-sur-Mer.

Bottom: Three zoea larvae measuring 2 to 3 mm: a squat lobster in the middle, surrounded by two shrimp.

Collected in the marshes of South Carolina, USA (left); off the coast of Roscoff, France (middle); and in the bay of Shimoda, Japan (right). Middle photo by Noé Sardet, Parafilms, Montreal.

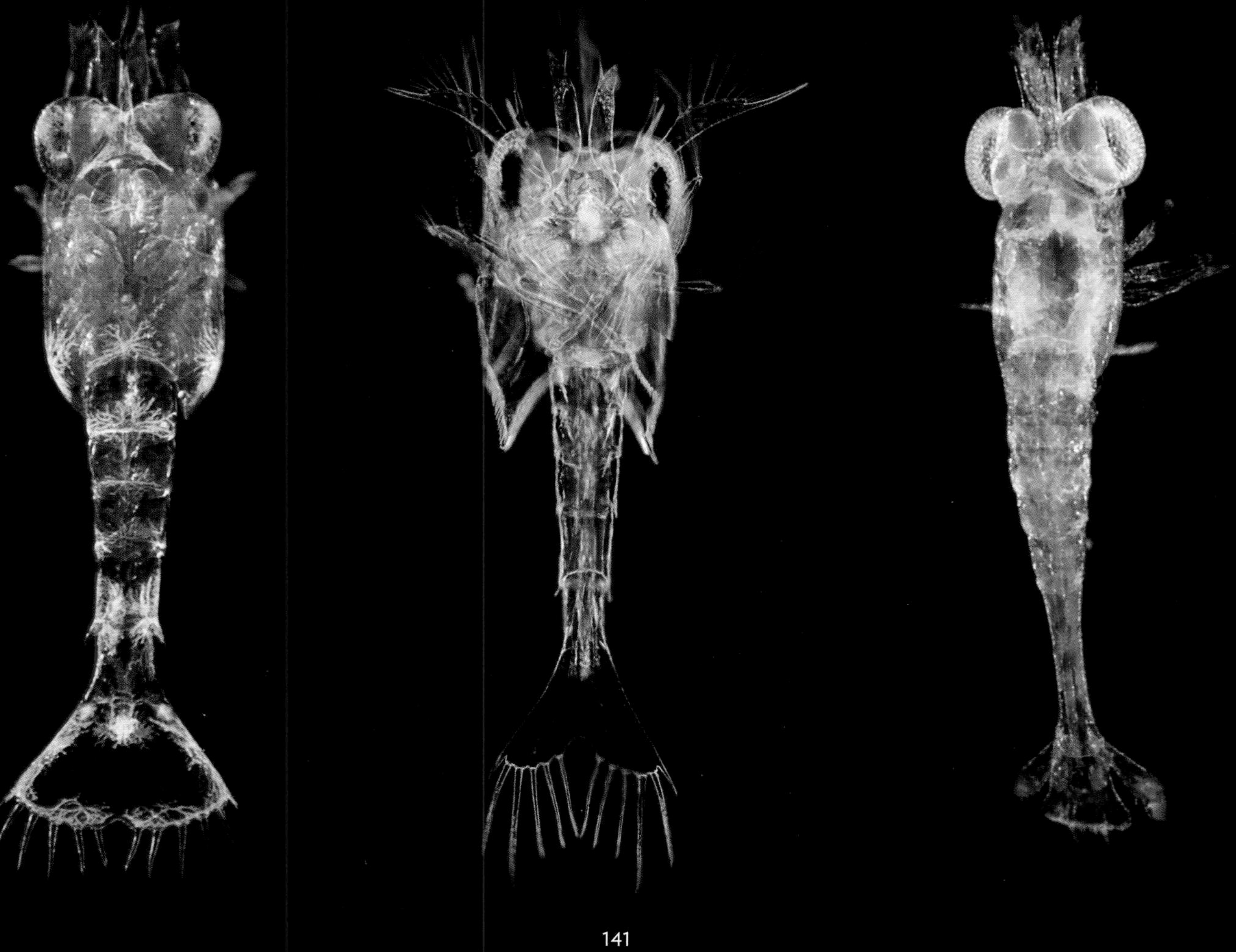

ALIMA LARVA OF THE MANTIS SHRIMP

The alima larva of the mantis shrimp *Squilla* sp. has huge compound eyes at the end of long stalks, and a thorny cephalothorax. Measuring 3 to 4 mm, the alima pass through nine stages of development. Like the adults, larvae of mantis shrimp are fierce predators, using their leg-shaped jaws, maxillipeds, to capture and cut up all kinds of prey.

COLLECTED AND PHOTOGRAPHED AT BARUCH MARINE FIELD LABORATORY WITH DENNIS ALLEN, UNIVERSITY OF SOUTH CAROLINA.

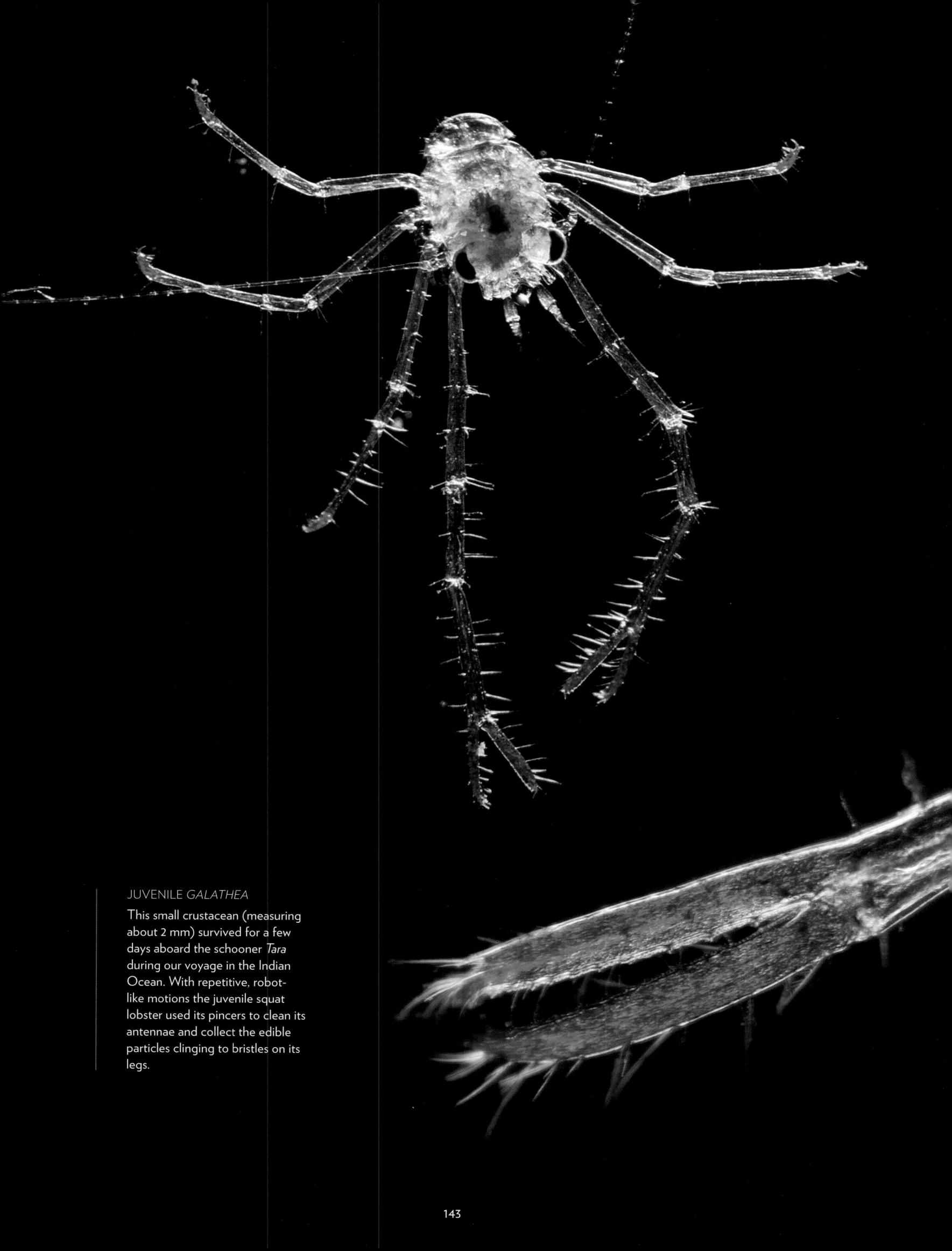

JUVENILE *GALATHEA*

This small crustacean (measuring about 2 mm) survived for a few days aboard the schooner *Tara* during our voyage in the Indian Ocean. With repetitive, robot-like motions the juvenile squat lobster used its pincers to clean its antennae and collect the edible particles clinging to bristles on its legs.

COPEPODS TO AMPHIPODS

VARIATIONS ON A THEME

Copepod crustaceans are the most abundant organisms among the zooplankton, numbering more than 14,000 species. These arthropods range in size from 0.2 to 10 millimeters and occupy nearly all marine and freshwater ecosystems. Some species are free-living, others are symbiotic, and some parasitize other organisms. Take for example the rainbow-hued parasitic male *Sapphirina* that travels piggyback on single or chained salps. Female *Sapphirina* lay eggs inside salps and devour their hosts at the end of their life cycle.

Apart from very rare exceptions, copepods are distinctly male or female, and they mate. Females release pheromones that attract a male (or succession of males) that then grabs the female by the abdomen and deposits a bag containing the sperm (a spermatophore) near the female's genital opening. In many species of copepods, fertilized eggs develop into embryos within egg sacs that the female carries around until the nauplius larvae hatch out.

Copepods are an essential link in the food chain. They generally feed on protists, and are in turn eaten by multicellular organisms like shrimp, chaetognaths, fish, and marine mammals. Each calanoid copepod grazes on 10,000 to 100,000 diatoms or dinoflagellates every day. Some species of copepods create currents with their appendages and buccal parts, bringing prey to the mouth for ingestion. Other copepod species remain almost motionless, suddenly pouncing on prey that emit specific chemical signals. Most copepods migrate to the ocean surface during the day to feed on phytoplankton, then sink into the depths at night, far away from predators.

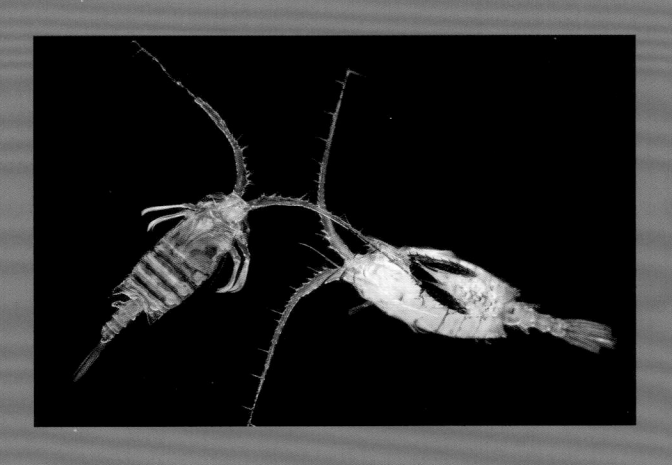

Besides the countless copepods, plankton teem with thousands of other crustaceans. Euphausiids, pelagic shrimp living in swarms known as krill, are a favorite food of baleen whales. Whales can impact the plankton in many ways, including preying on the zooplankton and fertilizing the phytoplankton with their iron-rich feces. Ostracods, sometimes called "seed shrimp," often shelter within calcified shells. Cumaceans and caprellids are filiform crustaceans that live in estuaries and marshes, and are only episodically planktonic. Finally, other crustaceans such as the hyperiid amphipods live in close association with jellyfish, siphonophores, pyrosomes, or salps. Amphipods play an important role in recycling the abundant gelatinous matter produced by these macroplankton.

SAPPHIRINA COPEPOD, PARASITIC AND IRIDESCENT
This male *Sapphirina* copepod reflects and diffracts light through tiny plates situated in the epidermal cells covering its surface. Depending on the orientation of the animal, its flat body switches from fully transparent to brightly colored. In the bay of Villefranche-sur-Mer, *Sapphirina* are most abundant when their hosts, the salps, proliferate. On board a collecting vessel, chains of salps beneath the surface can be detected easily by flashes of light coming from the parasitic copepod. Female *Sapphirina* have a very different morphology (see page 149).
PHOTO BY SHARIF MIRSHAK, PARAFILMS, MONTREAL.

Copepods

Copepods of four genera: *Centropages*, *Sapphirina*, *Pleuromamma*, and *Copilia*. Those on the right belong to the order Calanoida, while those on the left are Poecilostomatoida.

Opposite page: The ventral side of *Temora longicornis*. The body and appendages (approximately 1 mm in size) are covered by a cuticle made of chitin (ochre color). The blue regions are rich in resilin, an autofluorescent elastic protein present in the joints of many arthropods.

Confocal microscopy by Jan Michels, University of Kiel, Germany.

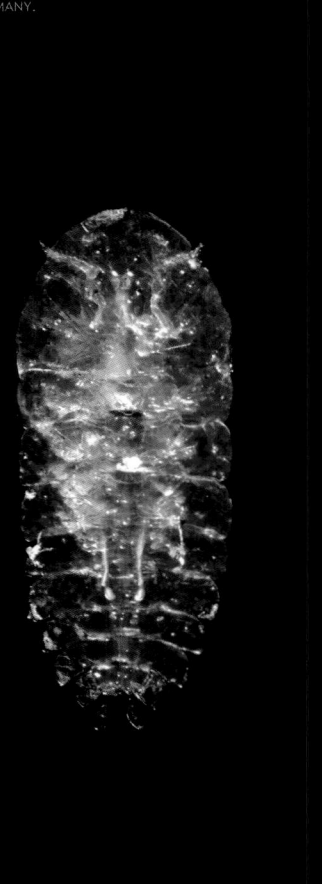

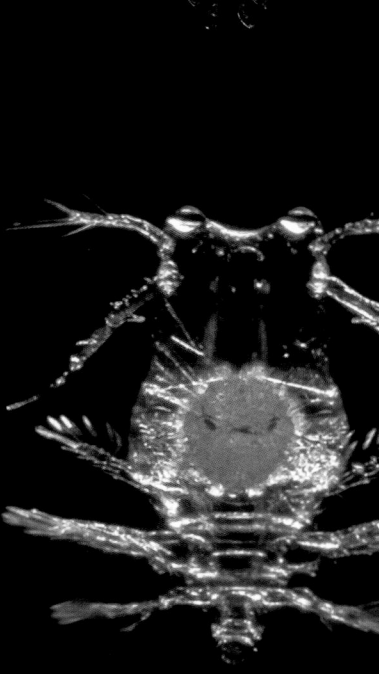

MATING AND EGGS

Male copepods use their legs or modified antennae to hold onto females while depositing their spermatophore, a bag full of sperm, near the female's genital orifice. The modified appendages in males, and the genital segments in the females, are distinctive characters used to determine species. Many copepods carry their embryos in special sacs.

This page, right: A variety of egg sacs collected in a 100-micron mesh plankton net in the bay of Shimoda (Japan).

Below: Two copepods of the genus *Oncaea* sp. mating, viewed from above and in profile.

Opposite page: Three species of female copepods bearing egg sacs: *Euchaeta* sp. (top), *Sapphirina* sp. (bottom), and a bluish harpacticoid copepod (right).

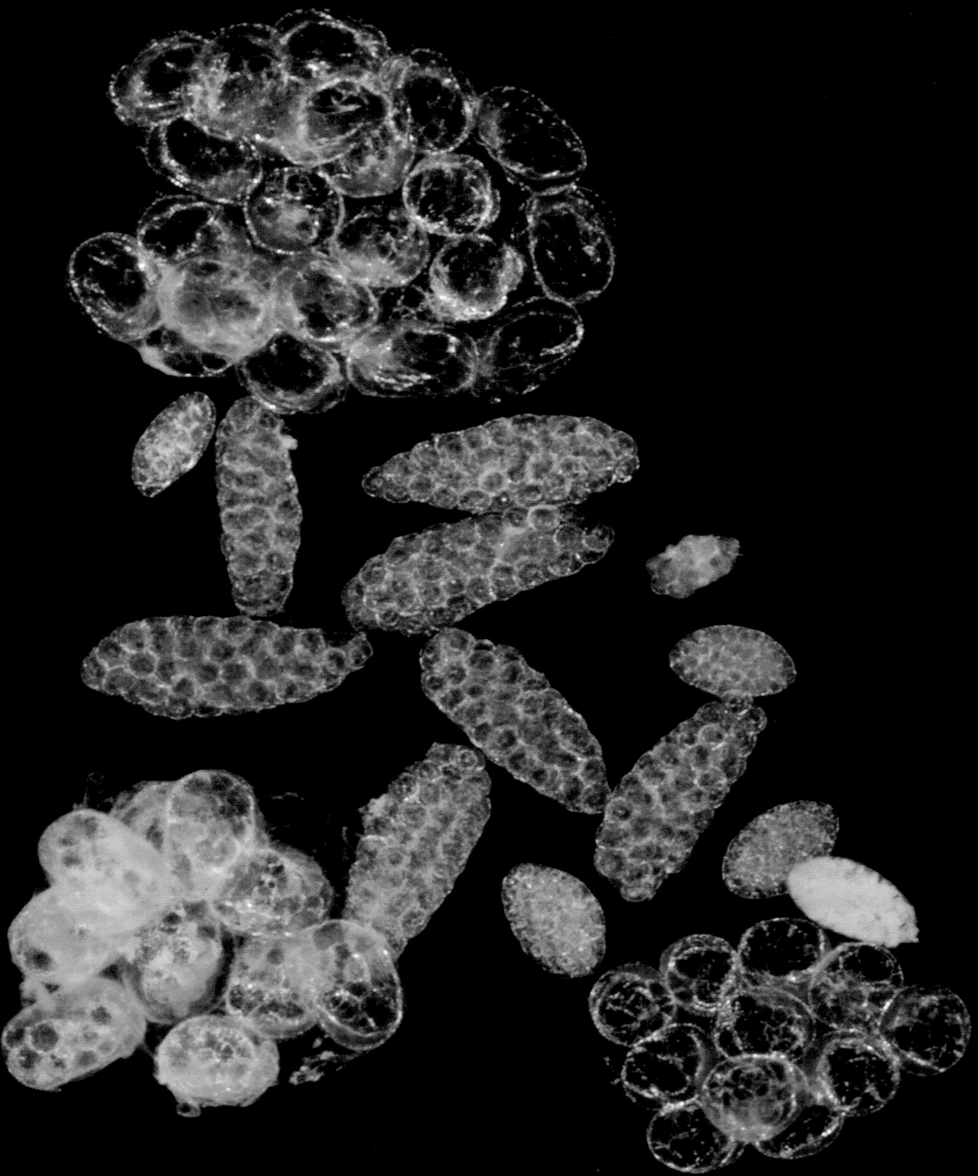

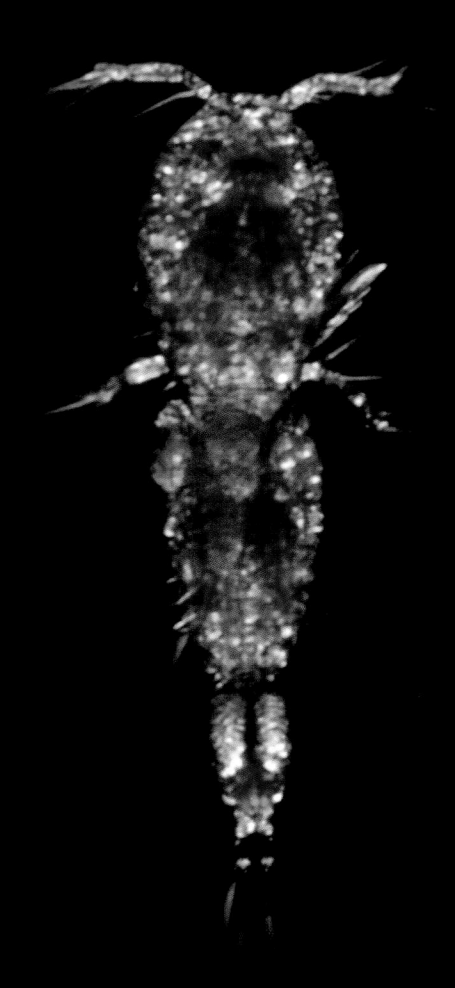

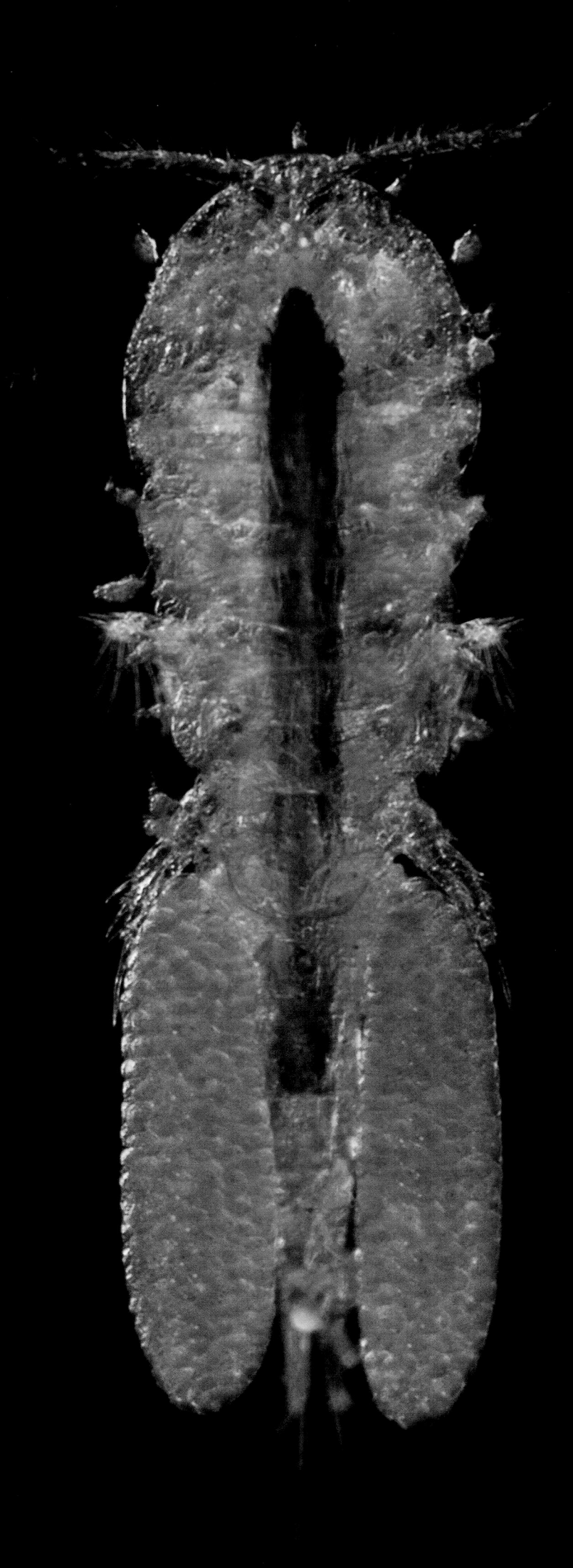

Penaeidae, Euphausiids, and Mysids

Penaeid shrimp larva *Solenocera crassicornis* (ventral view).
COLLECTED IN THE BAY OF VILLEFRANCHE-SUR-MER.

Penaeid shrimp larva *Solenocera crassicornis* profile view).

COLLECTED IN THE BAY OF VILLEFRANCHE-SUR-MER.

Euphausiid shrimp *Meganyctiphanes norvegica*, a main component of krill.

COLLECTED IN THE BAY OF VILLEFRANCHE-SUR-MER.

Two mysids: a juvenile *Brasilomysis castroi*, and the head of *Chlamydopleon dissimile*.

COLLECTED IN THE MARSHES OF SOUTH CAROLINA.

Ostracods and Amphipods

Right: Two *Vibilia armata*, parasitic hyperiid amphipods, found inside salps where they develop and reproduce.
COLLECTED IN THE BAY OF VILLEFRANCHE-SUR-MER.

Below: Twenty ostracods and two amphipods of the genus *Oxycephalus*. *Vargula hilgendorfii* ostracods are known as "sea fireflies." When disturbed, they emit a blue light generated by a luminescent protein related to that of terrestrial fireflies.
COLLECTED AT NIGHT IN TOBA BAY (JAPAN) USING A NET FITTED WITH A FLASHLIGHT.

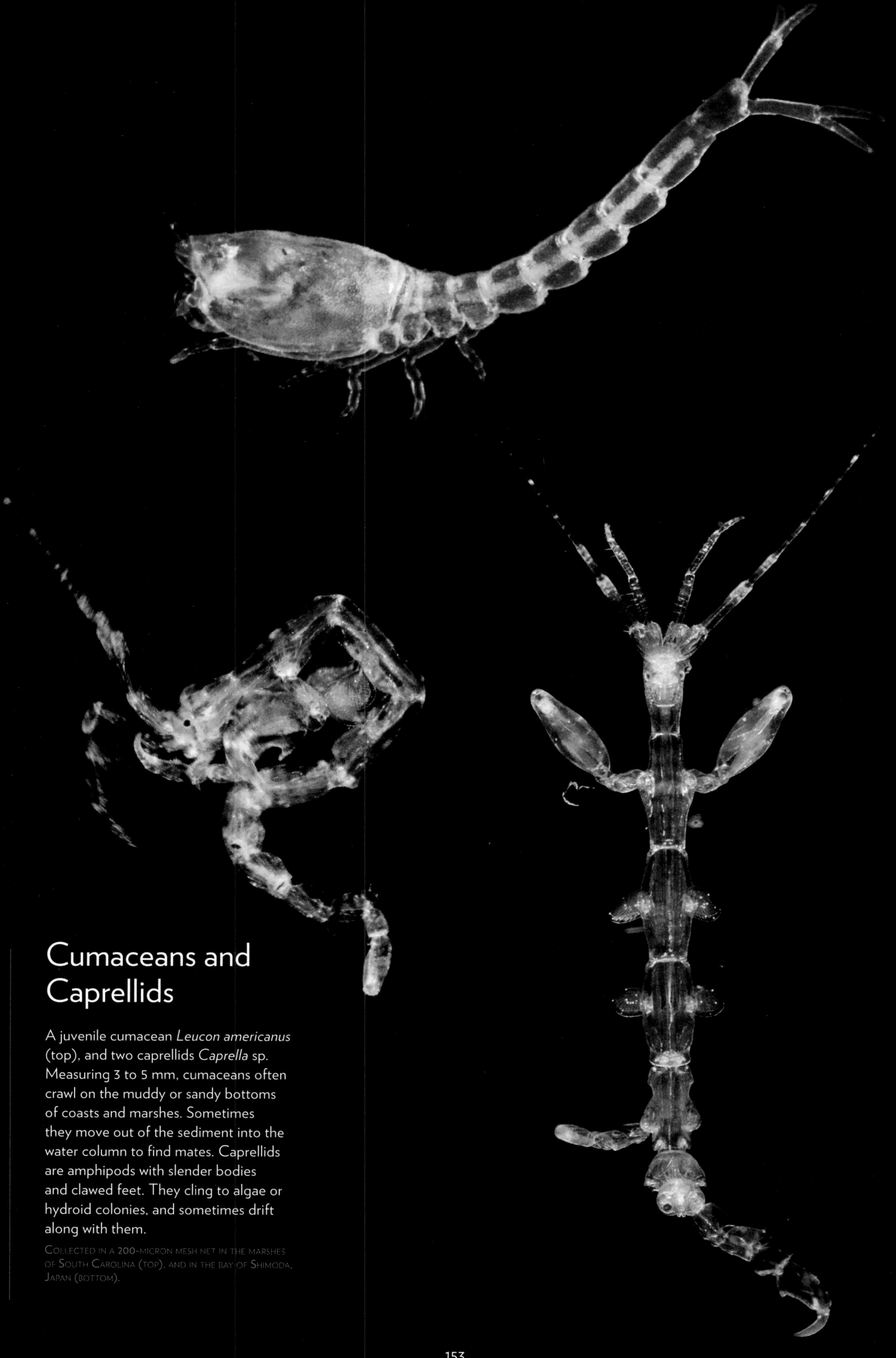

Cumaceans and Caprellids

A juvenile cumacean *Leucon americanus* (top), and two caprellids *Caprella* sp. Measuring 3 to 5 mm, cumaceans often crawl on the muddy or sandy bottoms of coasts and marshes. Sometimes they move out of the sediment into the water column to find mates. Caprellids are amphipods with slender bodies and clawed feet. They cling to algae or hydroid colonies, and sometimes drift along with them.

COLLECTED IN A 200-MICRON MESH NET IN THE MARSHES OF SOUTH CAROLINA (TOP), AND IN THE BAY OF SHIMODA, JAPAN (BOTTOM).

PHRONIMA

MONSTER IN A BARREL

Of the more than 5,000 species of amphipod crustaceans, most are benthic, buried in the seabed or exploring rocky bottoms. But several hundred amphipod species are pelagic, floating with the plankton. Many of these are hyperiid amphipods, living as parasites on specific gelatinous organisms. Each hyperiid species is specialized in capturing a specific gelatinous animal (a siphonophore, salp, jellyfish, or pyrosome).

Phronima prefers to parasitize and prey on salps. Yet this amphipod has a most unusual story. The cartoonist Moebius was inspired by *Phronima*'s form, and his drawings were used to create the monster in the Hollywood blockbuster *Alien*. Like all hyperiid amphipods, *Phronima* has a large head and oversized eyes, but in addition, it is equipped with two impressive claws. Like a hermit crab, the female *Phronima* is a squatter. But instead of occupying a ready-made snail's shell, *Phronima* builds a barrel using the envelope of its gelatinous host.

Phronima catches its prey, eats some, and recycles the rest to make its cellulose tunic. As *Phronima* grows, it builds ever larger barrels, moving inside and out, but always holding on with its spiky legs. Protected by the barrel, *Phronima* moves quickly, gathering small particles with its hind legs covered in long bristles. It only leaves the house to capture large prey. In this case, *Phronima* nibbles the catch at the edge of the shelter, then pulls it inside to finish the meal.

The female *Phronima* takes excellent care of her offspring, an extremely rare behavior among

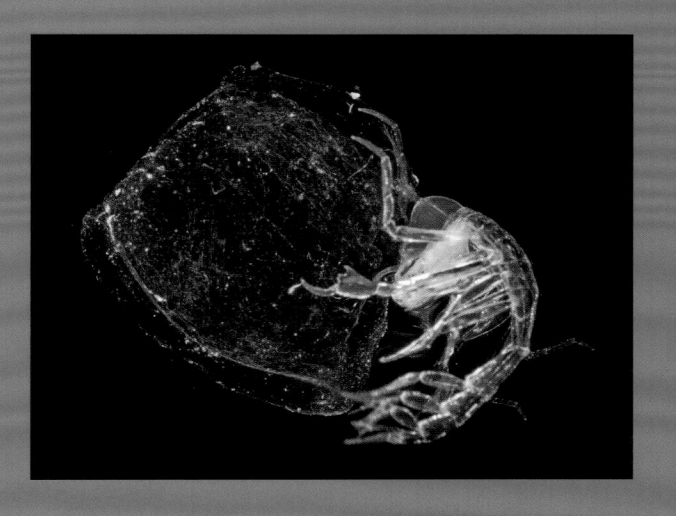

crustaceans. She lays and incubates eggs in a marsupial-like pouch, then raises and feeds the young inside the barrel. For this reason, *Phronima* is commonly called the "pram bug." Once the young have developed into miniature *Phronima*, they finally leave the nursery. Easy prey for fish, jellies, chaetognaths, and annelids, only a few of them will survive. An essential link in the food chain, *Phronima* eat and recycle gelatinous planktonic animals and are a favorite food for many of those same carnivores in the sea.

Plankton Chronicles website
Phronima

PHRONIMA IN ITS BARREL

Phronima feed on planktonic gelatinous animals and use their envelopes to make barrel-shaped houses. This *Phronima* sp. measures about 25 mm. Gripping the inside of its barrel with two front and two hind legs, it moves its tail to swim with the currents.
COLLECTED DURING THE WINTER IN THE BAY OF VILLEFRANCHE-SUR-MER.

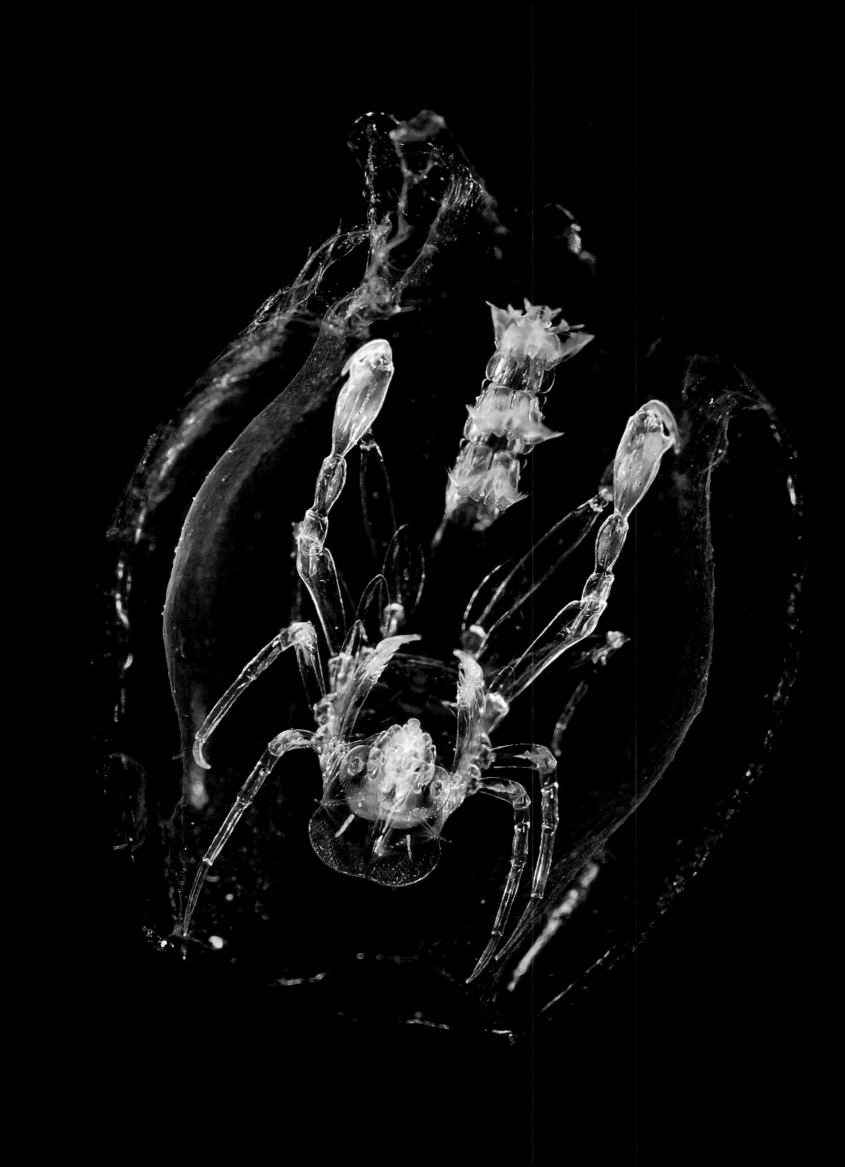

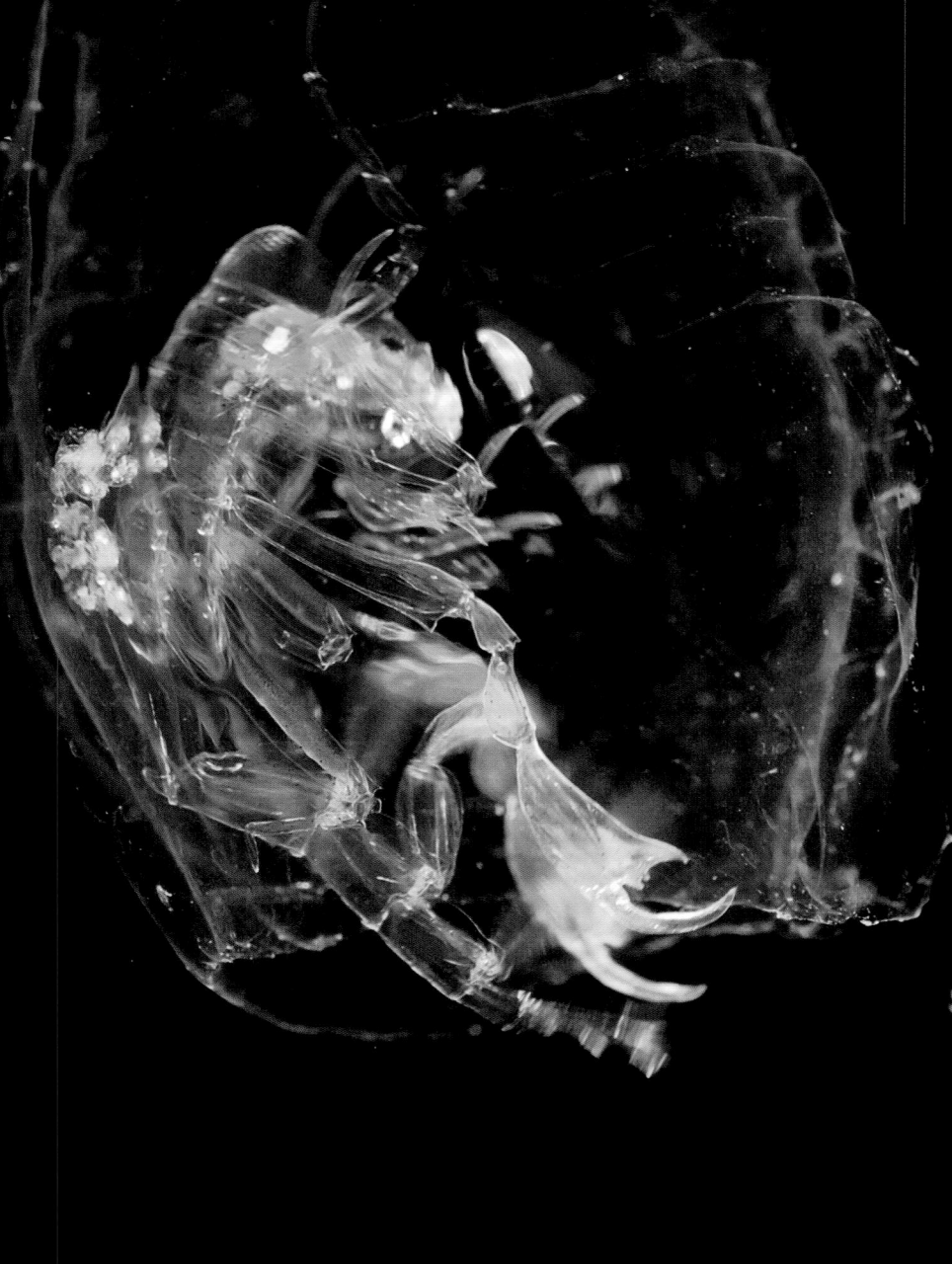

Phronima females spend much time in their barrels caring for their young. Larvae are visible here bunched on the inside surface of the barrel. In contrast *Phronima* males leave the barrel at the slightest disturbance.

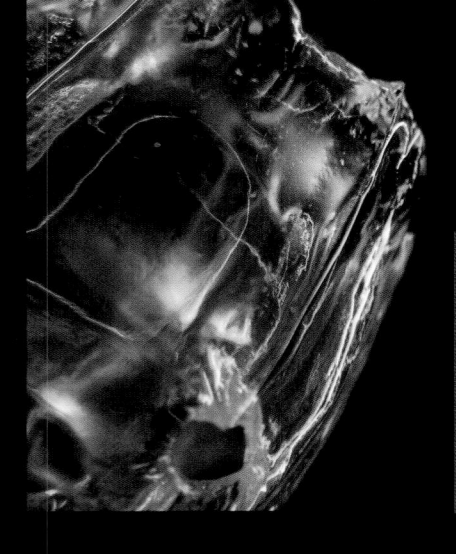

GELATINOUS BARRELS

Phronima eat the insides of salps and pyrosomes and use their emptied envelopes to construct barrels. During successive molts and periods of growth, *Phronima* build bigger barrels.

YOUNG *PHRONIMA*
Collected in the bay of Villefranche-sur-Mer
and during the crossing of the Indian Ocean
by the *Tara Oceans Expedition*.

A JUVENILE AMPHIPOD

Above left: The huge compound eyes of *Phrosina semilunata* with four dark-red retinas.

Above right: The digestive system appears as a reddish-brown mass below the compound eyes.

Collected in the bay of Villefranche-sur-Mer. Photo by Christian Sardet and Sharif Mirshak, Parafilms, Montreal.

PANORAMIC VISION WITH FOUR EYES

Phronima are carnivores. With their claws and sharp mouthparts, they shred and ingest gelatinous organisms. Four large compound eyes give them remarkable panoramic vision essential for hunting prey and escaping predators.

Opposite page: Close-up of the head of *Phronima sedentaria* with four large compound eyes containing dark-red retinas. The two medial eyes have elongated crystalline cones that extend the whole length of the head.

Right: Another view of the two medial eyes, comprised of long crystalline cells covering the entire head and ending in the facets visible here.

Collected in the bay of Villefranche-sur-Mer. Photo by Christian Sardet and Sharif Mirshak, Parafilms, Montreal.

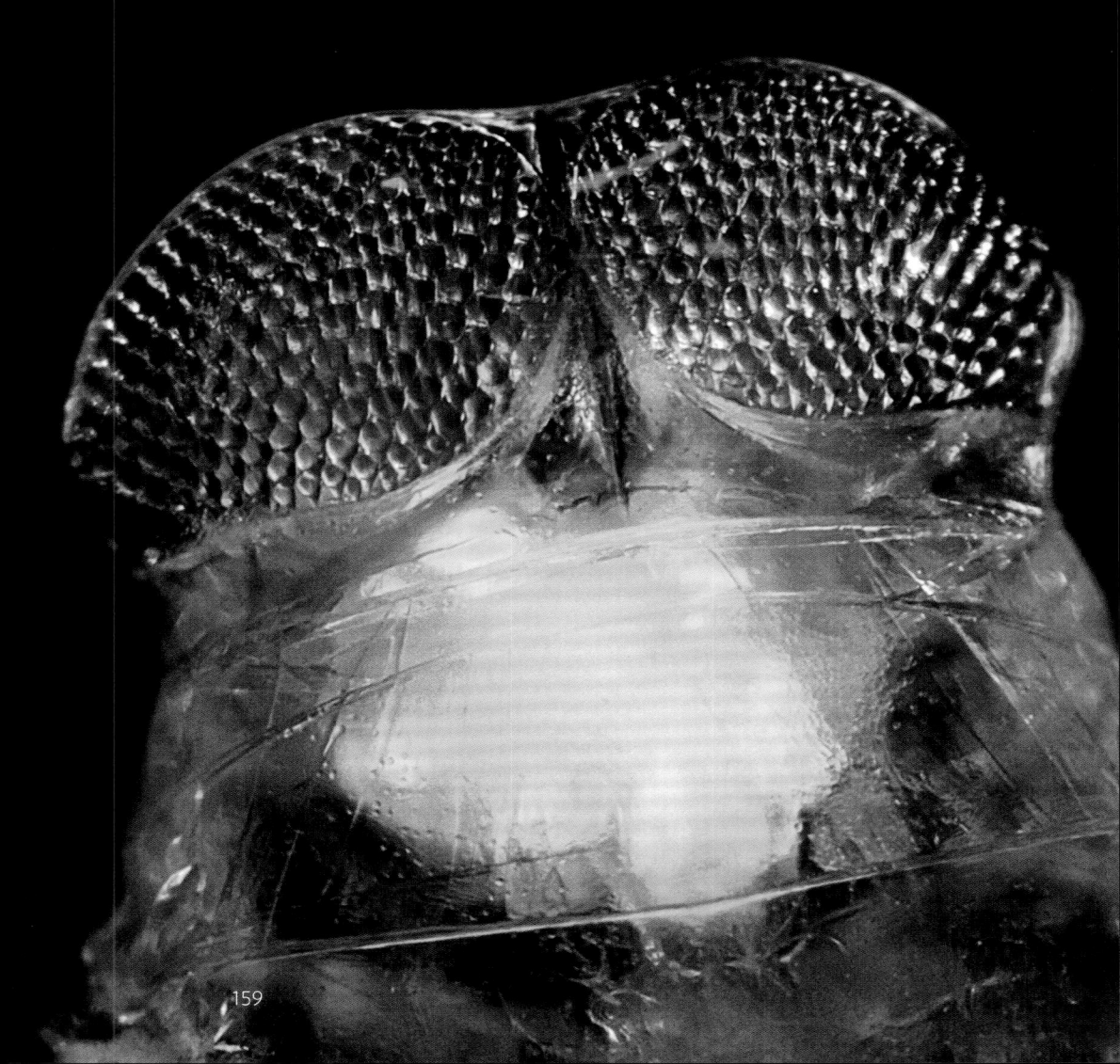

Claws and Camouflage

Phronima sedentaria use two large claws to catch prey and defend themselves. Like the rest of the animal, these claws can change color by means of pigmented cells, called chromatophores, located in the cuticle, above powerful muscles.

Opposite page: When chromatophores are fully spread they appear star-shaped and the whole animal takes on a reddish tint. Powerful muscles are visible inside the claw.

This page: When chromatophores contract into tiny brown spots, the animal loses its red color and becomes transparent inside its barrel. In this way the *Phronima* resembles a salp or a pyrosome, rarely eaten by other carnivores.

Collected in winter in the bay of Villefranche-sur-Mer. Photo by Christian Sardet and Sharif Mirshak, Parafilms, Montreal.

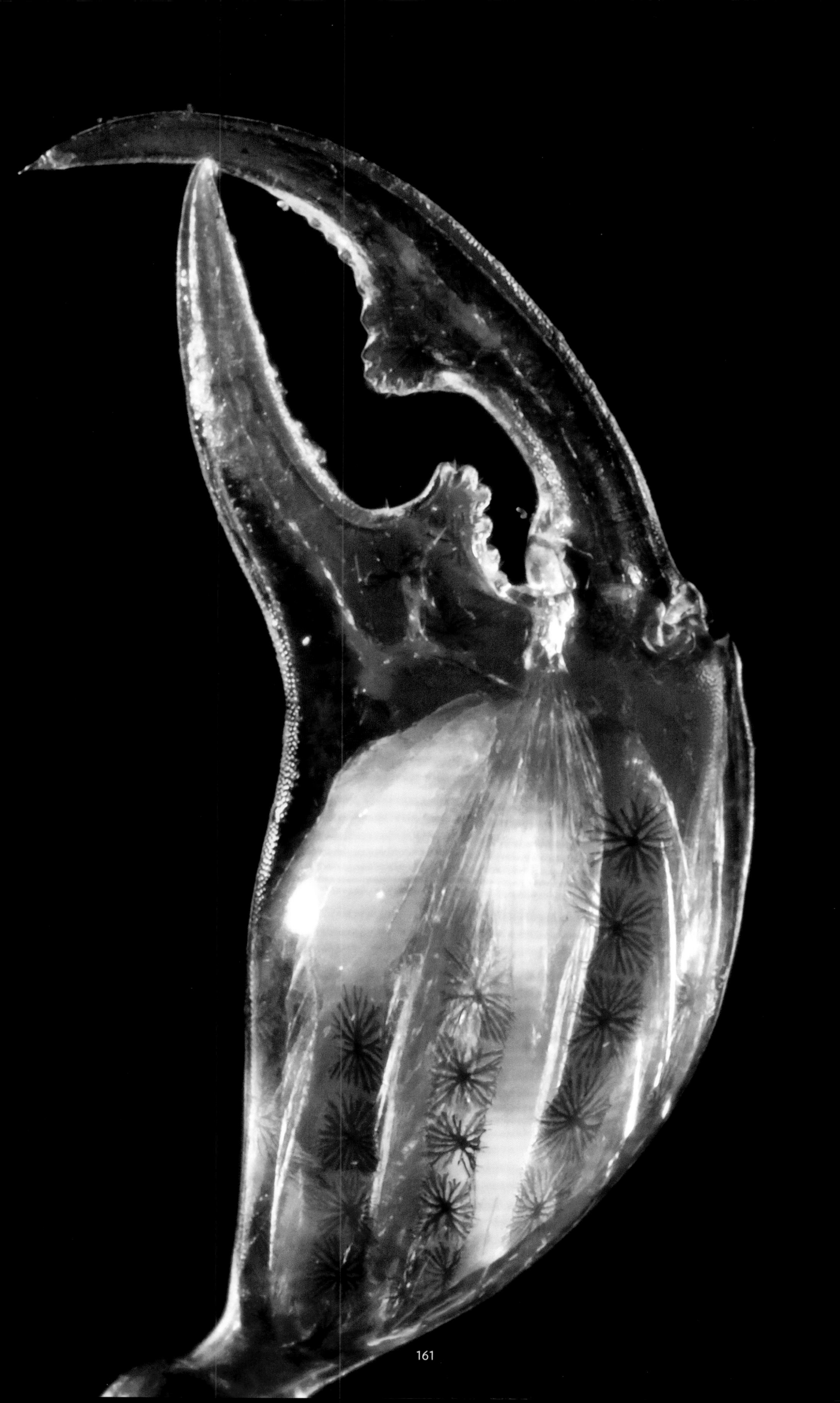

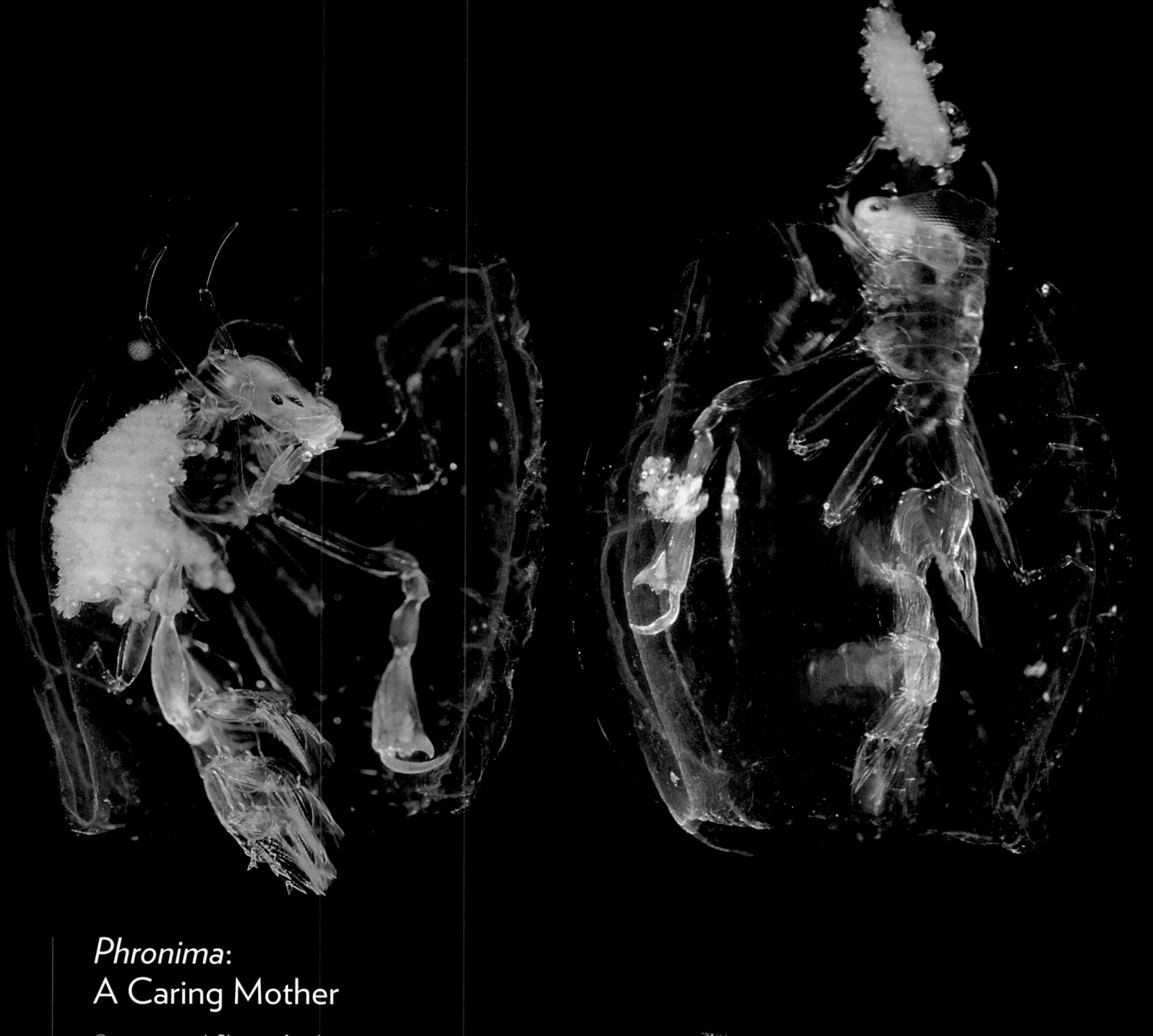

Phronima:
A Caring Mother

Opposite page: A *Phronima* female incubates embryos in a marsupial-like pouch.

Photo by Noé Sardet, Parafilms, Montreal.

This page, above left: A female *Phronima* holds her progeny against the wall of the barrel. She feeds the young for two to three weeks, making sure they stay bunched together inside. Above right: a group of young *Phronima* escaped from the barrel, but were caught by their mother and brought back inside.

JUVENILE SOLIDARITY

Phronima juveniles are able to move around on the wall of the barrel, but they tend to stay grouped, with heads facing outward, as if held together by a mutual force of attraction.

PTEROPODS AND HETEROPODS

MOLLUSKS THAT SWIM WITH THEIR FEET

Most snails and slugs move slowly using a slimy muscular foot. These creatures belong to the class of gastropods, mollusks whose origins date back over 500 million years. Some of the gastropods crawling on the seabed slowly adapted to the open ocean and evolved into planktonic pteropods and heteropods.

The progressive transformation of the foot into a single fin (the heteropods) or two fins (the pteropods) allowed these mollusks to colonize the water column. At present more than a hundred species of planktonic mollusks have been identified. Pteropods drift with the currents but can also perform quick movements. They seem to fly through the water using their fins as wings. For this reason pteropods are commonly called "sea butterflies."

Certain pteropods such as *Limacina*, and heteropods such as *Atlanta*, construct coiled shells of calcium carbonate (aragonite) characteristic of their crawling gastropod relatives. Others, like the pteropods *Cavolinia* and *Creseis*, evolved protection in the form of symmetrical, cone-shaped, or globular shells. A few millimeters to several centimeters in size, the heteropods and pteropods often display vibrant colors. They are either males or females and reproduce by mating. Individuals of some species switch sex, first developing as male and later changing to female. Certain females like *Firola* drag long filaments containing clutches of embryos. Others like *Creseis* or *Cavolinia* release transparent pouches full of rapidly dividing eggs that drift with plankton.

Planktonic mollusks live at different depths, but most migrate to the surface at night. Although widespread in warm seas, some species such as *Clione*, nicknamed "sea angel," form large swarms in freezing northern waters. Sea angels are naked marine mollusks, that is, lacking shells, in the group called gymnosomes. Traditionally known as "whale food" by fishermen, *Clione limacina* was the first species of gymnosome to be described, as early as 1676.

The gymnosome pteropods *Clione* and *Pneumodermopsis* are voracious predators. Moving their fins frantically, gymnosomes charge at high speed to catch their prey, including shell-covered pteropods, the thecosomes. With or without shells, heteropods like *Atlanta* or *Firola* are also voracious carnivores. They locate and capture prey thanks to a pair of large, complex eyes. Most thecosome pteropods have a different feeding strategy, trapping small organisms and bacteria in a net of mucous deployed at the opening of the shell near their mouth. Like all mollusks, heteropods and pteropods gnaw their prey using an abrasive tongue called a radula. The radula is a kind of ribbon equipped with a double row of teeth that are constantly renewed.

We may wonder what the future holds for shelled heteropods and pteropods. Not only are they prey for gymnosomes and many kinds of fish, but the shells of thecosome mollusks are becoming increasingly fragile. This is due to increased ocean acidification, a direct consequence of elevated CO_2 levels resulting from human activities.

Plankton Chronicles website
Pteropods and heteropods

THREE PLANKTONIC MOLLUSKS

Creseis conica, the largest organism in this photo (about 1 cm), is a thecosome pteropod with a symmetrical cone-shaped shell. To the right is *Pneumodermopsis paucidens*, a naked pteropod, or gymnosome. To the left is *Atlanta peronii*, a heteropod with a round, spiraled shell.

Plankton collected in winter in the bay of Villefranche-sur-Mer.

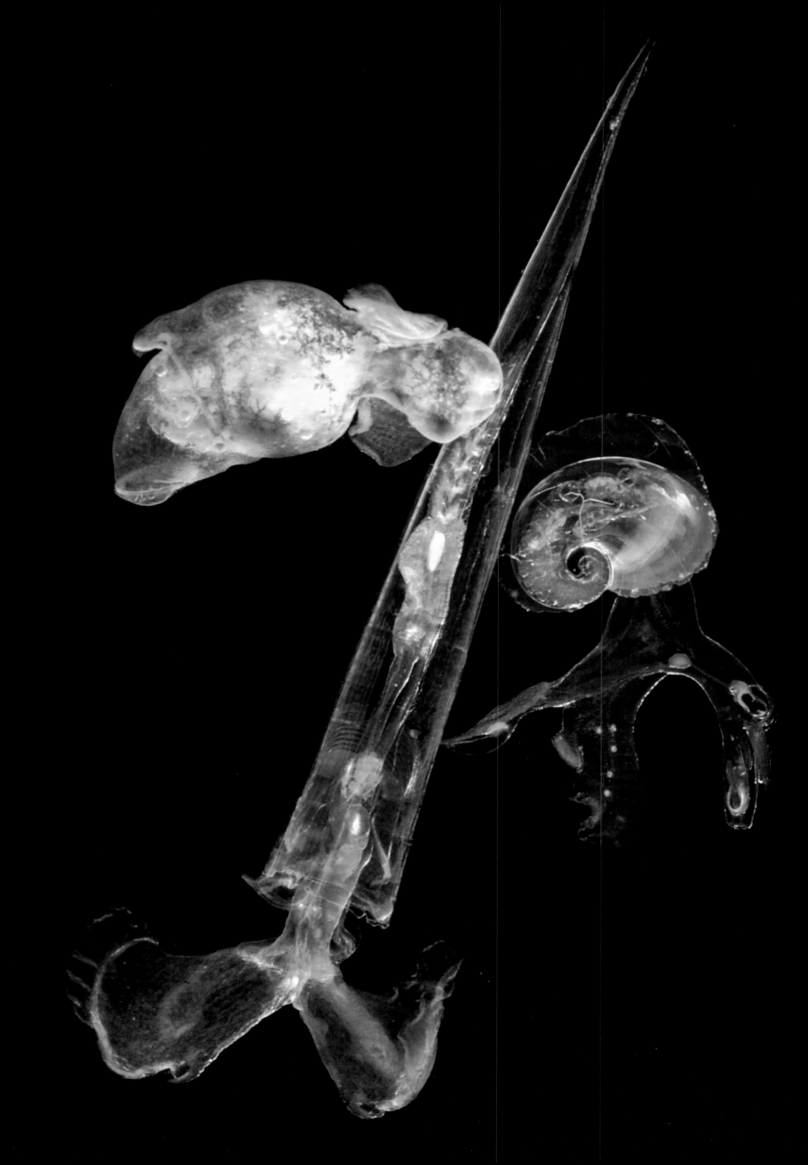

Sea Angels Are Formidable Predators

The naked pteropod *Clione limacina*, or "sea angel," is a torpedo-like creature a few centimeters long. Furiously flapping its fins, it speeds through the water hunting its favorite prey, the coil-shelled thecosome pteropod *Limacina helicina* (lower left corner). On contact, *Clione* immediately ejects six buccal cones, grabs the prey, then eats it slowly with its raspy tongue. *Clione* roam the cold polar waters where they can reach high densities comparable to the tiny shrimp that constitute krill. Sea angels are themselves a major food for marine mammals.

PHOTO TAKEN BY ALEXANDER SEMENOV IN THE WHITE SEA, SOUTH OF THE BARENTS SEA (RUSSIA).

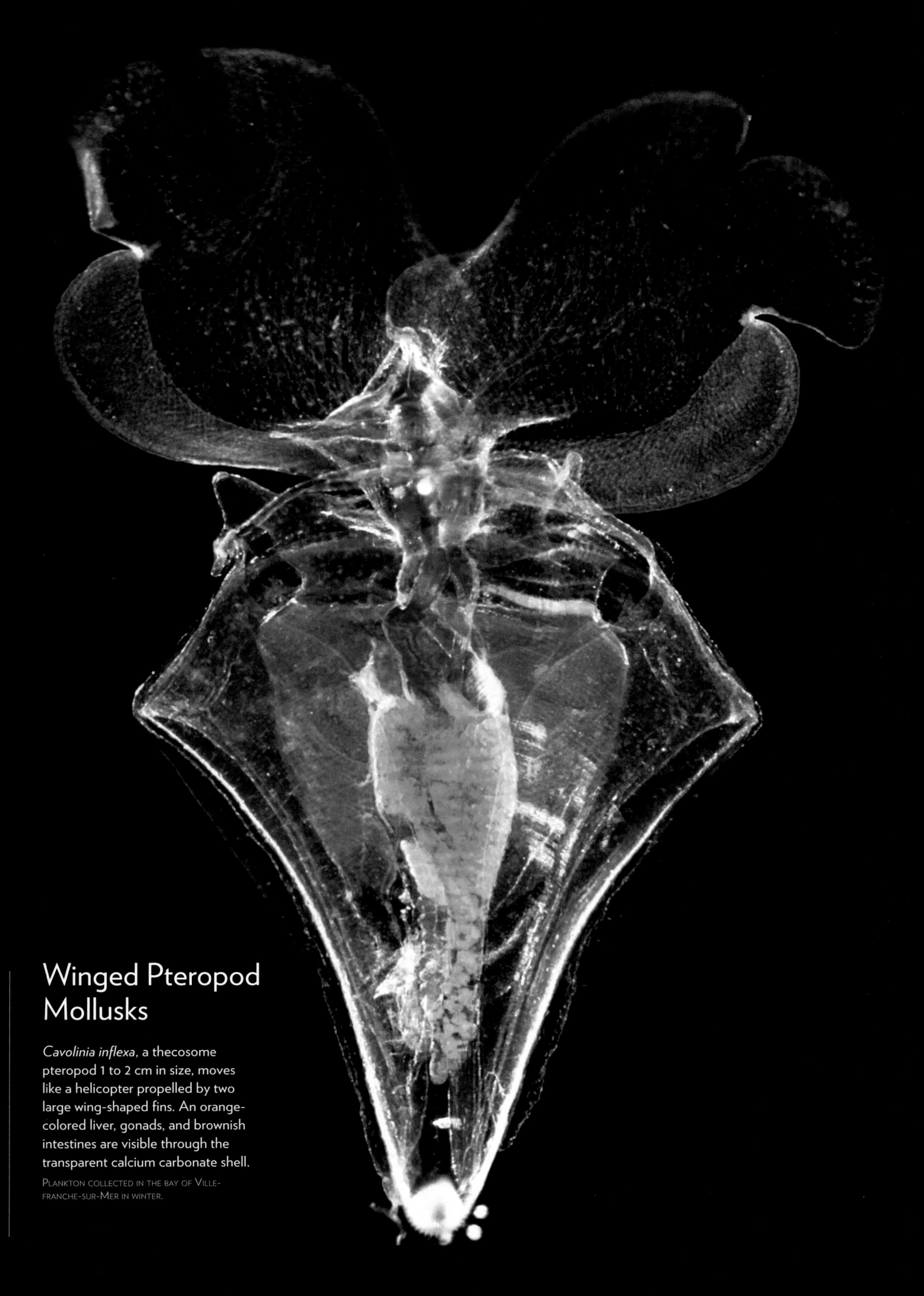

Winged Pteropod Mollusks

Cavolinia inflexa, a thecosome pteropod 1 to 2 cm in size, moves like a helicopter propelled by two large wing-shaped fins. An orange-colored liver, gonads, and brownish intestines are visible through the transparent calcium carbonate shell.

PLANKTON COLLECTED IN THE BAY OF VILLE-FRANCHE-SUR-MER IN WINTER.

Three Pteropods with Shells

The pteropods *Styliola subula* (left), *Creseis acicula* (middle hiding inside its shell) and *Creseis conica* (right) are all thecosomes, a category derived from the Greek *theque* ("casing"). Through their transparent calcium shells, hepatic glands and digestive organs are visible, colored orange, yellow, or green, depending on what the mollusk eats.

PLANKTON COLLECTED IN THE BAY OF VILLE-FRANCHE-SUR-MER IN WINTER.

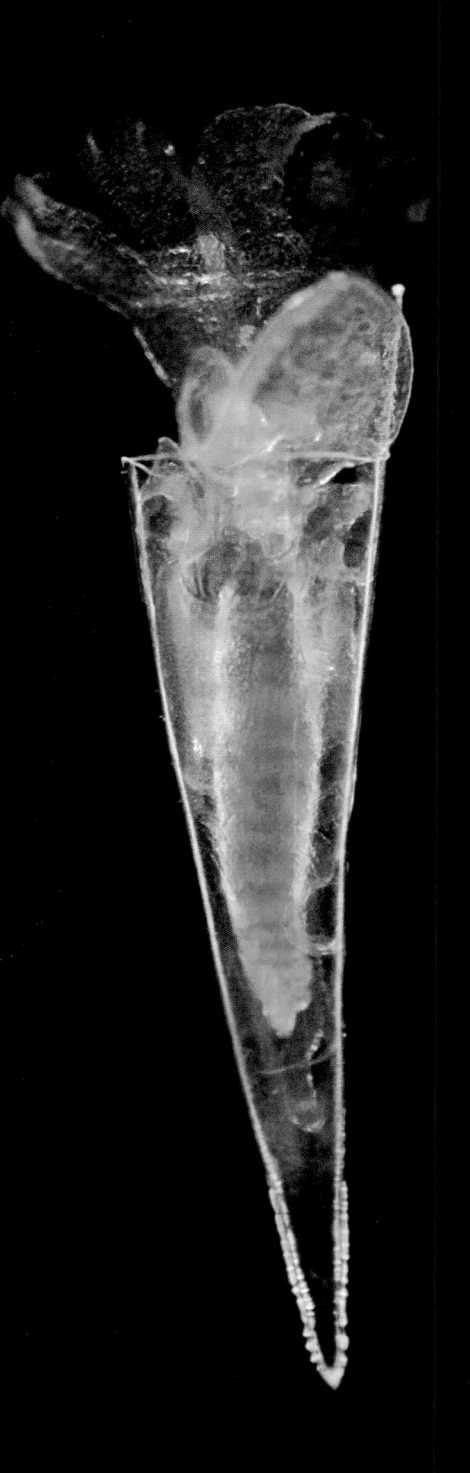

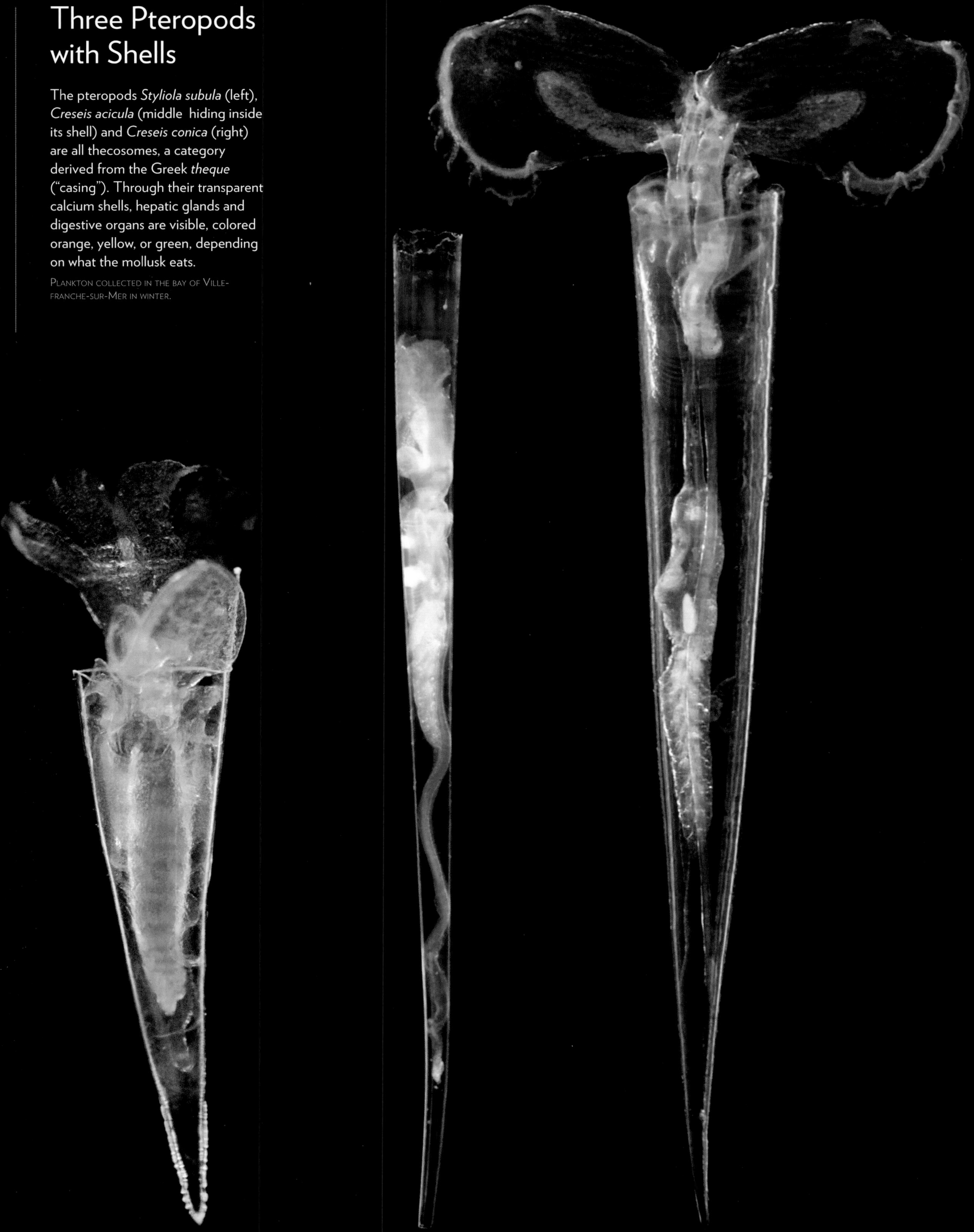

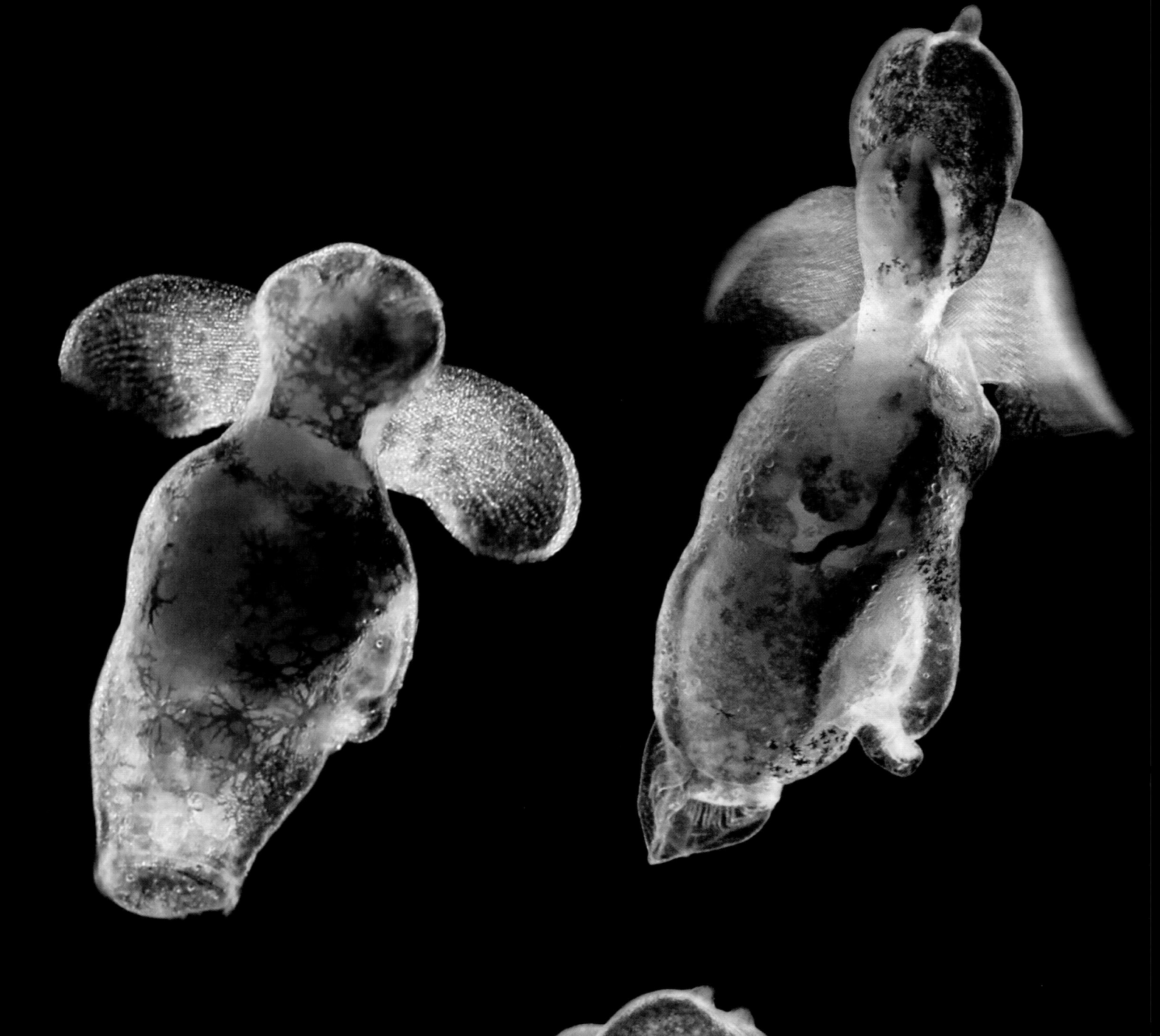

Quick as Torpedoes

Among the fifty known species of gymnosome pteropods, *Pneumodermopsis paucidens* is the most common in the waters of the French Riviera. Flapping their fins, these gastropod mollusks swim very quickly. Their elastic skin is dotted with mucus cells, as well as pigment cells called chromatophores. Contracting or spreading these chromatophores changes the appearance of the gymnosome. At the slightest sign of danger, *Pneumodermopsis* rolls into a compact brown ball.

A Gymnosome Devouring a Thecosome

The gymnosome *Pneumodermopsis paucidens* is a carnivore. Using delicate suckers, it attaches itself to the shell of its favorite prey, the thecosome pteropod *Creseis*. *Pneumodermopsis* inserts its long proboscis deep into the *Creseis* shell and sucks out its tissues. After its meal, the voracious gymnosome stretches itself out, digesting and defecating, before beginning the hunt for new victims.

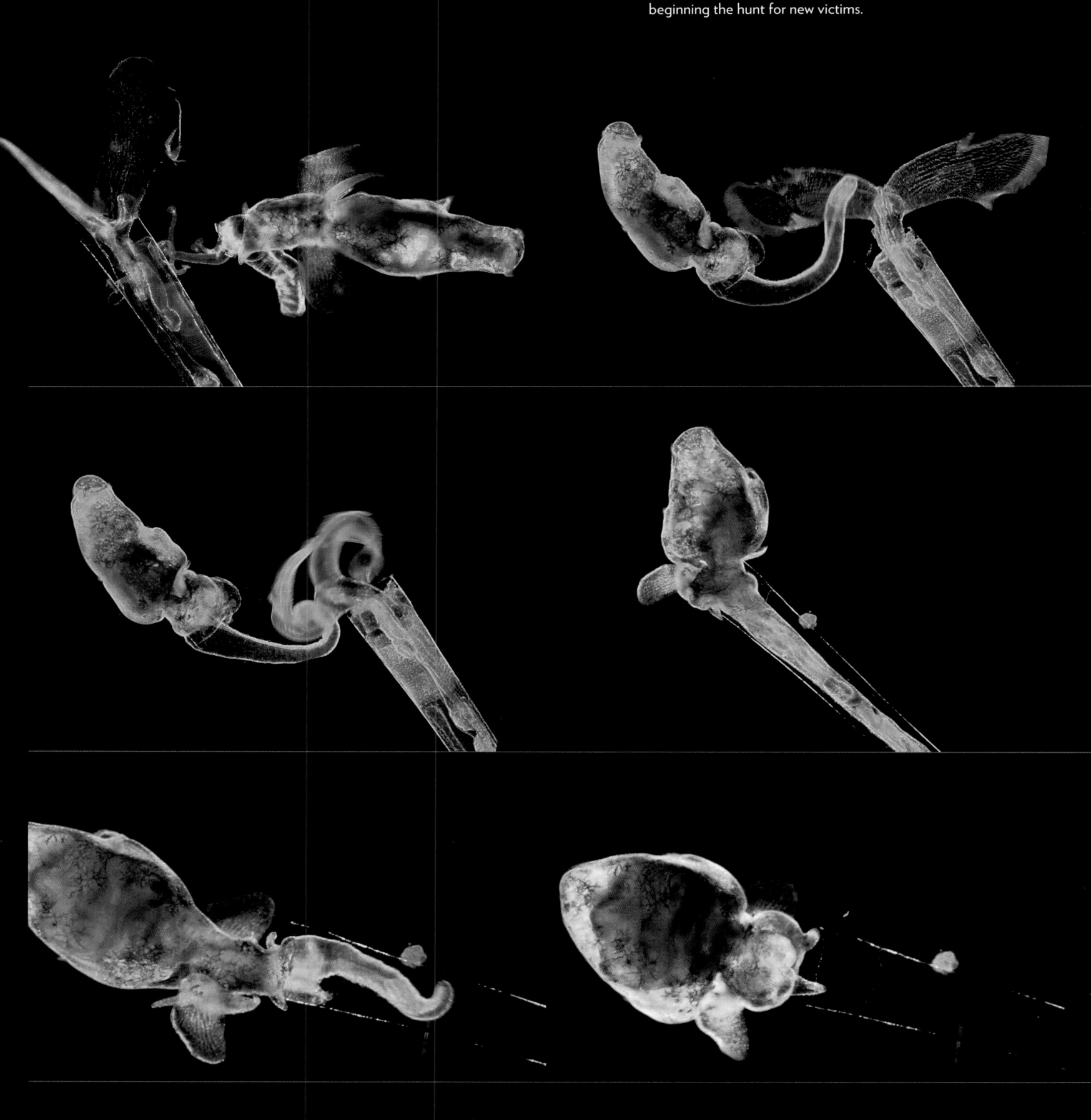

The Beautiful, Transparent *Atlanta*

Atlanta peronii is a common species in the bay of Villefranche-sur-Mer. A heteropod mollusk with a coiled shell, it possesses a flattened foot that serves as a single swimming fin. Gills, digestive tract, heart, gonads, and genitals are easily visible through the transparent shell. In the male (top photo, this page), the penis appears behind the head, and a reddish suction cup can be seen at the edge of the fin. This suction cup is used to hold the female (top photo, opposite) in position during mating. *Atlanta* are carnivorous predators. Peering into the darkness with their large mobile eyes, they catch prey and chew them with a raspy tongue situated at the extremity of a long snout.

WE COLLECTED THE PURPLE-COLORED *ATLANTA* (CLOSE-UP BELOW) IN THE INDIAN OCEAN DURING THE *TARA OCEANS EXPEDITION*.

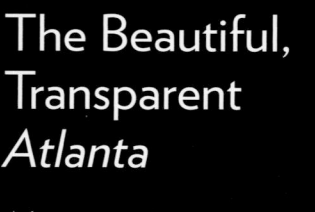

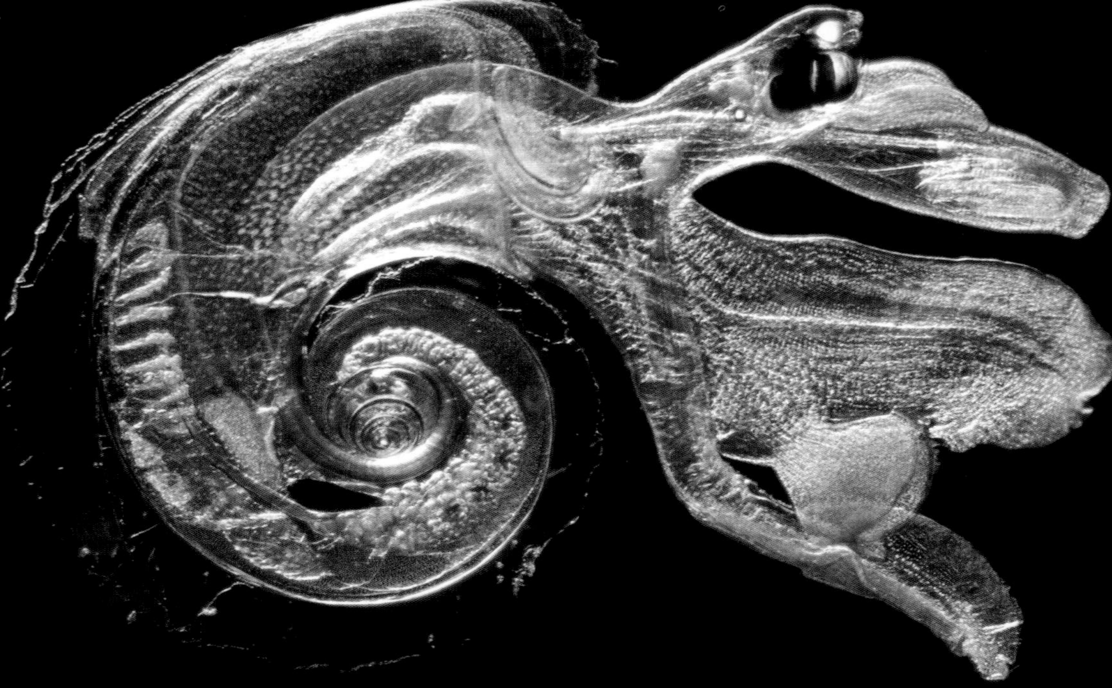

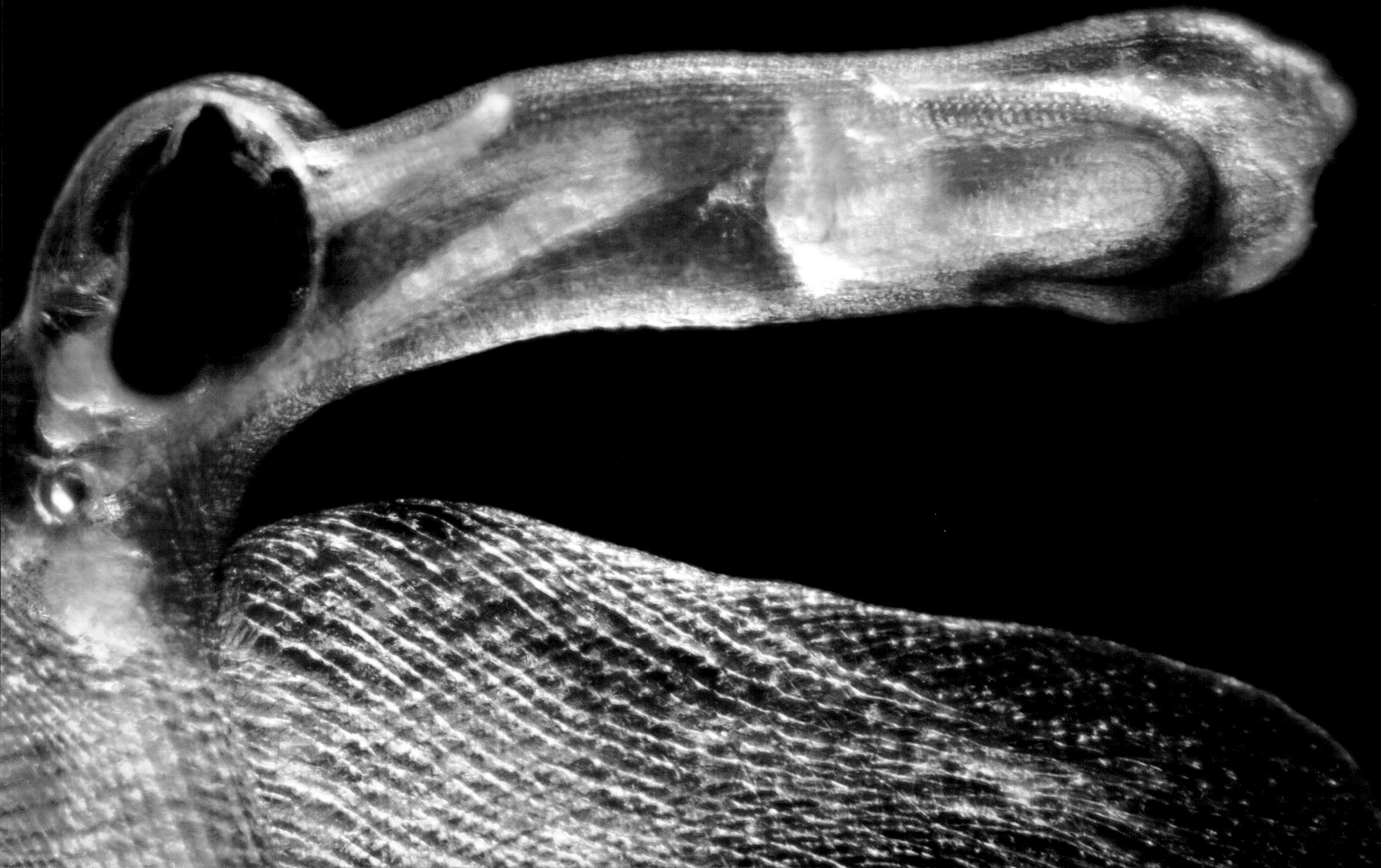

Two *Atlanta peronii* heteropods hide inside their calcareous shells. A pair of eyes is visible in each.

Firola,
the Sea Elephant

Firola is one of the largest planktonic mollusks, measuring 10 to 30 cm. Divers call it "sea elephant" because its long proboscis resembles an elephant's trunk. The species name is *Pterotrachea* or *Firola coronata.* With a single fin characteristic of heteropods, *Firola* swims in an inverted position. The adult *Firola* has no shell, lost after the larval stage during metamorphosis. At the head of its cylindrical body is a long snout, a mouth, and two remarkably mobile eyes. Like all heteropods, *Firola* is carnivorous. This particular specimen is easily identified as male since only males have a suction cup situated at the edge of the fin. The cup serves to immobilize the female during mating. Female *Firola* are often observed towing a long filament filled with embryos.

A WATCHFUL EYE

Firola detects and stalks its prey with two large mobile eyes capable of panoramic vision. The eyes have ovoid crystalline lenses and garnet-red retinas, and are innervated by two ganglia, opaque white masses at the center of a nerve network. Halfway between the eye and the ganglia one can see a small white mass, an organ called the otolith that confers balance and orientation.

A RASPY TONGUE

Like other gastropod mollusks, *Firola* has a raspy tongue called a radula. The radula is covered by series of chitinous teeth and functions as a grater. Rows of teeth are continuously and rapidly renewed.

CEPHALOPODS AND NUDIBRANCHS

BEAUTIFUL COLORS AND CAMOUFLAGE

Cephalopods and nudibranchs are evolutionarily quite distinct. Cephalopods constitute their own class within the phylum Mollusca, while nudibranchs are aligned with the class Gastropoda. The name *cephalopod* comes from the Greek *kephale* ("head") and *podos* ("foot"). This is the most intelligent group of mollusks and includes octopus and squid. Cephalopods have well-developed nervous and visual systems, and are fast swimmers that brave the currents. Nudibranchs owe their name to the fact that their gills are exposed on the surface of their bodies (from the Latin *nudus* ("naked") and the Greek *branchia* ("gill"). Also known as "sea slugs," most nudibranchs crawl on the seabed.

Few nudibranchs are planktonic, but one of the most notorious is *Glaucus atlanticus*, a predator of *Physalia*, *Porpita*, and *Velella*, the cnidarians that sail along the ocean surface (see page 122). After eating its gelatinous prey, *Glaucus* appropriates its victim's stinging cells and stores them in pockets called cnidosacs, to be used to immobilize future prey. For most nudibranchs, the stinging cells and their toxins constitute an effective defense against would-be predators as well. Nudibranchs can also curl themselves into tight balls to thwart attackers.

In other respects cephalopods and nudibranchs are alike. Certain cephalopods and all nudibranchs have evolved without the external shells that serve as shelter for most mollusks. Both have developed extraordinary uses of color for camouflage and communication. Plankton nets sometimes bring to the surface beautiful nudibranch and cephalopod embryos and

juveniles that have the same colorful attributes as their parents, but with fewer pigment cells.

Cephalopods exhibit the most spectacular changes in color in order to hide, or to stand out. Chromatophores, special skin cells filled with pigments, allow cephalopods to rapidly change their coloration and patterning to blend into the environment or signal to others. Fast color changes are mediated by nerve signals and muscle contractions that control the size and opacity of pigment cells. Cephalopod juveniles in the plankton possess a few large chromatophores. When these spread out, the cells reflect light, turning the skin vivid colors. In contrast, when chromatophore cells contract into tiny dots, the skin loses its colors. True chameleons of the ocean, cephalopods also use their repertoire of color changes to communicate with one another.

PLANKTONIC JUVENILES

Opposite page: Juvenile *Octopus vulgaris* (top).
PHOTO BY STEFAN SIEBERT, BROWN UNIVERSITY, USA.
Young squid of the family Gonatidae (middle).
PHOTO BY KAREN OSBORN, SMITHSONIAN NATIONAL MUSEUM OF NATURAL HISTORY, WASHINGTON DC, USA.
Juvenile squid *Loligo vulgaris* (bottom right).
PHOTO BY SHARIF MIRSHAK, PARAFILMS, MONTREAL.
Young nudibranch, *Flabellina* sp. (bottom left).
COLLECTED IN THE BAY OF VILLEFRANCHE-SUR-MER.

176

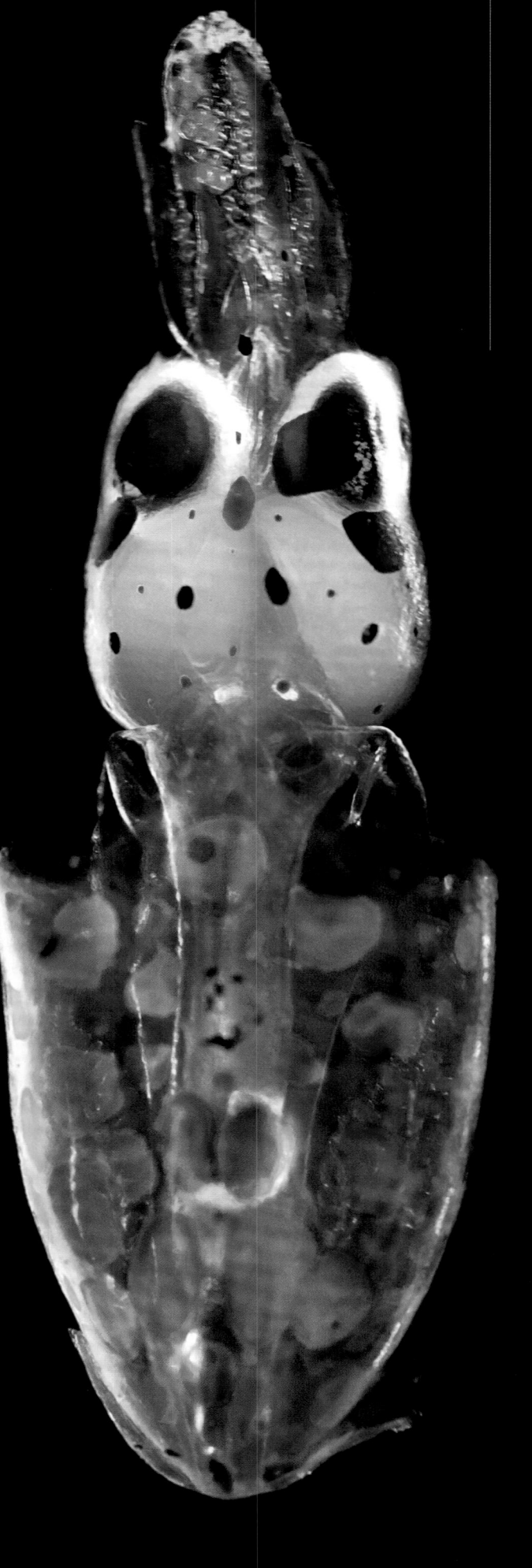

Cephalopods, Chameleons of the Sea

This newly hatched *Loligo vulgaris* squid (left, measuring 4 to 5 mm) already displays dozens of red and yellow pigment cells (chromatophores) that modulate the color of its skin. Below, tentaculate squid *Planctoteuthis* sp. larvae demonstrate how contracted (left) or dilated (right) chromatophores change their color for camouflage and communication.

Photos by Sharif Mirshak, Parafilms, Montreal (left), and Karen Osborn, Smithsonian National Museum of Natural History, Washington DC, USA (center and right).

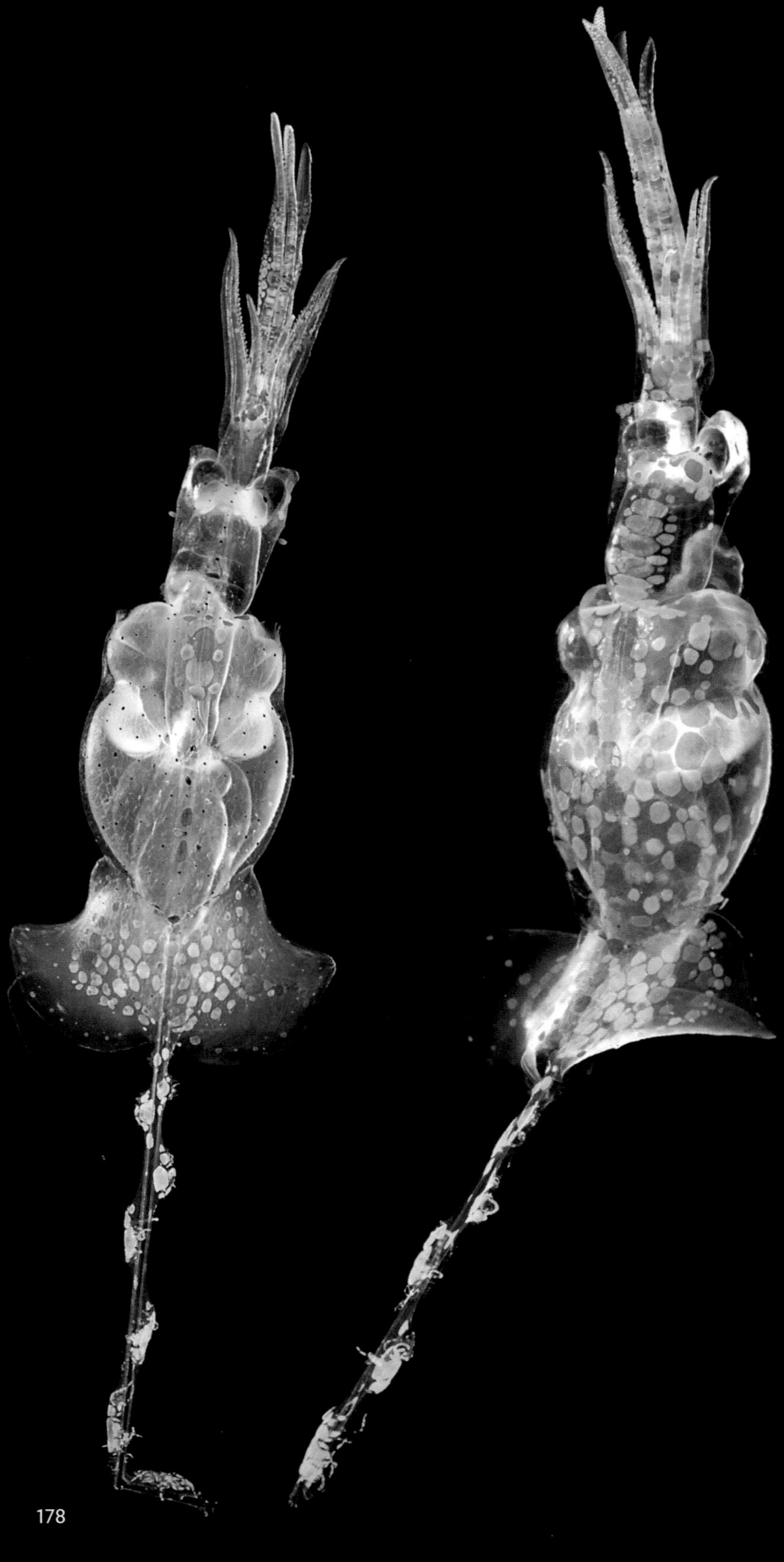

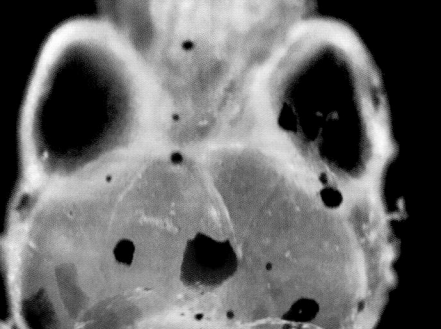

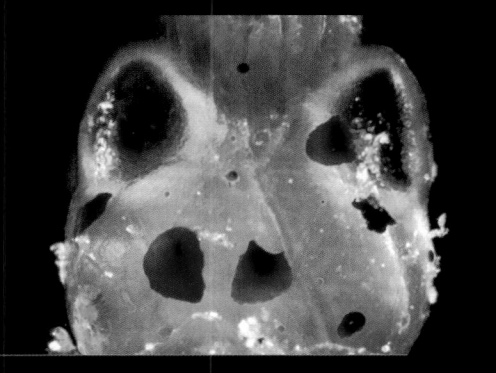

MODULATING COLORS WITH PIGMENT CELLS

Cephalopod chromatophores are the most complex pigment cells in the animal kingdom. They contract or relax, concentrating or spreading out thousands of pigment granules under the skin. Muscle and nerve cells control chromatophore activity, changing the animal's color and opacity in waves. Such variations in appearance are used for camouflage, and also for communication during mating and other interactions.

PHOTOS BY STEFAN SIEBERT, BROWN UNIVERSITY, USA.

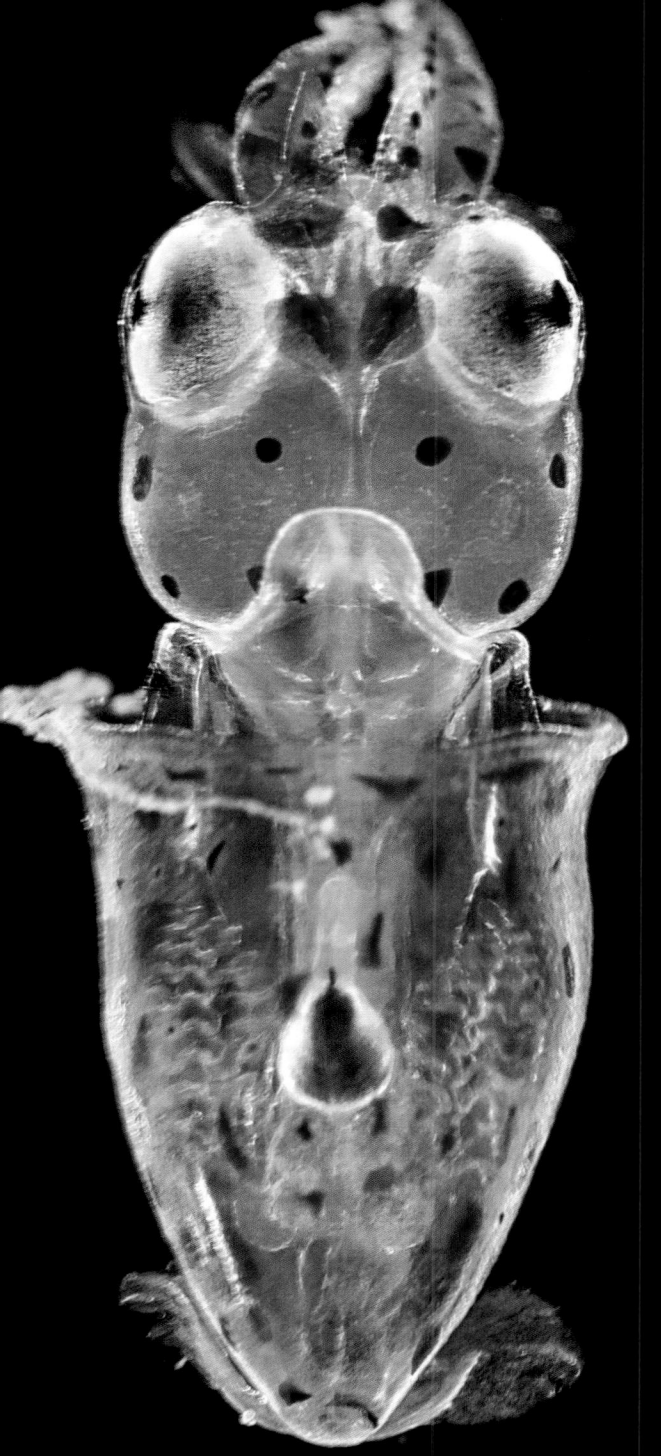

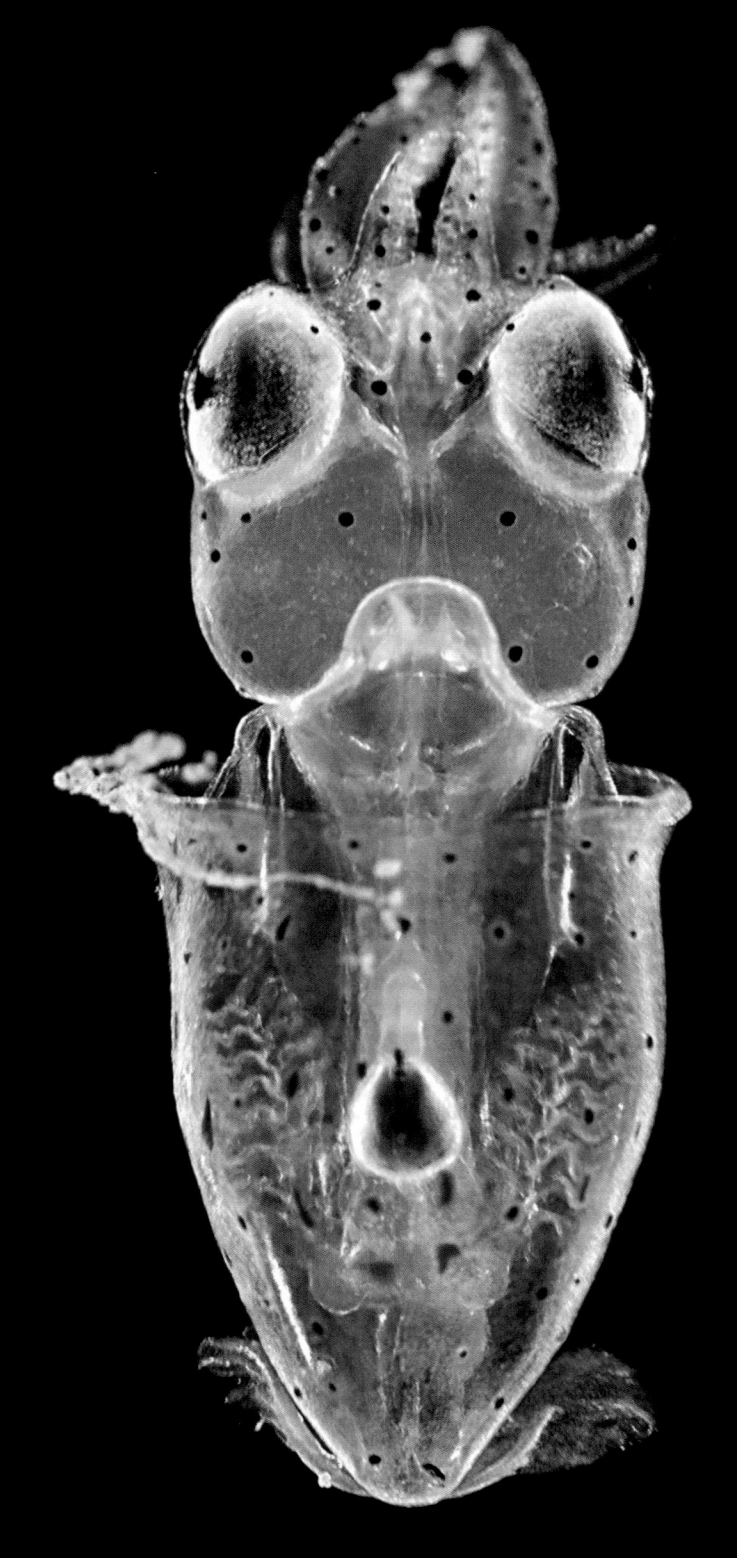

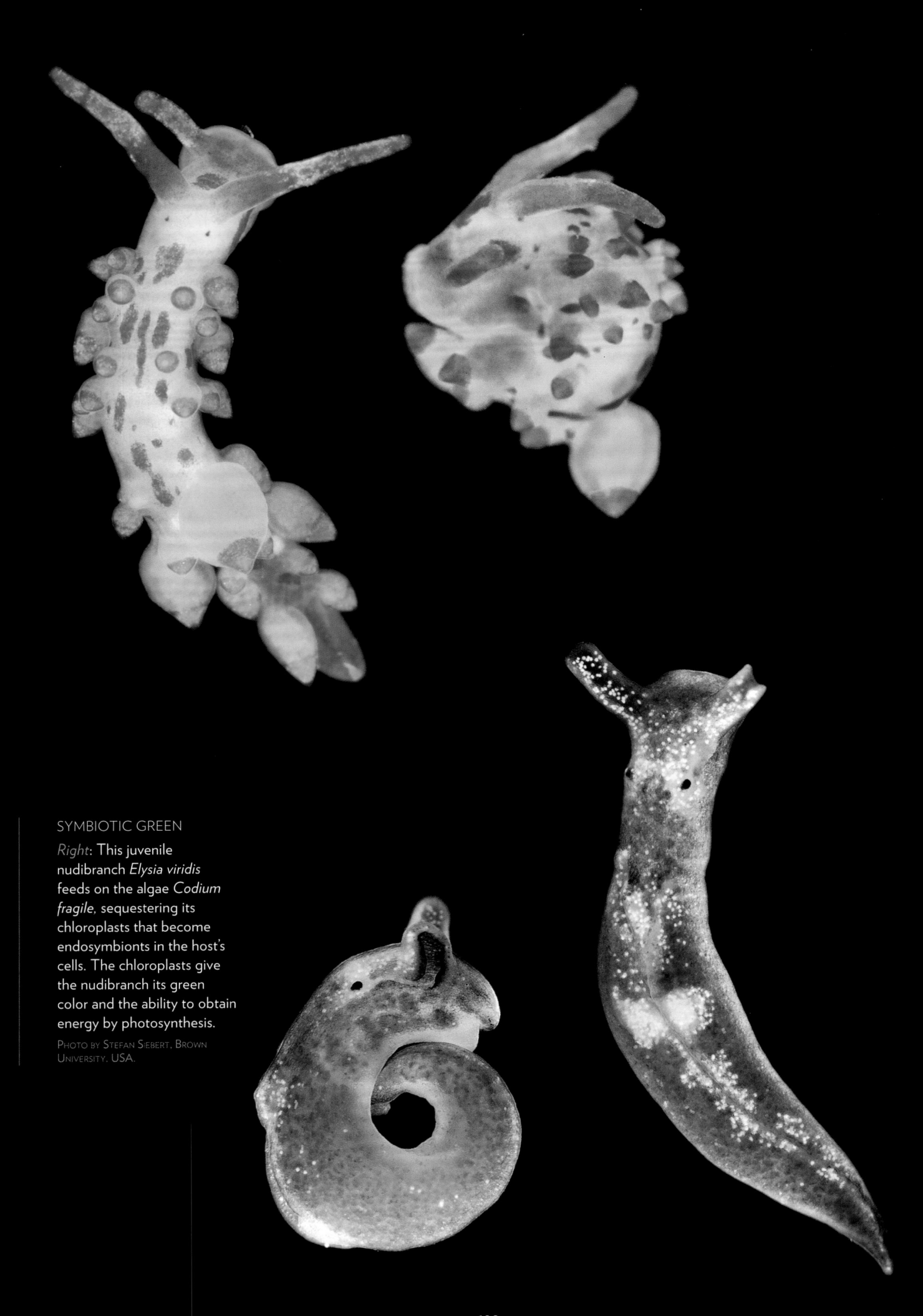

SYMBIOTIC GREEN

Right: This juvenile nudibranch *Elysia viridis* feeds on the algae *Codium fragile*, sequestering its chloroplasts that become endosymbionts in the host's cells. The chloroplasts give the nudibranch its green color and the ability to obtain energy by photosynthesis.

PHOTO BY STEFAN SIEBERT, BROWN UNIVERSITY, USA.

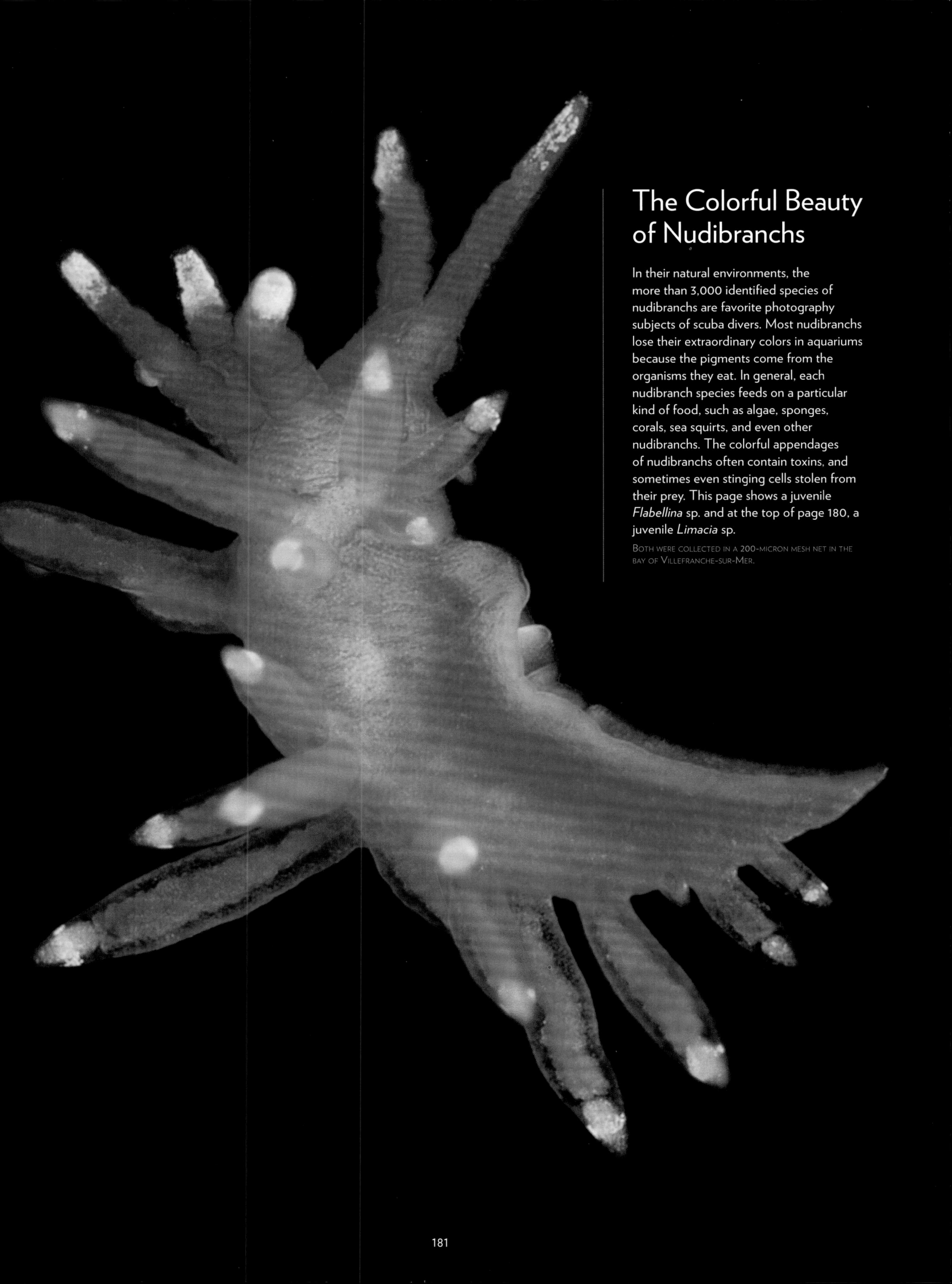

The Colorful Beauty of Nudibranchs

In their natural environments, the more than 3,000 identified species of nudibranchs are favorite photography subjects of scuba divers. Most nudibranchs lose their extraordinary colors in aquariums because the pigments come from the organisms they eat. In general, each nudibranch species feeds on a particular kind of food, such as algae, sponges, corals, sea squirts, and even other nudibranchs. The colorful appendages of nudibranchs often contain toxins, and sometimes even stinging cells stolen from their prey. This page shows a juvenile *Flabellina* sp. and at the top of page 180, a juvenile *Limacia* sp.

BOTH WERE COLLECTED IN A 200-MICRON MESH NET IN THE BAY OF VILLEFRANCHE-SUR-MER.

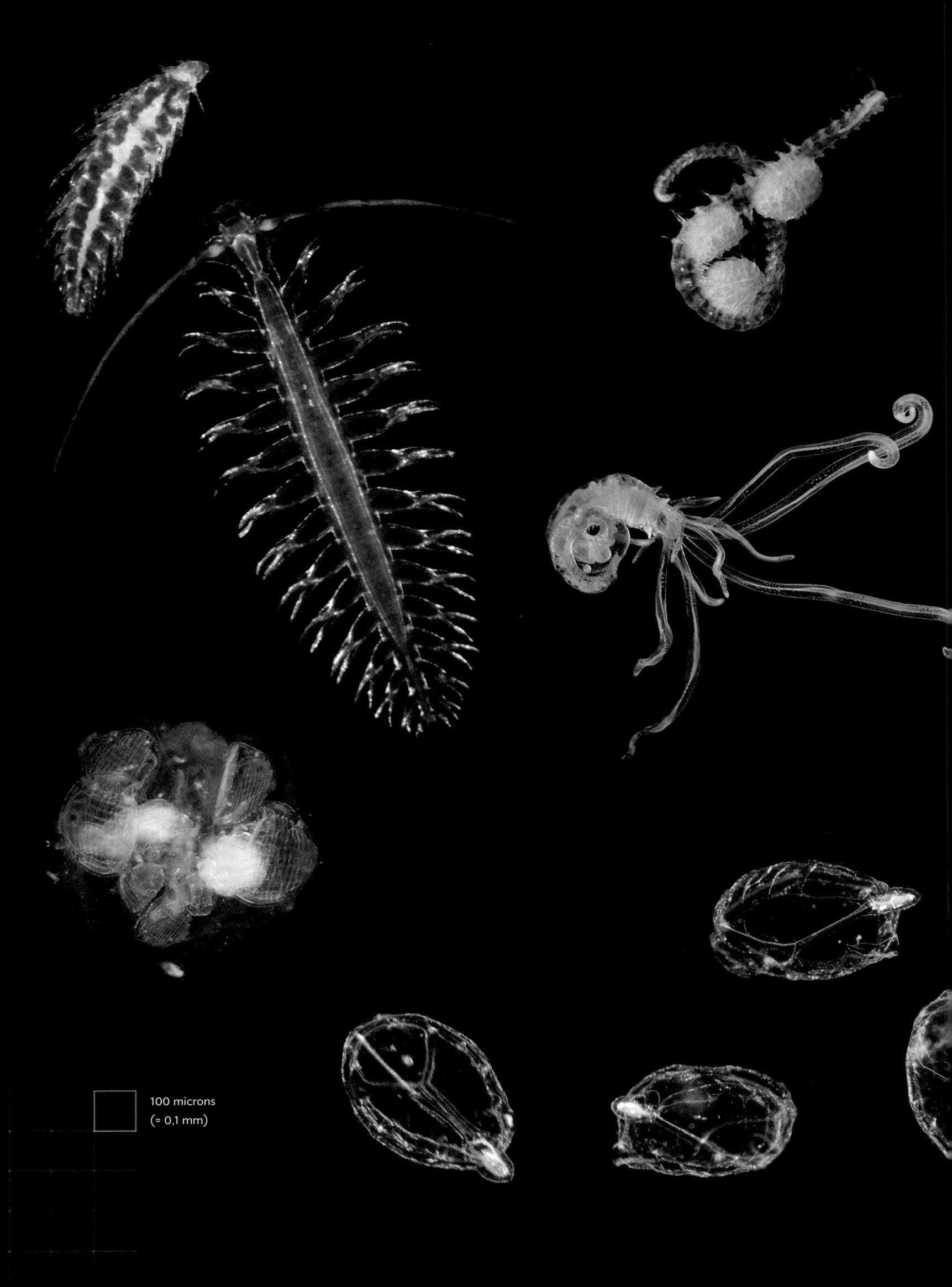

100 microns
(= 0,1 mm)

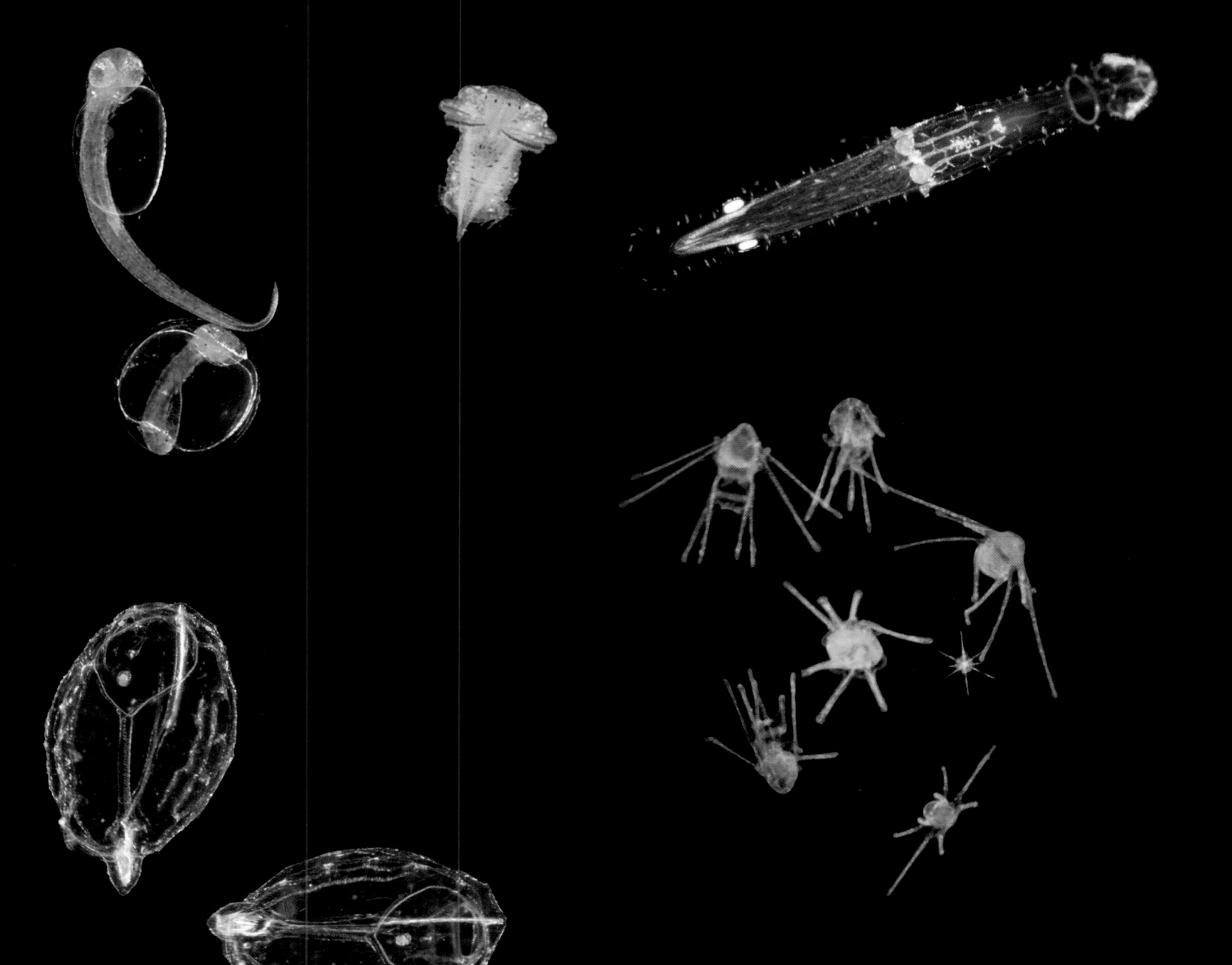

WORMS and TADPOLES
Arrows, Tubes, and Nets

CHAETOGNATHS
ARROWS IN THE OCEAN

With an angular head, beady eyes, and ferocious behavior, chaetognaths seem like miniature crocodiles. Their common name is "arrow worm," owing to their elongated bodies. Arrow worms have rows of teeth around the mouth and sharp, hook-like bristles on each side of the head. These appendages give chaetognaths their scientific name, from the Greek *chaeto* ("bristle/spine") and *gnathos* ("jaw"). In a split second the jaws grab and devour small copepods or larvae. Chaetognaths can also inject neurotoxic venoms to immobilize their prey. When food is scarce, chaetognaths become cannibals, gobbling up their smaller relatives.

Eat and be eaten: chaetognaths are an important link in the food chain of the world ocean. They consume small planktonic animals, and are themselves an abundant source of food for squids, jellies, and fish. About 200 species of chaetognaths have been identified. The smallest, *Mesosagitta minima*, measures only a few millimeters. The biggest, *Pseudosagitta lyra*, grows to more than 5 centimeters in length. Chaetognaths have colonized all surveyed ecological niches in the ocean, including deep hydrothermal vents where new species have recently been discovered.

Arrow worms have an amazing way to reproduce. Each individual is a hermaphrodite, daily producing both male and female gametes. Large quantities of sperm and oocytes (eggs) are stored in huge gonads that sometimes occupy the entire body. The animal is so transparent that oocyte growth and sperm maturation can be readily observed in the living animal. Chaetognaths mate following an unusual courtship ritual. In the species

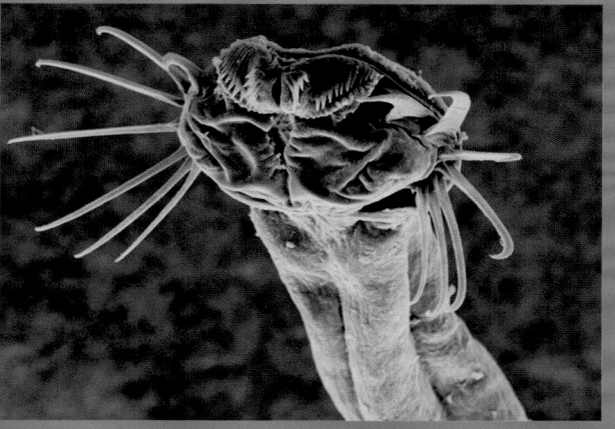

PHOTO BY YVAN PEREZ, UNIVERSITÉ D'AIX-MARSEILLE

Spadella, two individuals wriggle into a head-to-tail position, touching each other to exchange the spermatozoids contained in seminal vesicles near their tails. Deposited behind the head of their partner, the sperm migrate down the body, finally entering the female reproductive organ and fertilizing the oocytes. Fertilized eggs are soon released into the marine environment, and the embryos grow rapidly within their protective casings. Miniature chaetognaths hatch in a day or two, joining their adult relatives adrift in the plankton.

SPADELLA, A HEARTY CHAETOGNATH

Spadella cephaloptera is not truly pelagic, and it is quite stocky compared with the more slender planktonic species. *Spadella* lives near meadows of *Posidonia* algae where it rests on the grass-like stalks. Even headless *Spadella* are able to survive and continue to produce gametes. Because it's so easy to raise and maintain, *Spadella* is a species of choice for laboratory studies of chaetognaths.

BENTHIC PLANKTON COLLECTED IN EARLY SUMMER NEAR MARSEILLE BY YVAN PEREZ, UNIVERSITÉ D'AIX-MARSEILLE, FRANCE.

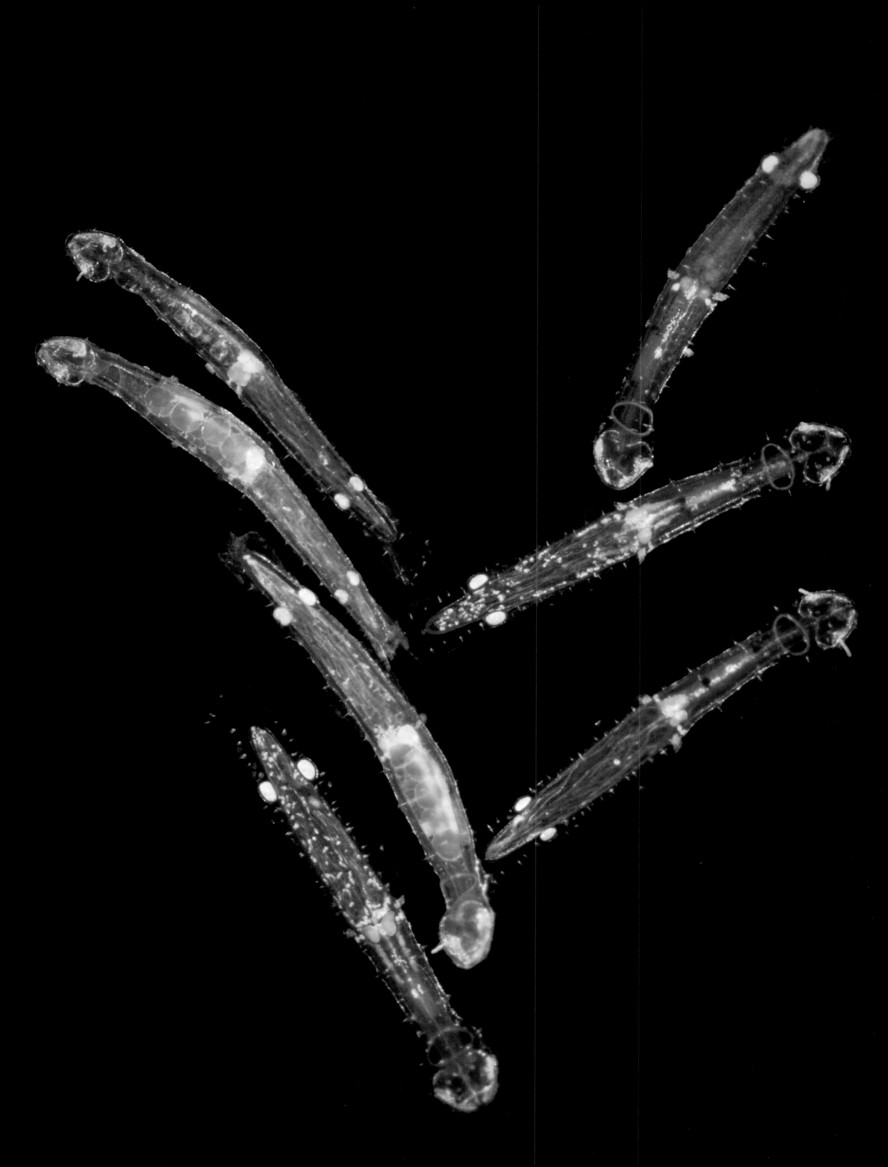

Spadella cephaloptera
COLLECTED IN THE MEDITERRANEAN SEA NEAR
MARSEILLE, FRANCE.

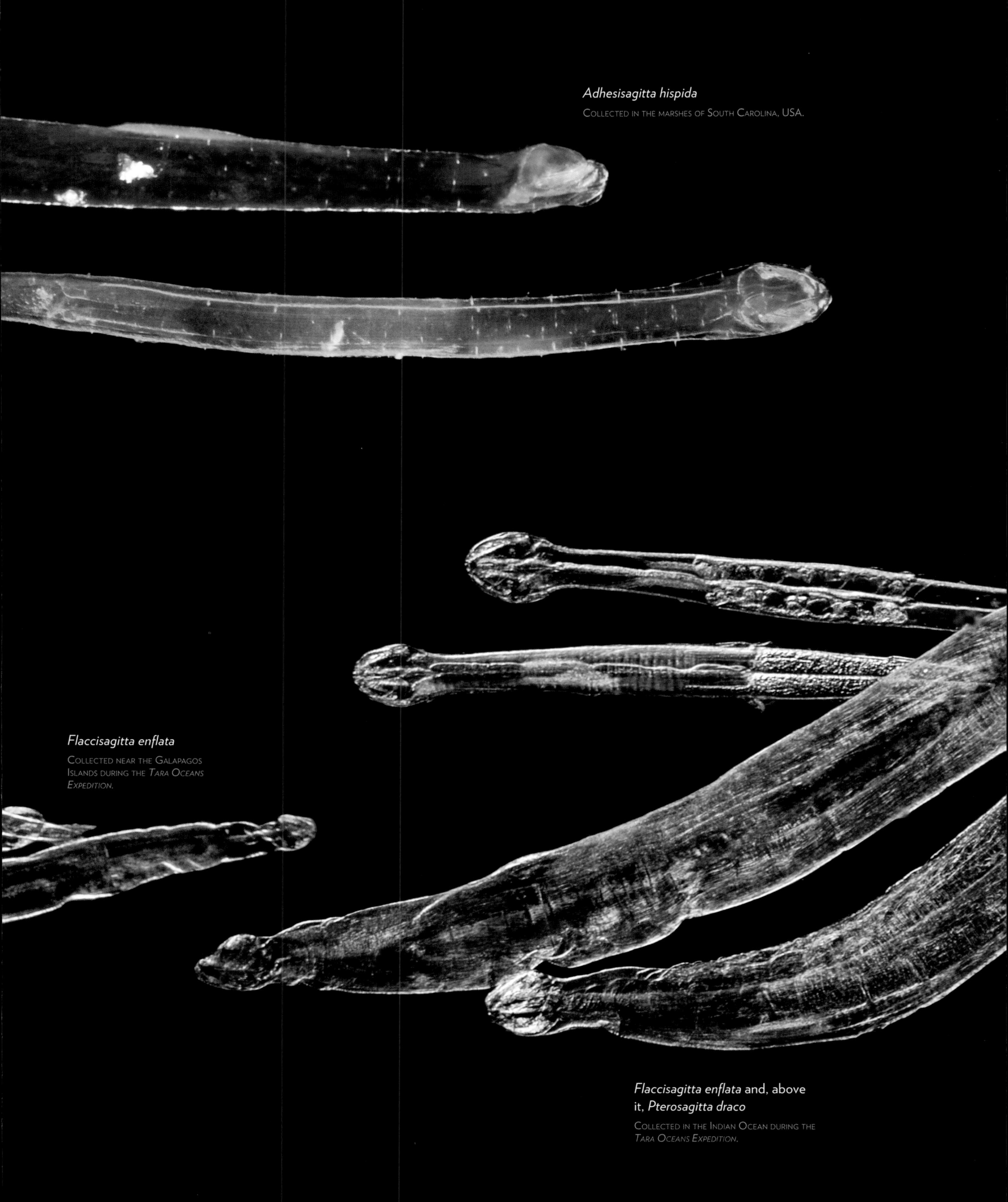

Adhesisagitta hispida
COLLECTED IN THE MARSHES OF SOUTH CAROLINA, USA.

Flaccisagitta enflata
COLLECTED NEAR THE GALAPAGOS
ISLANDS DURING THE *TARA OCEANS*
EXPEDITION.

Flaccisagitta enflata and, above
it, *Pterosagitta draco*
COLLECTED IN THE INDIAN OCEAN DURING THE
TARA OCEANS EXPEDITION.

Carnivores and Occasionally Cannibals

With their two sets of hooks and rows of chitinous teeth, chaetognaths catch, tear apart, and devour whole copepods and crustacean larvae. They are fierce predators capable of engulfing prey almost as big as themselves, such as small fish, or more rarely, cephalochordates such as *Amphioxus* seen on the opposite page.

Lower left: Cannibalism is not uncommon among chaetognaths. They detect the movements of their prey, including other arrow worms, with numerous sensory cilia distributed along their body.

Lower right: A chaetognath gulps down the pink cadaver of a copepod.

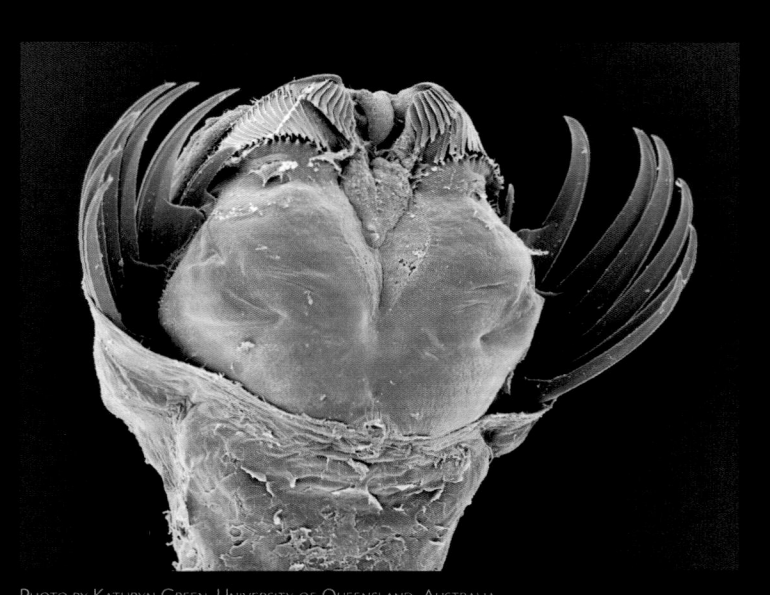

PHOTO BY KATHRYN GREEN, UNIVERSITY OF QUEENSLAND, AUSTRALIA.

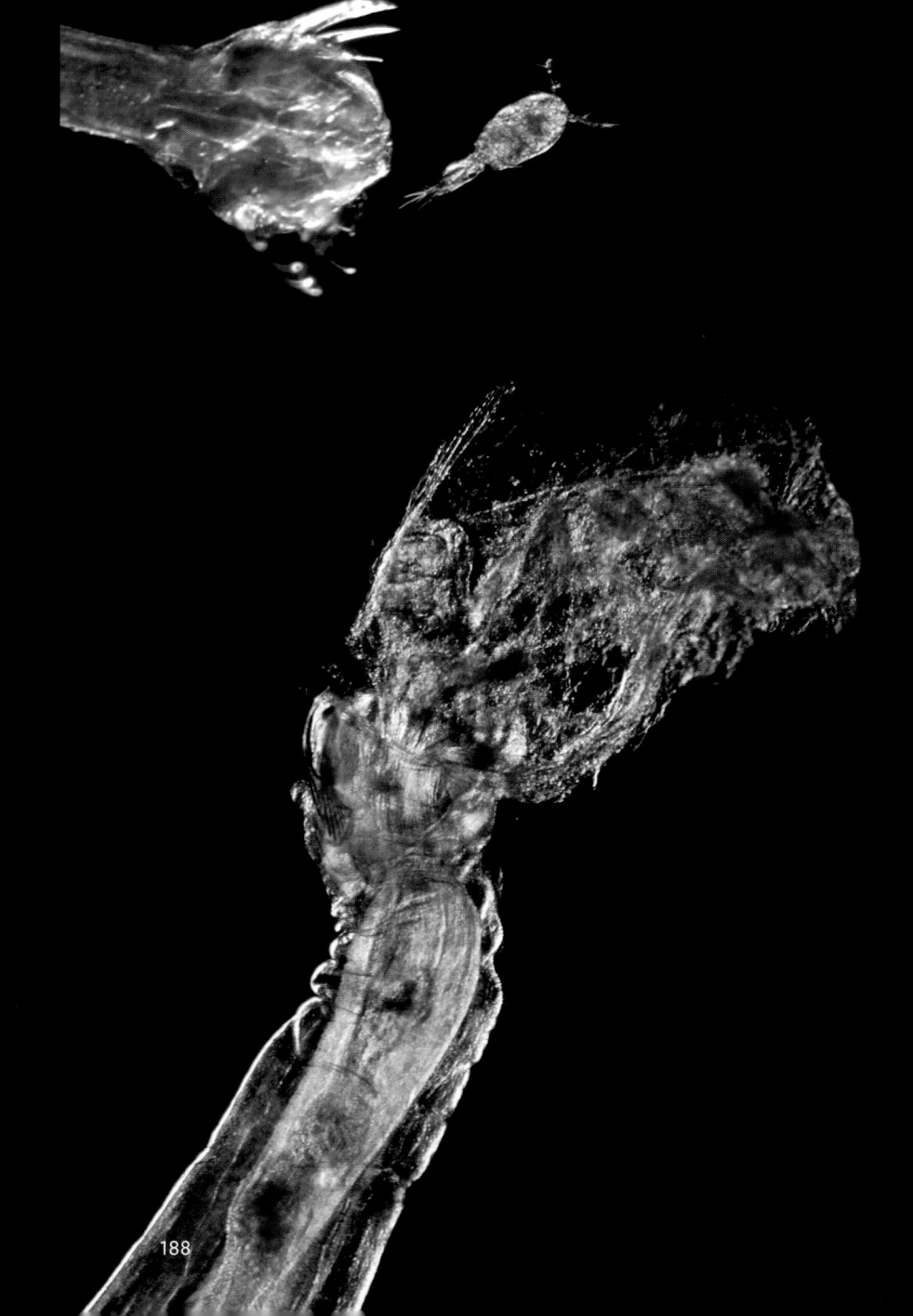

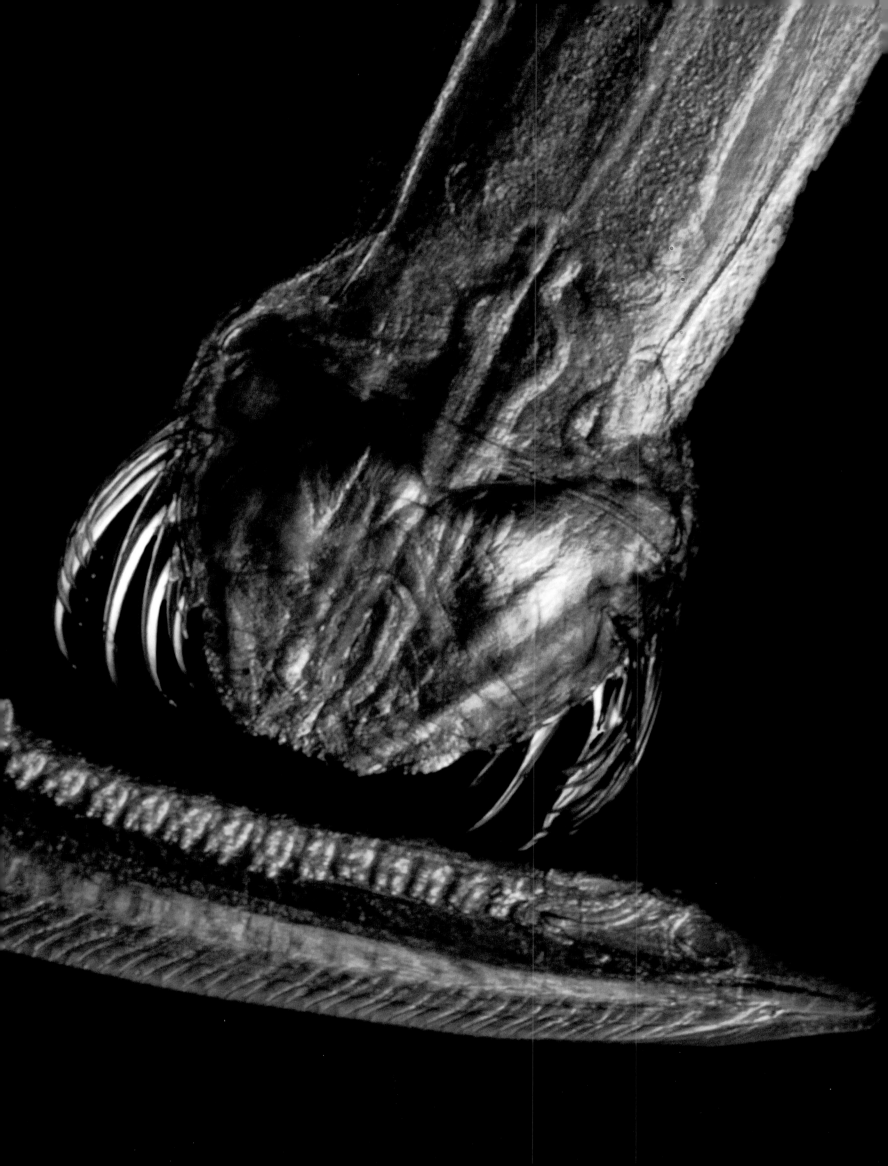

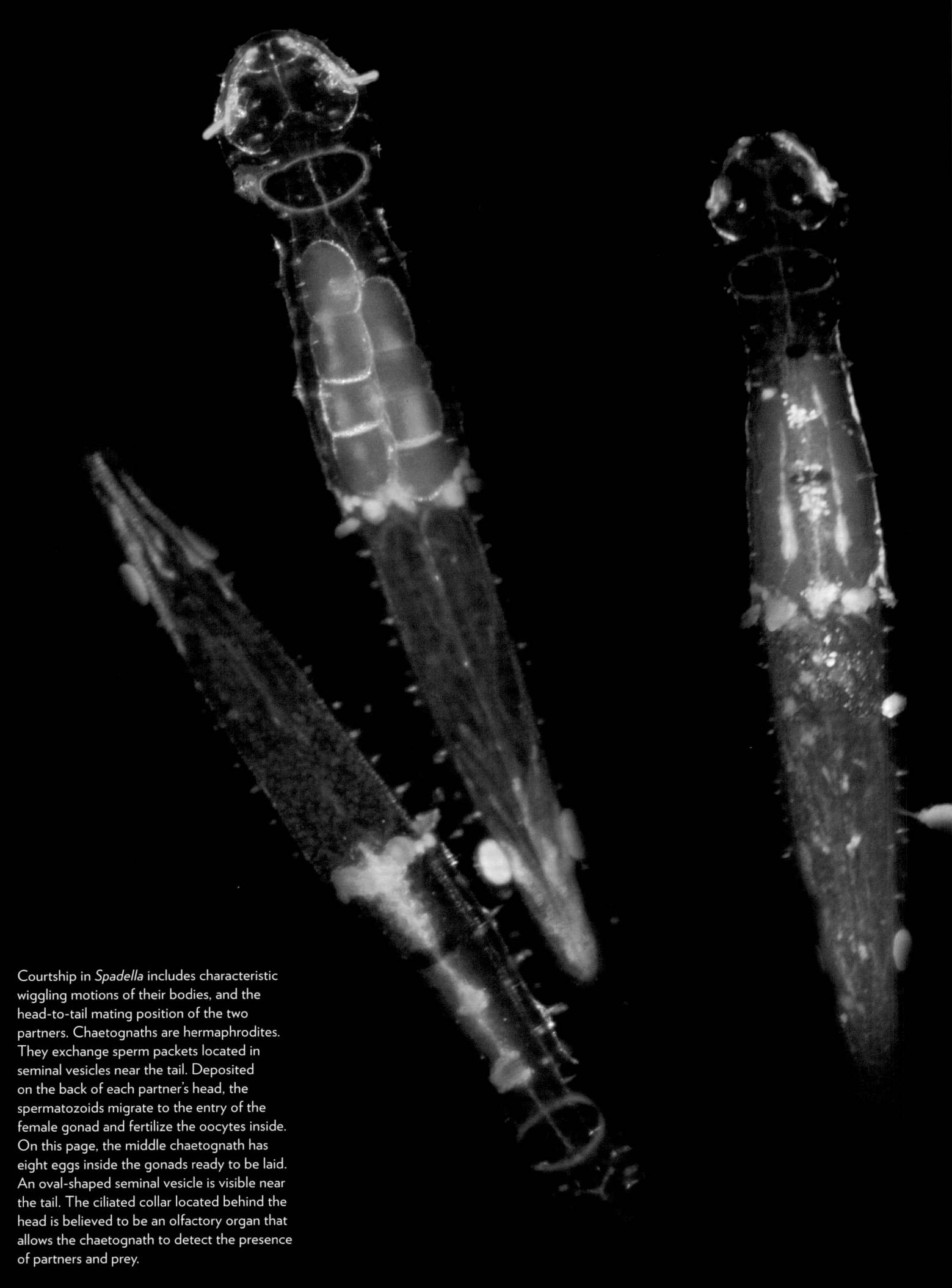

Courtship in *Spadella* includes characteristic wiggling motions of their bodies, and the head-to-tail mating position of the two partners. Chaetognaths are hermaphrodites. They exchange sperm packets located in seminal vesicles near the tail. Deposited on the back of each partner's head, the spermatozoids migrate to the entry of the female gonad and fertilize the oocytes inside. On this page, the middle chaetognath has eight eggs inside the gonads ready to be laid. An oval-shaped seminal vesicle is visible near the tail. The ciliated collar located behind the head is believed to be an olfactory organ that allows the chaetognath to detect the presence of partners and prey.

Prolific Hermaphrodites

Within a pair of female gonads up front near the head, and a pair of male gonads situated near the tail, oocytes and sperm mature rapidly. Every few days, chaetognaths lay a batch of fertilized eggs, each encased in a protective shell. Chaetognath development occurs without metamorphosis or molting. Once an egg is laid, a miniature chaetognath develops inside the shell, ready to hatch in a day or two.

POLYCHAETE ANNELIDS

WORMS IN THE SEA

Whether we are talking about earthworms in the vegetable garden, or giant marine worms in the deep ocean abyss, all annelids consist of repeating ring-like segments. Most of the 12,000 species of known annelids crawl, burrow in mud or sand, or live in tubes they build for themselves. Annelids swim or creep by contracting transversal and annular muscles. The muscular tissues, along with a fibrous cuticle and hydrostatic liquid within the central cavity, or coelom, extending from the head to the anus, act as a strong, flexible skeleton.

Many species drift with plankton for only short periods, when they are embryos or larvae. But some annelids, primarily the polychaetes, are planktonic their entire lives. The name *polychaete* comes from the Greek for "many" and "hairs." Polychaetes are sometimes called bristle worms thanks to the countless hairs at the end of their multiple appendages (parapodia). Each segment or ring supports one pair of parapodia that the polychaete uses like oars to move through the water column.

Polychaetes are extremely diverse, numbering thousands of species found in nearly all aquatic and marine habitats. One the most unusual families is that of the Tomopteridae. Unlike most polychaetes, the annelids of the Tomopteridae family have no chaetae. They move quickly through the water beating their parapodia and deploy two long tentacles to catch prey, such as chaetognaths or fish larvae. In certain circumstances, *Tomopteris* emits flashes of yellow light and a cloud of particles, probably a defense mechanism. To escape predators, it sometimes

Photo by Alexander Semenov.

curls up into a tight ball and sinks. Another family, the Alciopidae, detect prey and predator with the help of a pair of huge eyes with lenses.

Other polychaetes such as *Nereis* and *Myriadina* usually crawl in sand and mud, but join the plankton for reproduction. At the approach of a full moon or giant tide, these annelids transform into pelagic reproductive plankton called epitokes. With enlarged eyes and parapodia modified for an awkward manner of swimming, thousands of epitokes gather in swarms, synchronously releasing male and female gametes for optimal fertilization. In the Pacific Islands, *Palolo* worms swarm at night near coastal reefs and are collected by islanders for traditional celebrations and festivals. The worms' terminal parts bearing the gametes are cooked in various ways, a famous local delicacy.

ANNELIDS OF SHIMODA BAY

One fall morning in stormy weather, we went out in a boat to collect plankton in Shimoda Bay with 20- to 200-micron mesh nets. Back in the laboratory, we sorted, cleaned, filmed, and photographed different polychaete annelids, including a bright green epitoke bearing eggs.

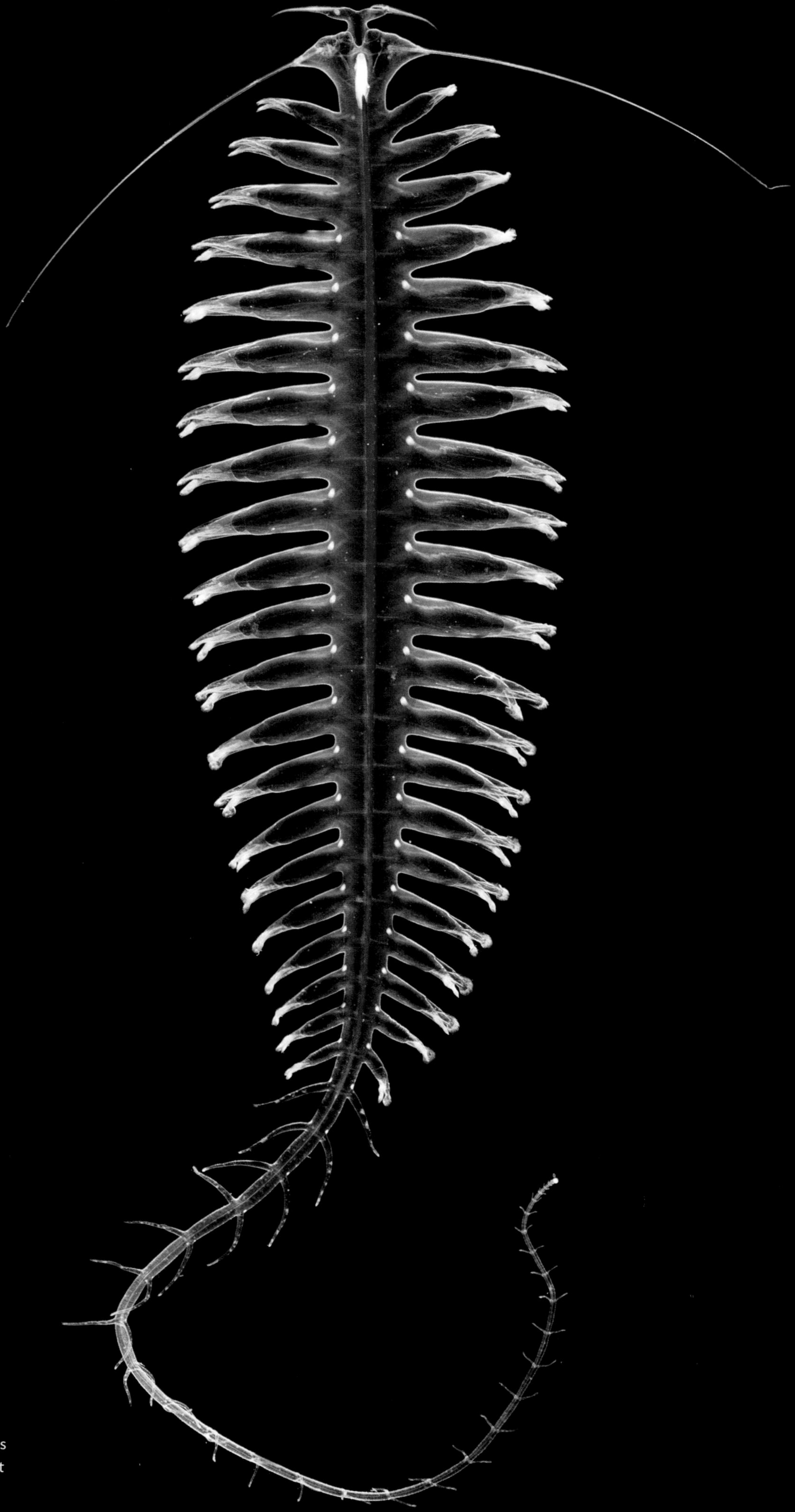

Tomopteridae

Polychaetes such as this Tomopteridae annelid *Eunapteris* sp. can measure just a few millimeters, or tens of centimeters. These fast swimmers have two long antennae and agile parapodia ideal for propelling them and catching zooplankton.

PHOTO BY KAREN OSBORN, SMITHSONIAN NATIONAL MUSEUM OF NATURAL HISTORY, WASHINGTON DC, USA.

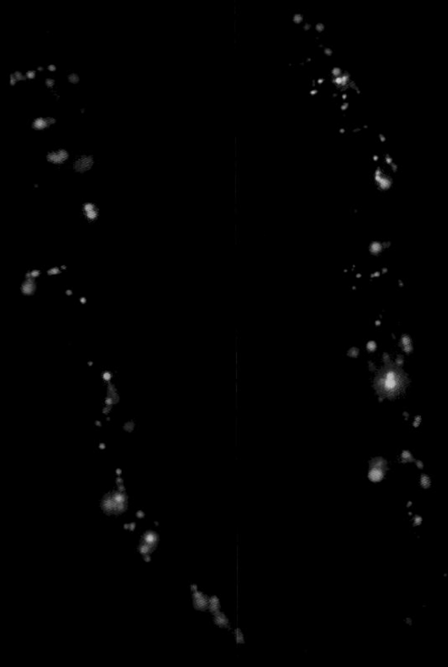

BIOLUMINESCENCE

The yellow-green light emitted by bioluminescent glands situated on the parapodia of a *Tomopteris helgolandica* was captured on film by making a contact print with the annelid.

PHOTO BY PER FLOOD, BATHYBIOLOGICA A/S.

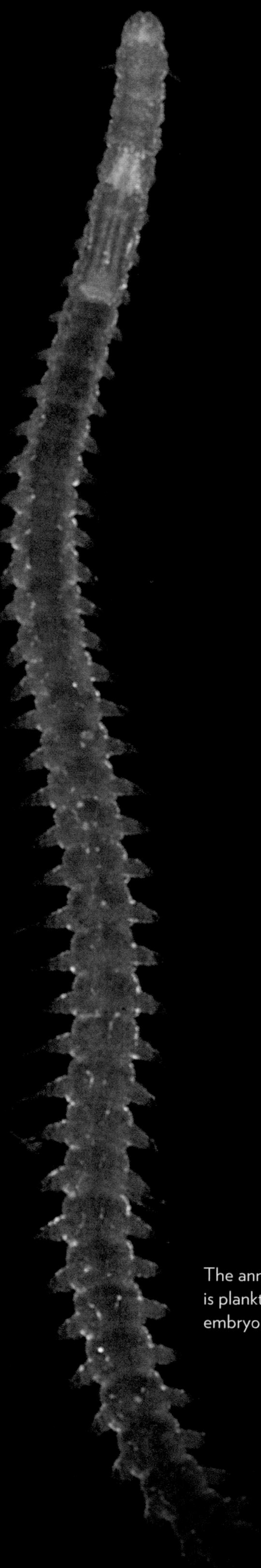

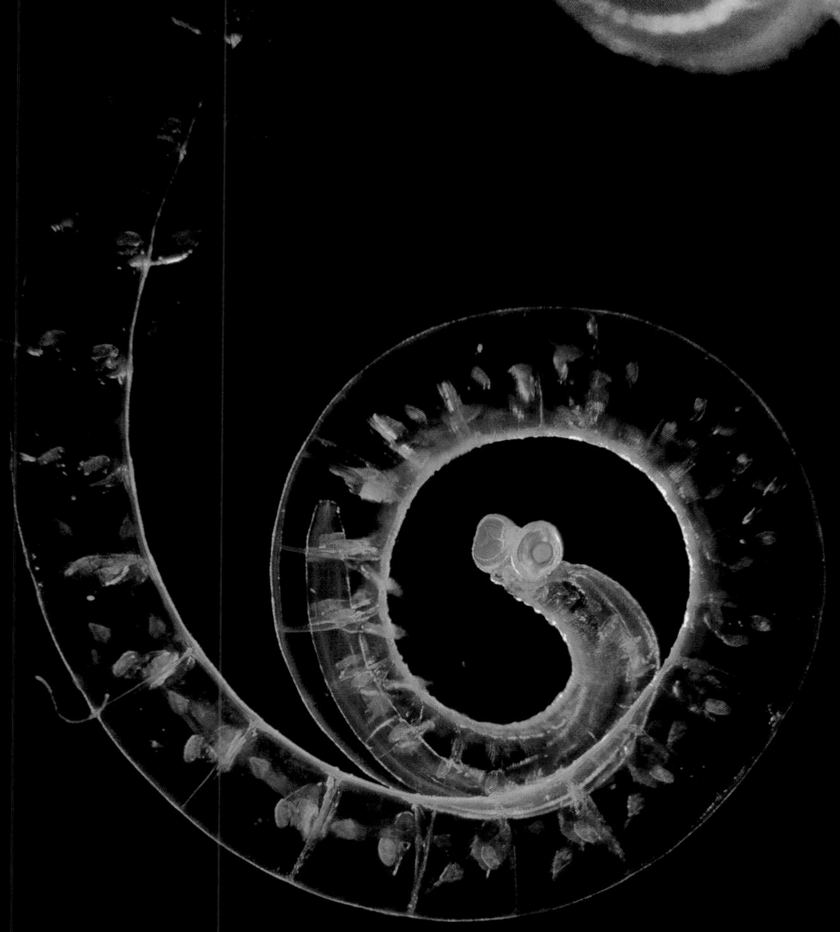

Complex Eyes of the Alciopidea

The undulating body of the polychaete *Vanadis* sp. is made up of about fifty segments, each possessing two yellow parapodia terminated by chaetae. The eyes of *Vanadis* measure about 1 mm in diameter. Equipped with a lens and giant red photoreceptors, they are apparently capable of forming an image. Often these worms float inconspicuously, waiting to catch the zooplankton passing by. But they can also swim rapidly to catch prey.

The annelid *Poecilochaetus* sp. is planktonic only during its embryonic and juvenile stages.

Strange Ways to Reproduce

A special feature of polychaete annelids, even those that crawl on the sea bottom, is their particular manner of reproduction, called epitoky. In rhythm with the season and lunar cycle, adult worms change form. Some species shed their rear ends containing gonads. These become ephemeral planktonic epitokes totally specialized in reproduction. Carrying eggs or sperm, the epitokes gather in swarms at sea. Male epitokes surround the females and shower them with sperm. After fertilization, embryos develop inside the brood pouch of the female epitoke. A few days later, she releases her larvae into the sea. Unable to feed, the female quickly perishes.

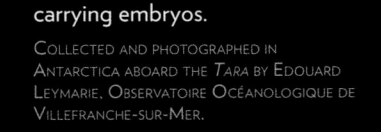

Female epitokes of *Nereis* sp. carrying embryos.

COLLECTED AND PHOTOGRAPHED IN ANTARCTICA ABOARD THE *TARA* BY EDOUARD LEYMARIE, OBSERVATOIRE OCÉANOLOGIQUE DE VILLEFRANCHE-SUR-MER.

LARVAE AND JUVENILES

Whether benthic or pelagic as adults, annelids are planktonic during their embryonic, larval, and juvenile stages.

This page: Two nectochaete larvae surround a blue mitraria larva of *Owenia* sp. Opposite page: Two juvenile annelids.

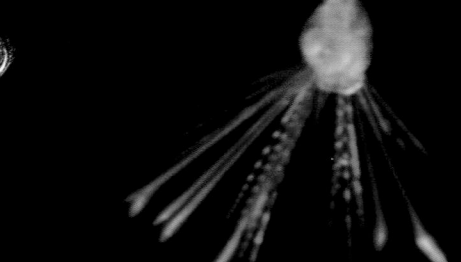
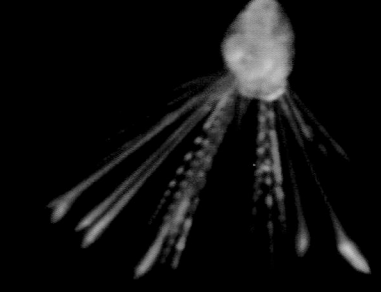
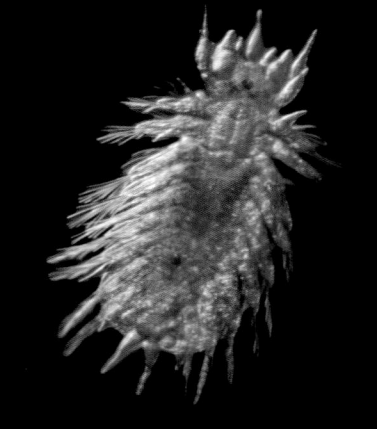

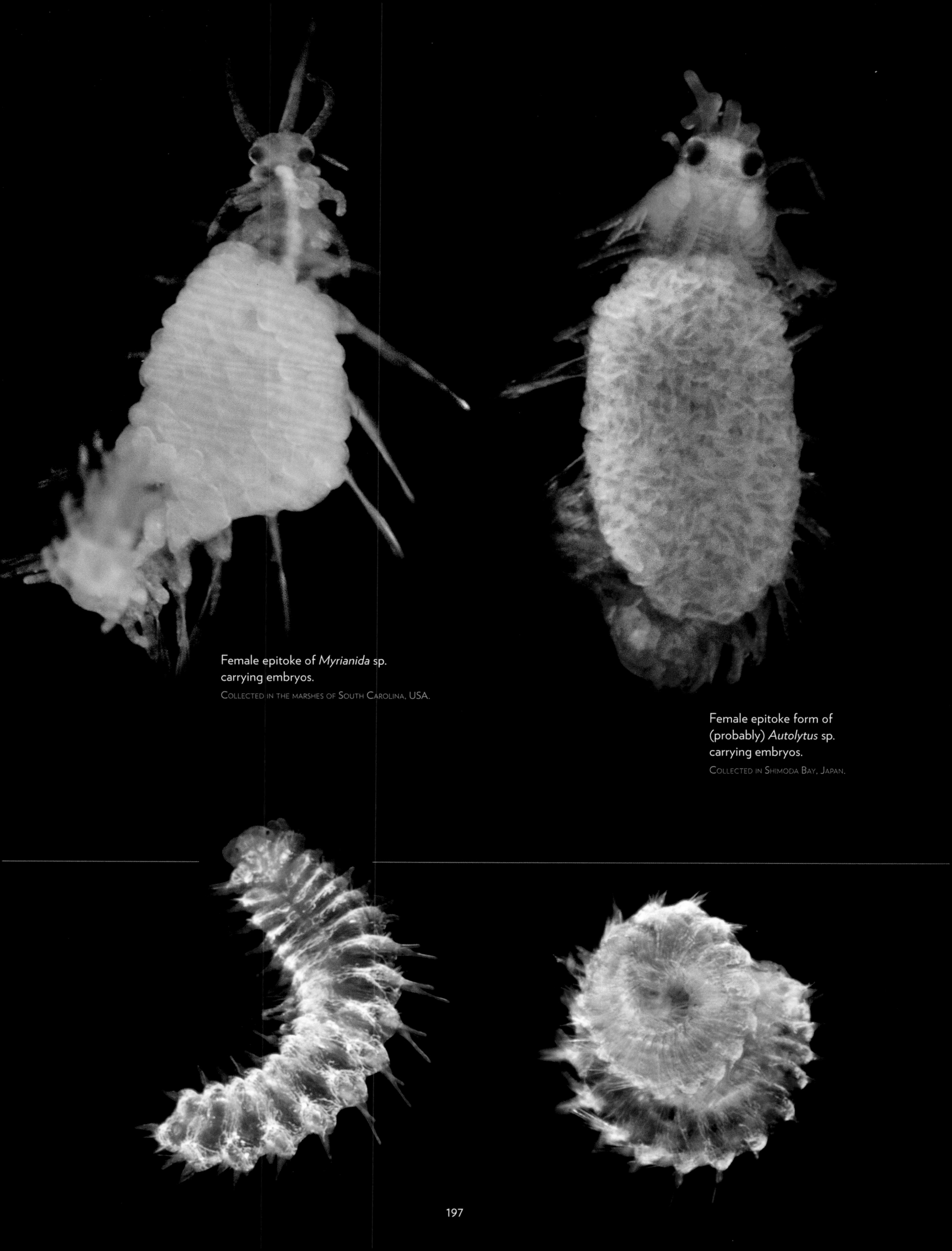

Female epitoke of *Myrianida* sp.
carrying embryos.

COLLECTED IN THE MARSHES OF SOUTH CAROLINA, USA.

Female epitoke form of
(probably) *Autolytus* sp.
carrying embryos.

COLLECTED IN SHIMODA BAY, JAPAN.

SALPS, DOLIOLIDS, AND PYROSOMES

HIGHLY EVOLVED GELATINOUS ANIMALS

Although they appear to be simple and primitive jellies, salps have a heart, a gill, and even the equivalent of a placenta. Salps, doliolids, and pyrosomes belong to the phylum Urochordata, which also includes ascidians (sea squirts) and larvaceans. They are among the closest invertebrate relatives of the vertebrates (like fish or humans). All members of the Urochordata are characterized by an embryonic structure known as a dorsal notochord, a flexible rod-shaped tissue that anticipates the spinal column of vertebrates. Salps, doliolids, and pyrosomes float freely their entire lifespan and possess a protective outer covering, a tunic, which gives the group their name: tunicates.

Salps can live as solitary individuals, or in colonies of identical individuals that form long chains. The most visible feature of transparent salps is their opaque, brownish "nucleus" containing the creature's viscera. Smaller salps measure a few millimeters, while giant specimens like *Salpa maxima* can reach 30 centimeters. Large or small, all salps move and feed by pumping phytoplankton-containing water through their tubular body using powerful ring-shaped belts of striated muscle. When microscopic algae proliferate, salps gorge themselves and respond to the abundance of food by reproducing explosively.

Solitary salps that reproduce asexually are called oozooids. They bud a stolon that segments into identical individuals, the blastozooids. These identical clones form chains several meters long consisting of hundreds of individuals that synchronize their swimming by means of electrical signals. Chained blastozooids mature and eventually split apart to reproduce sexually. Each individual grows an ovary. After fertilization, an embryo develops inside the organism, surrounded by a tissue that resembles a primitive placenta. The embryo will

Photo by Fabien Lombard, UPMC, Observatoire Océanologique de Villefranche-sur-Mer.

turn into a new oozooid. Thousands of individuals can be born from a single oozooid in just a few days. As a consequence, blooming salp populations sometimes cover hundreds of acres, producing huge quantities of fecal pellets that nourish other organisms in the water column. When salps have exhausted their food resources, they are infested by bacteria and viruses and become prey for crustaceans. Their remains sink into the depths, carrying a large quantity of organic carbon to the sea floor.

Relatives of salps, pyrosomes are bioluminescent tunicates that aggregate in colonies shaped like long socks. The colonies consist of many identical zooids sharing a common cellulose tunic. Pyrosomes feed on bacteria and microorganisms that filter through the gills of the zooids. Some pyrosomes are so large that a diver can fit inside the cavity in the middle of the colony.

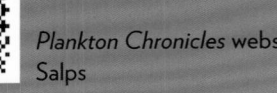

Plankton Chronicles website
Salps

198

A SALP OOZOOID

This salp, *Thalia democratica*, was collected in the bay of Villefranche-sur-Mer. At the top, the oral siphon opens onto a mucous net (invisible here) that collects microorganisms for food. In the center, ciliated branchial bars facilitate the flow of water through the animal. At the bottom, the digestive system called the nucleus and a nascent chain of blastozooids.

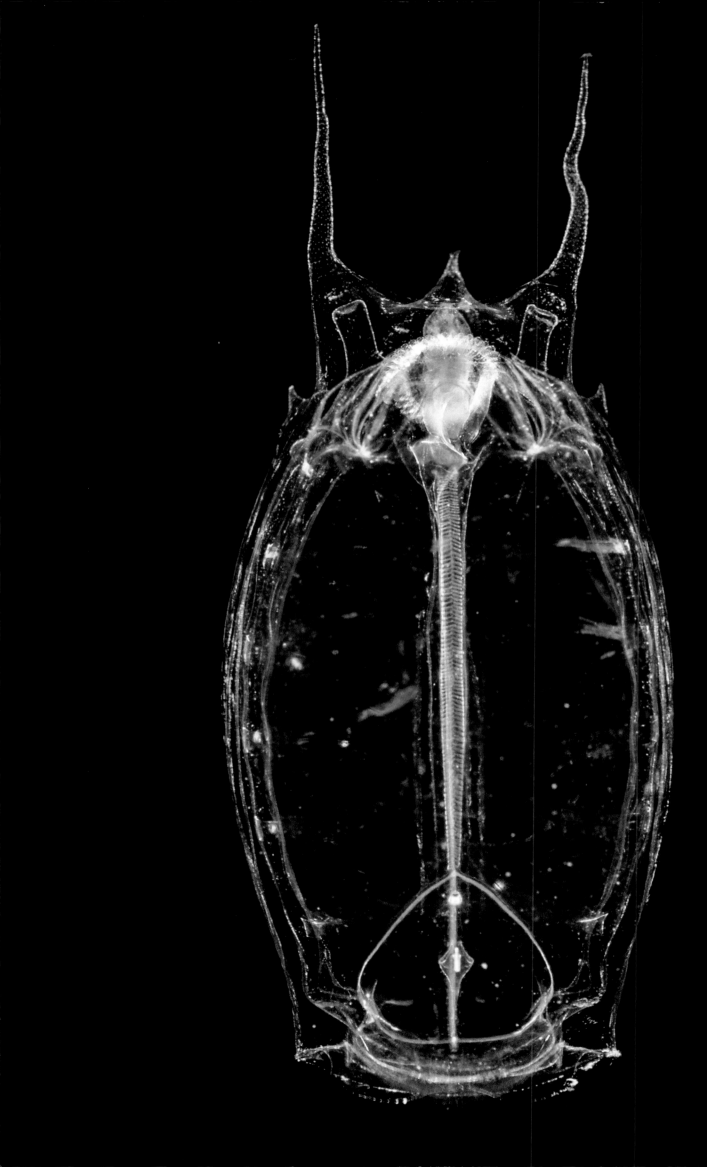

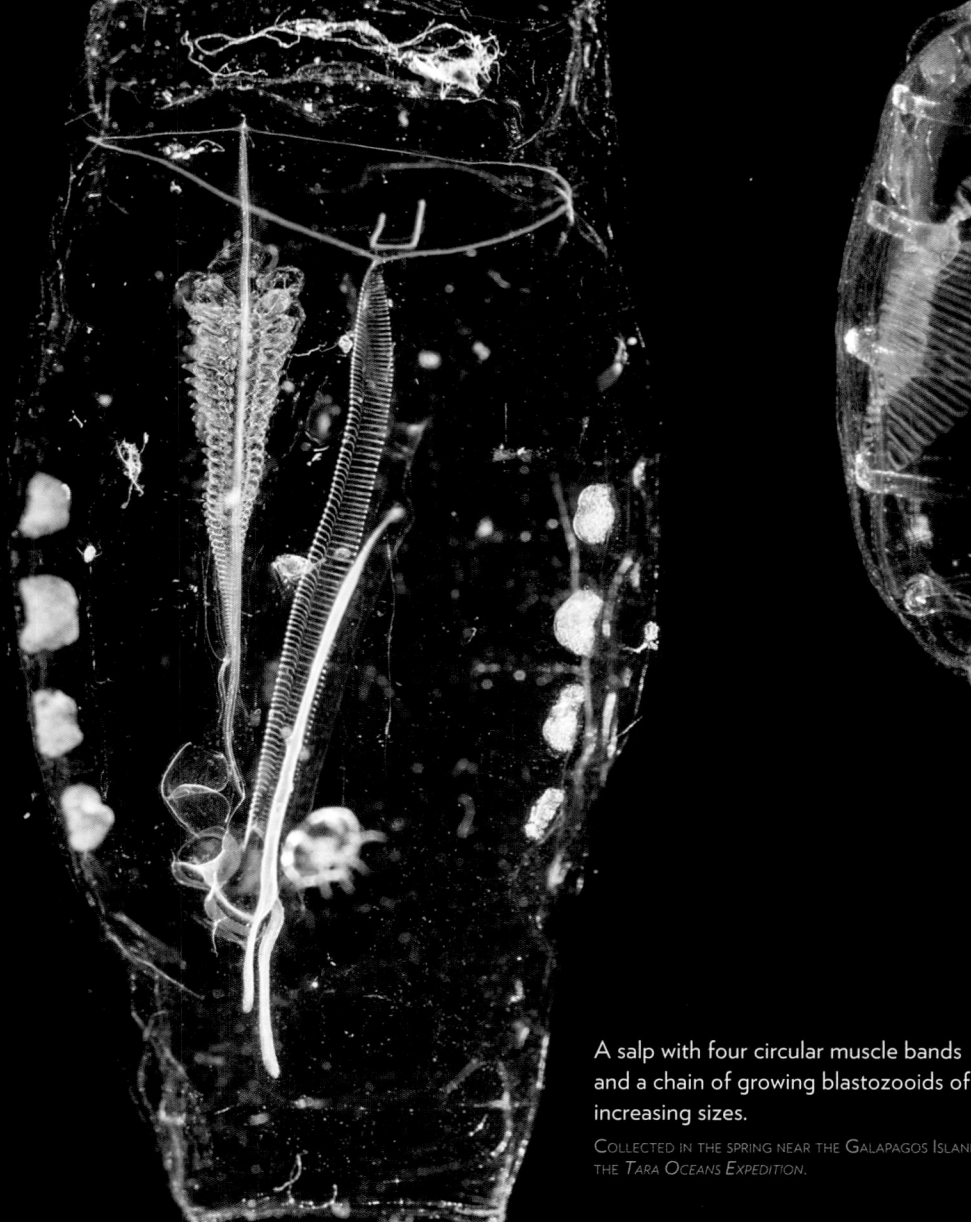

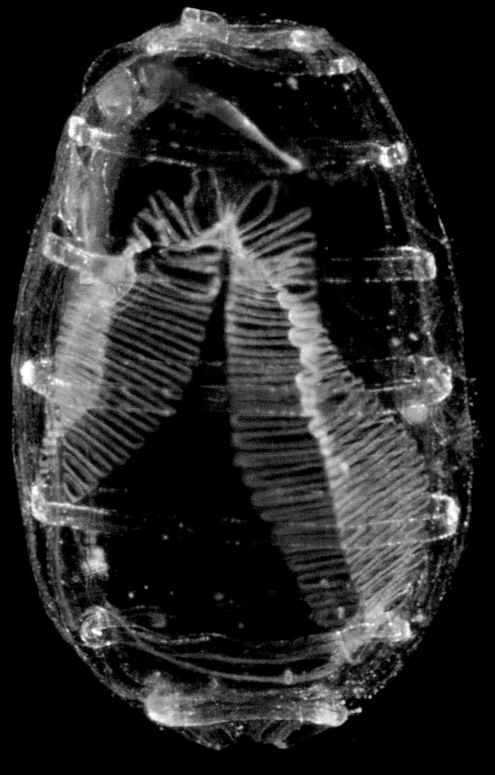

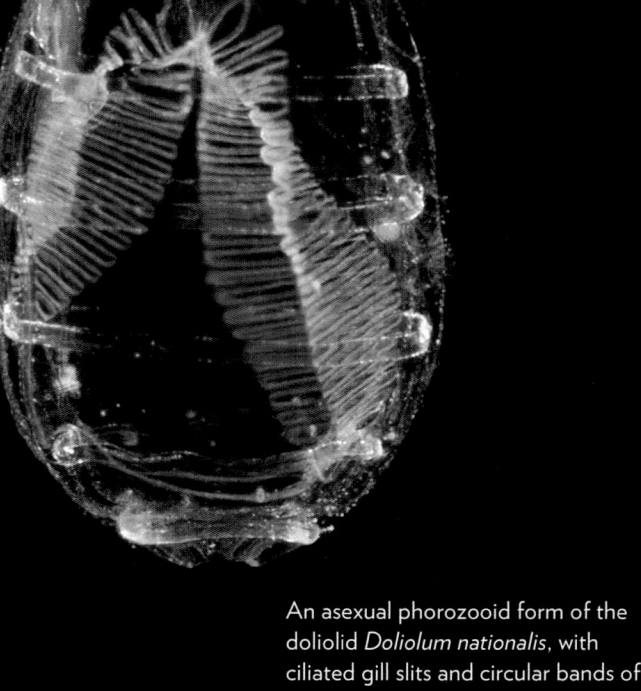

An asexual phorozooid form of the doliolid *Doliolum nationalis*, with ciliated gill slits and circular bands of muscle, developed from budding.

COLLECTED IN THE FALL IN TOBA BAY, JAPAN.

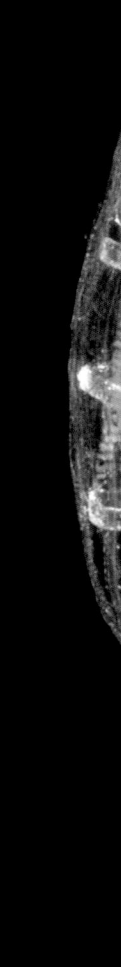

A salp with four circular muscle bands and a chain of growing blastozooids of increasing sizes.

COLLECTED IN THE SPRING NEAR THE GALAPAGOS ISLANDS BY THE *TARA OCEANS EXPEDITION*.

Nerves and Muscles

The salp nervous system consists of a central ganglion, a sort of primitive brain, with a cup-like structure of photosensitive red pigment cells forming the eye (left). A network of nerves radiates from the ganglion and controls striated muscles (right).

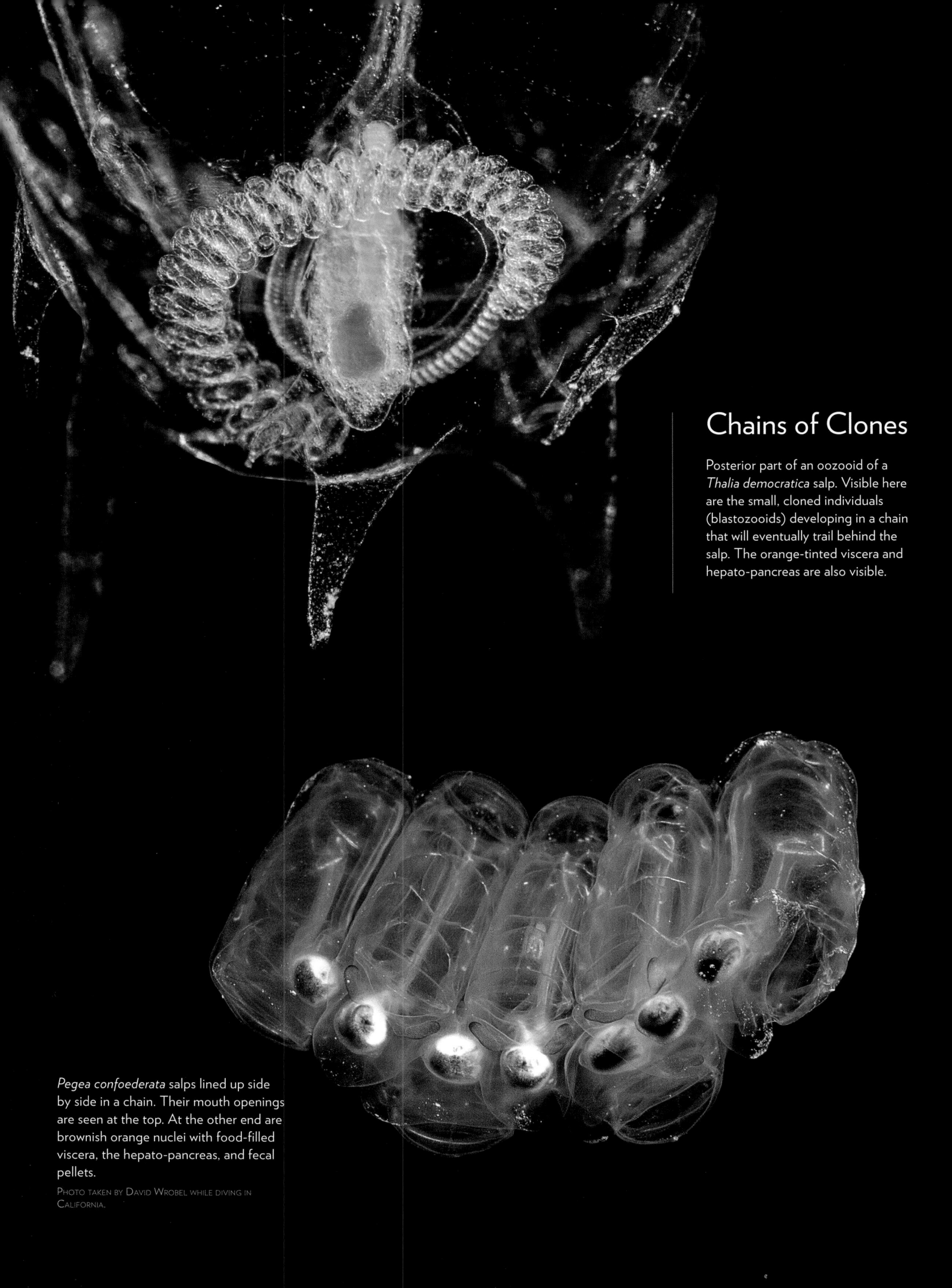

Chains of Clones

Posterior part of an oozooid of a *Thalia democratica* salp. Visible here are the small, cloned individuals (blastozooids) developing in a chain that will eventually trail behind the salp. The orange-tinted viscera and hepato-pancreas are also visible.

Pegea confoederata salps lined up side by side in a chain. Their mouth openings are seen at the top. At the other end are brownish orange nuclei with food-filled viscera, the hepato-pancreas, and fecal pellets.

Photo taken by David Wrobel while diving in California.

THIS SPECIMEN WAS COLLECTED IN SPRING BY DIVERS IN THE BAY OF VILLEFRANCHE-SUR-MER. PHOTO BY STEFAN SIEBERT, BROWN UNIVERSITY, USA.

Colonies of Pyrosomes

Pyrosome colonies contain hundreds of identical individual zooids organized in a cylinder open at one end. The zooids' ciliated gill slits are visible in close-up views (here to the right, and opposite page). These gill slits efficiently filter and concentrate microorganisms that comprise the pyrosome's food. The yellowish pyrosome on the left was collected during the *Tara Oceans Expedition* off the coast of Ecuador.

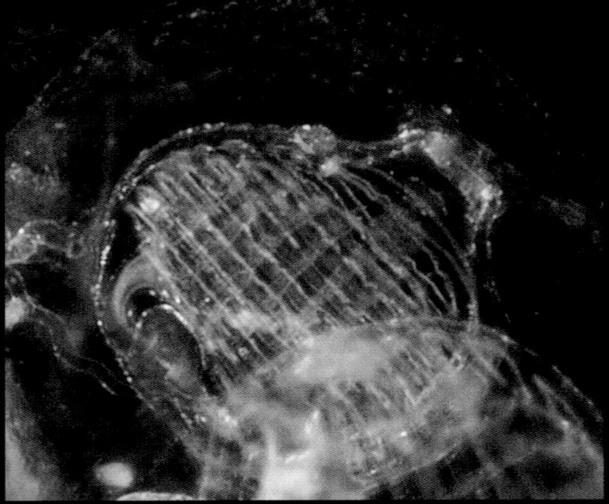

Opposite page: Close-up of the bioluminescent species *Pyrosoma atlanticum*. The name *pyrosome* comes from the Greek words *pyro* and *soma* meaning "fire" and "body." Each zooid contains a pair of organs that emit a pale blue light. The whole colony (seen above) measures about 3 cm. *Pyrosoma atlantica* was first described by the French naturalist Péron in 1803.

202

LARVACEANS

TADPOLES THAT LIVE IN A NET

Larvaceans, also known as appendicularians, look a lot like tadpoles, with an oval-shaped body and a long, flexible tail. Groups of large cells with complex, hand-shaped nuclei in the larvacean epithelium, secrete proteins and sugars that assemble themselves into a fine mesh net. The animal sits in the middle of this structure that serves both as a house and a filter for food. The tadpole-like larvacean whips its muscular tail, creating currents that channel bacteria, algae, protists, and small particles through the filters and into its mouth. Several times a day, the larvacean leaves its food-clogged house and swims around frantically, secreting a brand-new net with rapid movements of its tail.

The tail muscle and adjoining notochord of larvaceans are characteristic features of the urochordates, a phylum closely related to the vertebrates. Larvaceans retain their tadpole appearance throughout their entire life. In contrast, doliolids and ascidians appear as tadpoles for only a brief larval stage. A larvacean's existence is very short, a few days at most. Their embryos develop fast, with cells dividing every few minutes, and they become reproductive adults in only two or three days. Most species of larvaceans are hermaphroditic, but species like *Oikopleura dioica* are dioecious, with distinct males and females that carry gonads like helmets on their heads. They release sperm and eggs in the open sea for fertilization and die soon after. Fast growers with a short lifespan, larvaceans proliferate in huge swarms throughout the world ocean, supplying an important link in

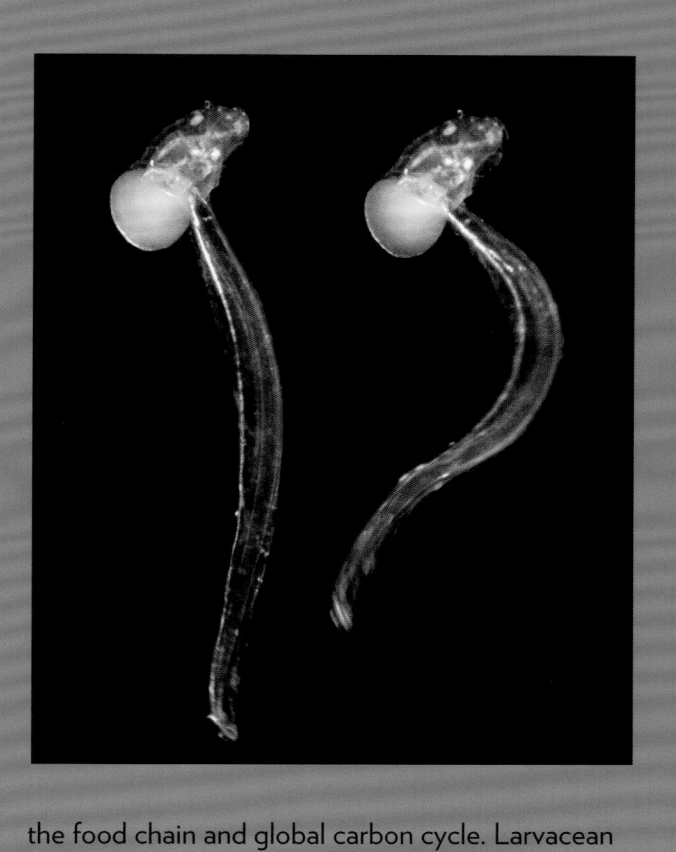

the food chain and global carbon cycle. Larvacean houses and the debris trapped in the nets make up a significant portion of the particulate matter known as "marine snow." The abandoned abodes sink to the ocean floor, taking with them remnants of bacteria and microorganisms collected by the larvacean. Through this process, CO_2 that was originally sequestered from the atmosphere by phytoplankton has been converted to living biomass and back again to a chemical form.

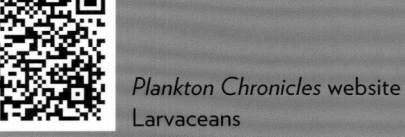

Plankton Chronicles website
Larvaceans

LARVACEAN TADPOLES BUILD HOUSES

Above: A naked larvacean, without its house, photographed with its tail in two different positions.

Opposite page: The larvacean *Oikopleura labradoriensis* appears here as a whitish oval in the middle of the filters of its net house. In the laboratory we use India ink and carmine particles to reveal the transparent structures.

COLLECTED AT FRIDAY HARBOR, USA, AND PHOTOGRAPHED WITH A FLASH THROUGH A MAGNIFYING GLASS BY PER FLOOD, BATHYBIOLOGICA A/S.

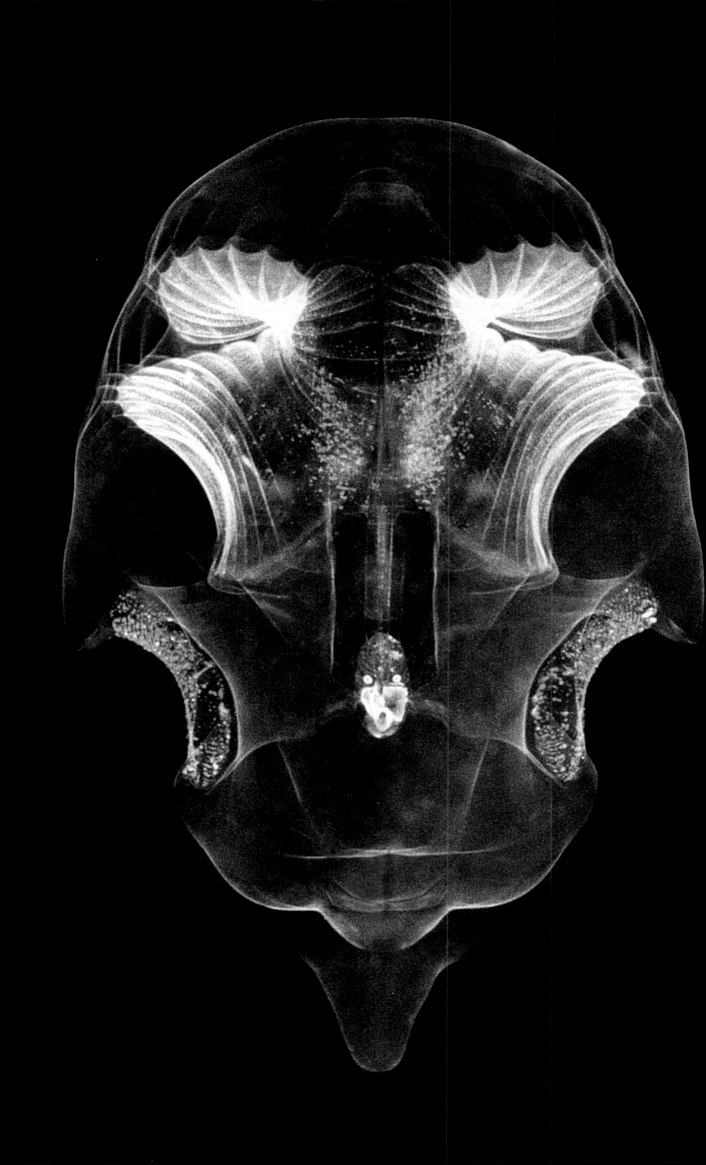

MALE AND FEMALE GAMETES

Unlike most hermaphroditic larvaceans, *Oikopleura dioica* is a dioecious species: individual animals are either male or female. Female gonads containing the eggs (left), and male gonads containing the sperm (right), look like helmets on top of the animal's head. Methods for culturing *Oikopleura dioica* as a laboratory model were mastered in the 1980s at the marine station of Villefranche-sur-Mer. Its life cycle from adult to adult is extremely fast, taking only one week, and its genome is the smallest sequenced for any animal. Might there be a link between the rapid life cycle and highly simplified genome of *Oikopleura dioica*?

COLLECTED AND PHOTOGRAPHED IN NORWAY BY PER FLOOD, BATHYBIOLOGICA A/S.

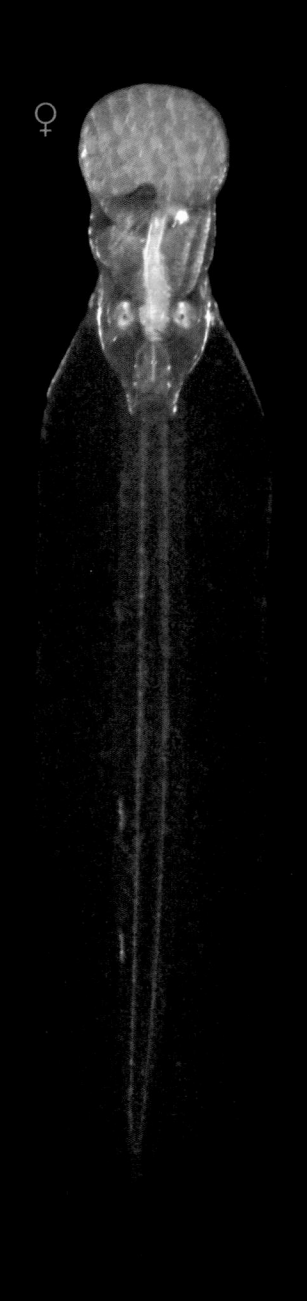

LARVACEANS BUILD THEIR OWN HOUSES AND FILTERS

Every four or five hours the larvacean secretes a new net and inflates its house (left), abandoning the previous one when it gets clogged. Fine mesh filters (center) take shape from the macromolecules secreted by groups of epithelial cells specialized in producing different parts of the house. The large nuclei of these cells containing multiple copies of the larvacean genome are revealed using a fluorescent molecule that binds to DNA.

THE TWO PHOTOS IN THE MIDDLE AND ON THE RIGHT ARE FROM ERIC THOMPSON, PHILIPPE GANOT, AND ENDY SPRIET, SARS LABORATORY, BERGEN, NORWAY.

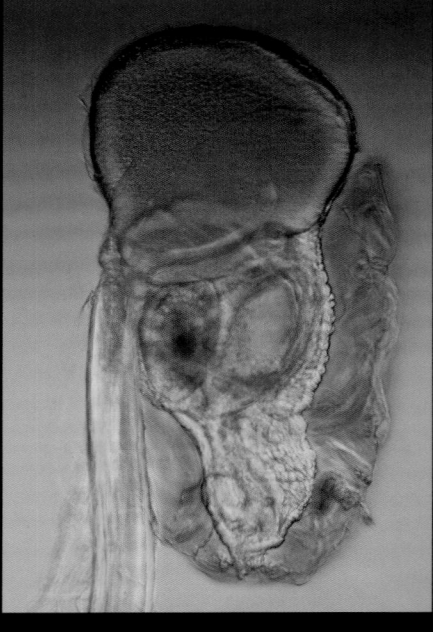

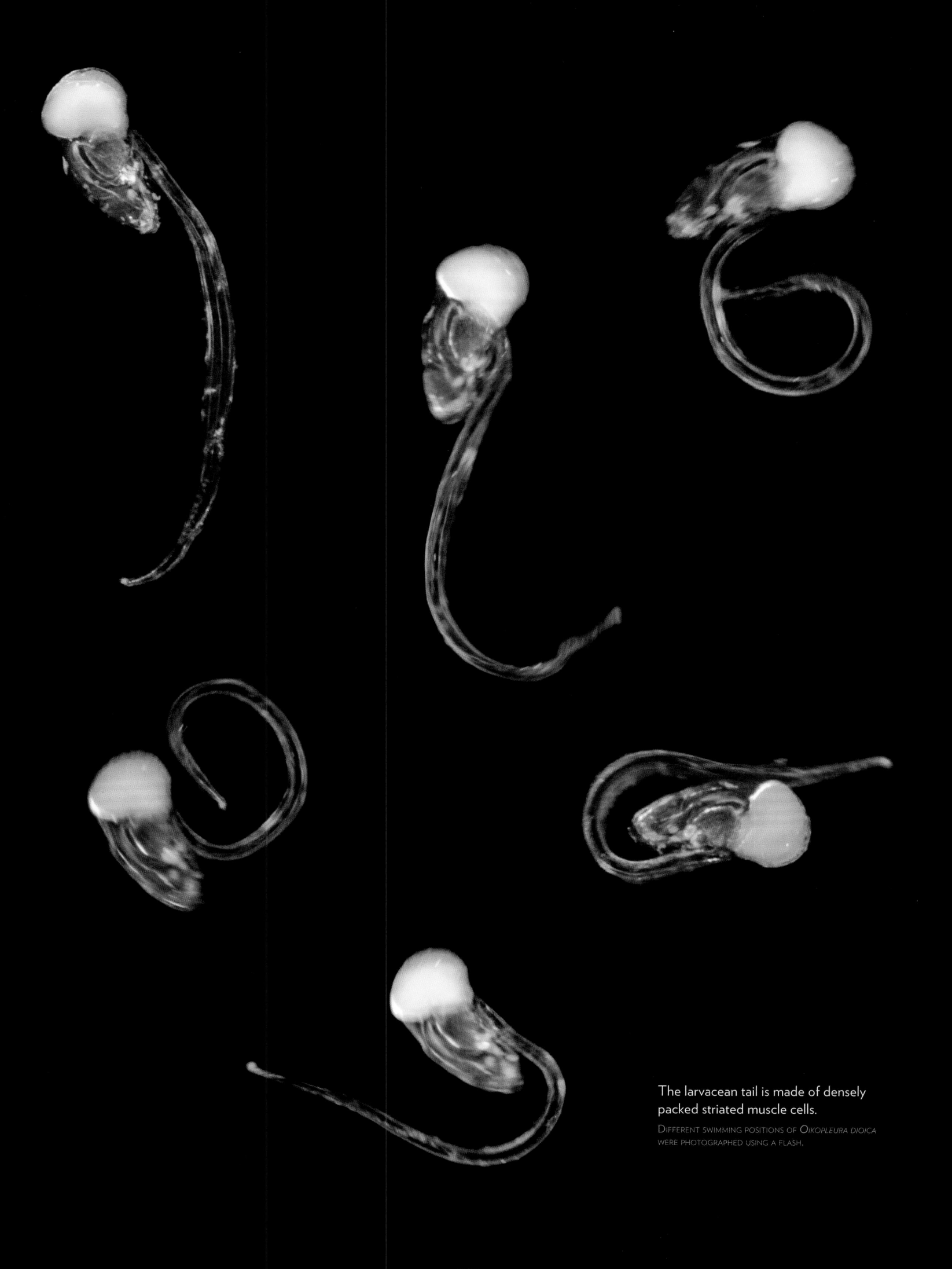

The larvacean tail is made of densely packed striated muscle cells.

Different swimming positions of *Oikopleura dioica* were photographed using a flash.

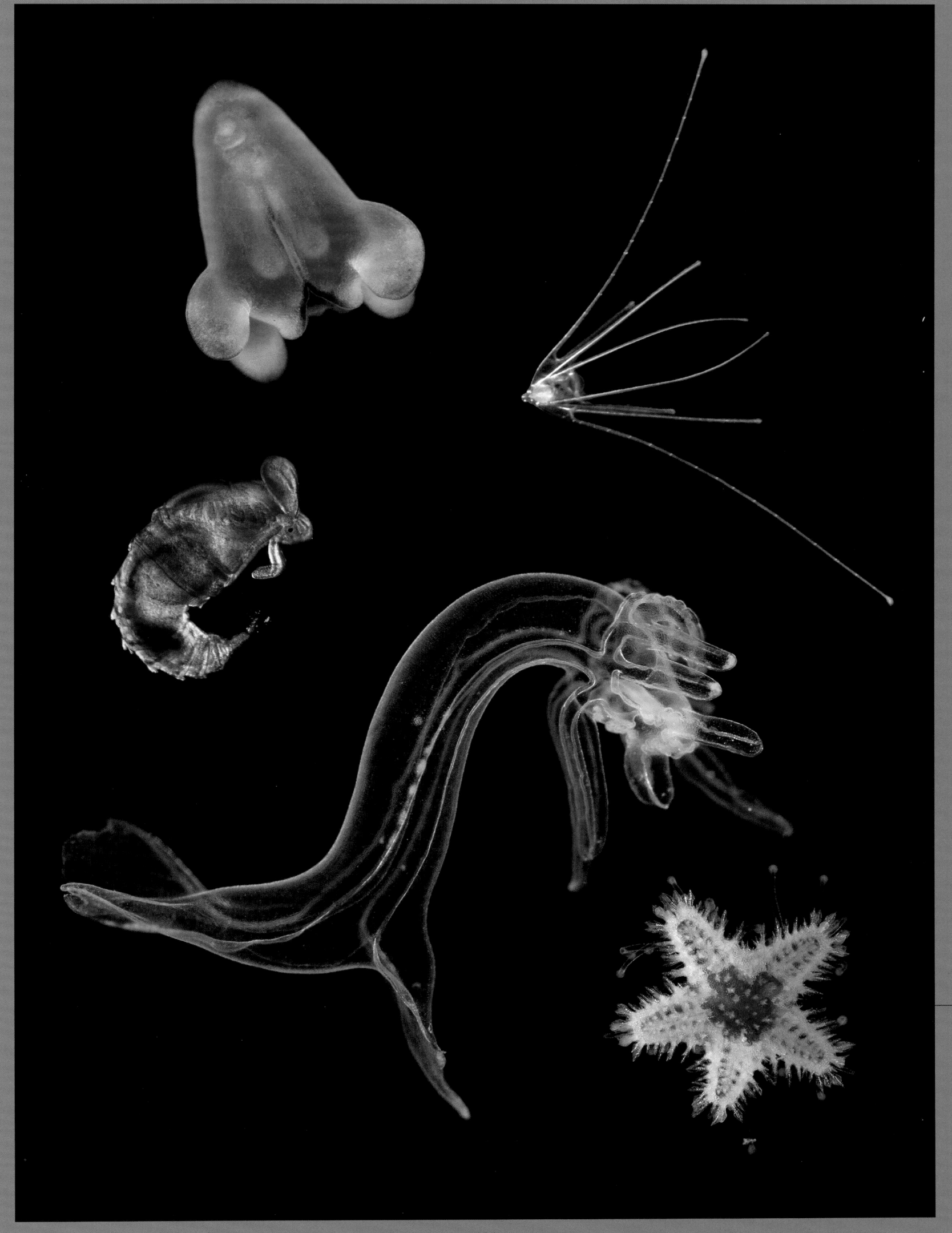

Embryos and Larvae

All kinds of animal embryos and larvae abound in the plankton. Spawned not only from adults that drift their entire lives in the currents, many come from benthic organisms living on the seabed and coastlines, such as urchins, anemones, corals, crabs, and shellfish. Great quantities of gametes also make up a portion of the plankton. Fish larvae often resemble tiny versions of the vigorous swimming adults, but in general, larval forms of sea creatures bear little resemblance to their parents. During the early days of marine research, larvae were often given names such as *zoea* or *planula* and were sometimes mistakenly considered separate species from the adult form. To become adults, larvae and juveniles must feed, grow, and (for many species) molt while drifting with the currents.

With a few exceptions, most echinoderms—for example, sea stars, urchins, sand dollars, and sea cucumbers—are exclusively benthic, moving slowly on the seabed, consuming mollusks or grazing on algae. But during their larval stages, echinoderms are planktonic. Urchins release millions of eggs and billions of sperm into the sea, most often when the moon is full or just before a storm. Fertilization in

the open sea gives rise to countless embryos, each developing into a pluteus larva that feeds on phytoplankton. After a few weeks of drifting, each pluteus larva metamorphoses, nurturing a miniature sea urchin inside. Gradually, the surrounding larval tissues are consumed by the nascent urchin, which hatches and starts exploring the rocks and seaweed with its tube feet.

Such transformations are commonplace for a majority of species that spend their larval life among the plankton. Many larvae and juveniles

drift and grow into adulthood as part of the zooplankton. Others eventually settle on the bottom, or swim freely in and out of the currents. If they survive, that is. Embryos and larvae are choice food for jellyfish, shrimp, fish, and other marine creatures. Only a tiny fraction of the huge quantities of eggs, embryos, and larvae drifting in the open ocean will reach adulthood. But a small percentage of the large initial production is enough to renew the life cycle and perpetuate the species.

WHICH LARVA IS WHICH?

From top to bottom: larva of *Cerianthus* sp. (sea anemone); pluteus larva of a brittle star (echinoderm); larva of *Chaetopterus* sp. (annelid worm); larva of *Luidia* sp. (sea star) that has just released a juvenile sea star.

PHOTO OF SEA STAR LARVA AND JUVENILE BY STEFAN SIEBERT, BROWN UNIVERSITY, USA.

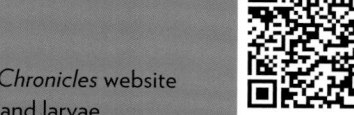

Plankton Chronicles website
Embryos and larvae

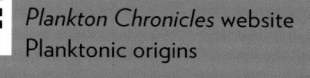

Plankton Chronicles website
Planktonic origins

From Sea Urchin Egg to Pluteus

1. Male urchins release billions of sperm that fertilize millions of eggs released by females in the open sea.

2. Eggs divide, becoming embryos.

3. The embryo develops into a pluteus larva in two days.

4. The pluteus larva feeds on green algae and grows.

5. In a few weeks, the pluteus larva transforms into a juvenile sea urchin.

6. The adult sea urchin, *Paracentrotus lividus*.

COLLECTED IN THE BAY OF VILLEFRANCHE-SUR-MER. THE PHOTO ON TOP IS FROM SHARIF MIRSHAK, NOÉ SARDET, AND FLAVIEN MEKPOH.

Diversity of Larval Forms and Behavior

Right: Two molluskan larvae swim rapidly using cilia.

Middle: Larva from a phoronid worm using its umbrella for movement.

Far right: A starfish larva swims with jerky motions using its long arms.

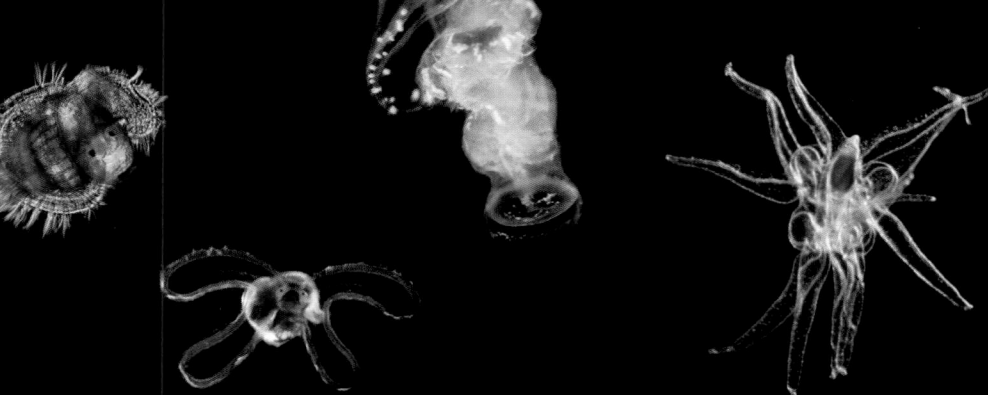

Fish in the Plankton

Before swimming freely with fully developed fins and tails, fish drift with the plankton as embryos, larvae, or juveniles.

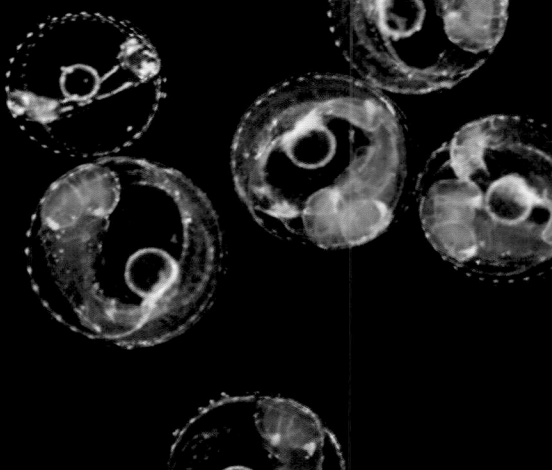

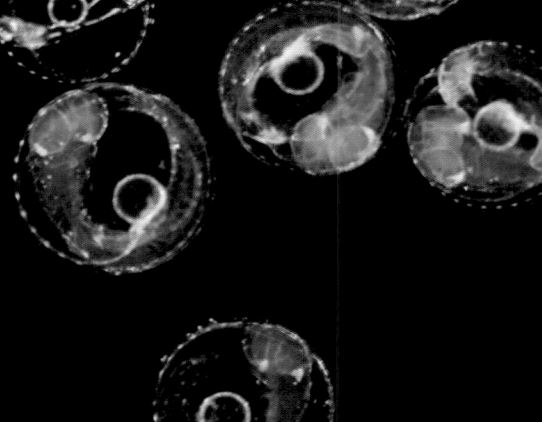

Sea Squirts: From Egg to Tadpole in a Single Day

Most tunicates are attached to the seabed and release large numbers of eggs and sperm in the open sea. After fertilization, embryos develop in a single day into simple planktonic tadpoles comprised of 3,000 cells.

EPILOGUE

The planktonic ecosystems of the world are estimated to contain over one million species. Less than a quarter of these, roughly 250,000, have been formally described. This book features 250 or so representative organisms, showcasing the beauty and variety of a subset of creatures that drift in the ocean. Of the vast remaining diversity, some may vary only slightly from their relatives portrayed in the preceding pages, while others surely take on different forms, behaviors, and functions. The plankton we know well often hail from abundant groups and from well-explored marine provinces and depths. There is still much to bring to light!

Essential to the health of the world's ocean and atmosphere, plankton constitute the bulk of the marine food chain, nourishing the fish and shellfish that hundreds of millions of people rely on for sustenance. Planktonic ecosystems are subject to tremendous stresses from rising atmospheric CO_2 and ocean acidification, overfishing, and various forms of pollution. Many scientists feel an imperative to better understand the diversity of the planktonic ecosystem and the relationships between plankton and the environment. These efforts include cataloging organisms around the globe and relating the biodiversity to geographical and temporal variation. Other researchers explore the interplay between environmental conditions and plankton migration, proliferation or decline.

Collecting, describing, and classifying the remaining majority of planktonic diversity constitute a potent challenge for current and future generations of scientists. The ocean and marine provinces are vast and ever changing, due to both natural fluidity and human activity. The occurrence and abundance of plankton also change over time and location. We are closer to the start than to the end of what there is to know. One begins to get a sense of the scale of the task—tracking countless types of creatures with dynamic life cycles in interconnected ecosystems, drifting in shallow waters and deep, moving with currents and tides, responding to day-night cycles, seasons, and interannual events such as El Niño Southern Oscillations.

So where to begin? Long-term academic and institutional research programs continue to produce an impressive body of sampling data and are complemented by research and surveillance networks of largely unaffiliated groups like NGOs. One novel approach fosters a participatory science movement, mobilizing volunteers to document plankton abundance from their private boats in easily accessible as well as remote waters. But broad descriptions of the planktonic organisms and their abundances, inferred from sampling on ocean-going vessels and from satellite views, do not adequately reflect the complexity of life in the ocean. We are just beginning to appreciate that our own genetic constitution as humans is dominated by the bacteria and viruses living in our guts and on our skin. The same holds true of individual planktonic creatures. Each represents a drifting ecosystem with a cohort of bacteria, viruses, parasites, and symbionts. These individuals form communities, connected by chemical signals, reproductive and survival relationships, and nutritional dependencies. Today, marine scientists take advantage of advanced imaging and genomic technologies from the biomedical realm. Endeavors such as the *Tara Oceans Expedition* and the *Global Ocean Sampling Expedition* generate mind-boggling quantities of genetic information. We can look forward to years of productive data mining and analyses, though the story that will be told from all this information will still be just the tip of the iceberg.

In addition to biodiversity and biogeography, we wish to better understand how planktonic habitats change, and the ways plankton responds to environmental variation. Both topics are domains of exciting and important research that asks, for example, what are the factors that cause phytoplankton blooms, and what are the consequences for other organisms and ocean chemistry? Populations of specific protists, particularly radiolarians, diatoms, dinoflagellates, and microalgae, can quickly proliferate and reach extraordinarily high densities if conditions of light, temperature, salts, and nutrients are favorable. Blooms of phytoplanktonic organisms are sometimes so gigantic that they color the sea and can be seen from satellites. After they have exhausted resources and been eaten by others, the phytoplankton disappear as fast as they appeared. These unicellular proliferations trigger successive blooms of zooplanktonic predators who themselves fall prey to bigger creatures, even whales and sharks. Causes and effects, however, are never so simple. The relationships between plankton communities and structures of food webs are governed by myriad sources of feedback and environmental inputs, as well as bursts of microbial activity that recycle organic matter.

What is the future of plankton in this fast-changing world? Ocean acidification resulting from excess

CO_2 absorption in large regions off the coasts of India and Panama is, in effect, a large-scale experiment. Some planktonic species will manage to acclimate while others, particularly those that rely on calcium for making shells or skeletons, may disappear. The same scenario applies to anoxic regions that spread along many coastal areas. These "creeping dead zones" occur mainly at the mouths of major river systems carrying agricultural fertilizers into the sea. Nitrates and phosphates stimulate phytoplankton growth on the surface and bacterial proliferation in the deep, resulting in the exhaustion of dissolved oxygen.

Marine phytoplankton account for roughly half of the total photosynthetic production on the planet. Monitoring unicellular photosynthetic organisms that fluctuate with different seasons in different locations is not an easy task. Some data suggest phytoplankton have significantly declined in the world ocean over the past century, though the interpretation of the results has proved quite controversial. Whether we are witnessing an actual global decline, or massive changes in plankton distribution, will require further study. Changes in geographic localization are thought to be due to a slight but steady rise in ocean surface temperature, increased stratification of water layers, and changes in complex wind and ocean circulation patterns.

Certainly many species will be forced to adapt. Studies tracking *Calanus finmarchicus* copepods over the last fifty years have found a decrease of these small crustaceans living in colder habitats, while a related species, *Calanus helgolandicus*, tended to increase in warmer climates. Cold-water predators such as cod and hake, feeding on the cold-water copepods, have experienced the detrimental effects of a decline in food, exacerbated by overfishing. On the other hand, some warm-water predators, such as jellyfish, are thriving. What planktonic organisms will benefit from increasingly disturbed marine eco-systems? Some scientists predict that jellies are best suited to survive a conceivable sixth mass extinction and will one day rule the seas. Only time will tell.

In the laboratory we extrapolate what can be learned from well-chosen model organisms representing various branches of the tree of life. In the field, there is much to be learned from observing the collective behavior of populations of organisms, such as the daily migrations of copepods in the water column. New and sophisticated observation buoys, gliders, and submersibles provide detailed descriptions of the fine structure of the ocean and increasingly track the organisms living in microhabitats. Some day these devices will report on the genes and chemical signatures of the organisms as well. Visionary vessels like *Sea Orbiter*, conceived to drift with the plankton, may become widely used platforms for observation and experimentation. These will broadcast in real time what this book has illustrated in still images: the irreplaceable beauty and diversity of planktonic life forms.

Christian Sardet and Rafael D. Rosengarten

ACKNOWLEDGMENTS

This book and the *Plankton Chronicles* project arose from many adventures, personal and professional.

The family adventure started when I was twelve years old. Grandfather Marc gave me a small microscope that I used to examine tiny animals in the ponds of Melle, a rural village in western France. I was fortunate to pursue this passion as a biologist and a researcher in France and the United States: Paris, Villefranche-sur-Mer, Roscoff, Woods Hole, Monterey, and Friday Harbor. With my son Noé, his associate in Montreal, Sharif Mirshak, and the CNRS, I started the *Plankton Chronicles* project in 2008. Over the years I benefitted from the loving support of my wife, Dana, my brother-in-law Ted, and all those who call me "brother plankton" or "uncle plankton." When it came to editing the English-language text, Rafael Rosengarten, my molecular biologist nephew, applied his expertise to the task as a scientist and writer. Mark Ohman contributed many helpful comments and a great prologue.

Professional adventures devoted to research about molecules and cells—in Lyon, Berkeley, and Gif-sur-Yvette—led to the creation and direction of a laboratory focused on the study of fertilization, embryos, and plankton at the Marine Station in Villefranche-sur-Mer. This was an ideal meeting ground for students, colleagues, and collaborators from all around the world. In the United States and Europe I was privileged to work with the people who had initiated a revolution in microscopic imaging. This prepared me for filming and photographing a plethora of fascinating cells, embryos, and organisms.

Another ongoing adventure, the *Tara Oceans Expedition*, took me on a worldwide exploration of plankton. This project was conceived with biologist colleagues Eric Karsenti and Gaby Gorsky and supported by the generous family of agnès b., Etienne Bourgois, Romain Troublé, and their extraordinary team. Many thanks to all the Taranautes—scientist colleagues and sailors—whom I frequently consulted about the arts of navigation and sampling and the intricacies of plankton.

I am indebted to Cedric Pollet, artist/photographer of trees, who introduced me to Antoine Isambert and Guillaume Duprat of Les Editions Ulmer. We immediately felt that together we could create a spectacular book. I am grateful as well to Christie Henry at the University of Chicago Press for her guidance and enthusiasm in publishing this English-language edition.

My gratitude extends to all the biologists who taught me so much: my long-time research companions on cells and embryos, Janet Chenevert, Evelyn Houliston, and the whole Biodev team who understood and encouraged me as I shifted my focus from cells and embryos to plankton; to zoologists Claude Carré and John Dolan, who furthered my education and filled in the gaps of a late vocation; to Karen Osborne, Jeremy Young, Marcus Weinbauer, Stefan Siebert, Casey Dunn, John Dolan, and Claude Carré, who generously contributed photos and important material; to all the colleagues who provided samples, images, and expertise—Sacha Bollet, Per Flood, Rebecca Helm, Christoph Gerigk, Anna Deniaud, David Luquet, Jean-Luc Prévost, Jean-Yves Carval, Sophie Marro, Marie-Dominique Pizay, Yvan Perez, Jeanine Rampal, Jean-Louis and Dominique Jamet, Fabien Lombard, Colette Febvre, Jean-Jacques Pangrazi, Martina Ferraris, Adriana Zingone, Philippe Laval, Eric Thompson, Matt Sullivan, Johan Decelle, Sebastien Colin, Christian Rouvière, Fabrice Not, Colomban De Vargas, Margaux Carmichael, Chris Bowler, Dennis Allen, Atsuko Tanaka, Lixy Yamada, Tsuyoshi Momose, Hiroki Nishida, and Kazuo Inaba.

I am grateful to everyone who helped bring this project to fruition, each in her or his own special way.

BIBLIOGRAPHY, WEBSITES

BOOKS

Arthus Bertrand, Y., and B. Skerry (2012) *Man and the Sea: Planet Ocean*. Goodplanet Foundation.

Bergbauer, M., and B. Humberg (2007) *La vie sous-marine en Méditerranée*. Vigot.

Blandin, P. (2010) *Biodiversité*. Albin Michel.

Boltovskoy, D., ed. (1999) *South Atlantic Zooplankton*. Backhuys Publishers.

Bougis, P. (1974) *Écologie du plancton marin*. Elsevier Masson.

Brusca, R. C., and G. J. Brusca (1990) *Invertebrates*. Sinauer Associates.

Burnett, N., and B. Matsen (2002) *The Shape of Life*. Sea Studios, Foundation and Monterey Bay Aquarium. Boxwood Press.

Carroll, S. B. *Endless Forms Most Beautiful*. W. W. Norton & Co.

Carson, R. (2012) *The Sea Around Us*. Oxford University Press.

Conway, D. V. P. , R. G. White, J. Hugues-Dit-Ciles, C. P. Gallienne, and D. B. Robins (2003) *Guide to the Coastal and Surface Zoo-plankton of the Southwestern Indian Ocean*. Vol. 15. Marine Biological Association of the United Kingdom.

Deutsch, J. (2007) *Le ver qui prenait l'escargot comme taxi*. Le Seuil.

Dolan, J. R., D. J. S. Montagnes, S. D. Agatha, W. Coats, D. K. Stoecker (2012) *The Biology and Ecology of Tintinnid Ciliates: Models for Marine Plankton*. Wiley-Blackwell.

Elmi, S., and C. Babin (2012) *Histoire de la terre*. Dunod.

Falkowski, P. G., and J. Raven (1997) *Aquatic Photosynthesis*. Blackwell Science.

Fortey, R. (1997) *Life*. Vintage Books.

Garstang, W. (1951) *Larval Forms and Other Zoological Verses*. Blackwell.

Gershwin L. (2013) *Stung! On Jellyfish Blooms and the Future of the Ocean*. University of Chicago Press.

Glémarec, M. (2010) *La biodiversité littorale, vue par Mathurin Méheut*. Éditions Le Télégramme.

Gudin, C. (2003) *Une histoire naturelle de la séduction*. Le Seuil.

Gould, S. J., ed. (1993) *The Book of Life*. W. W. Norton & Co.

Goy, J. (2009) *Les miroirs de Méduses*. Apogée.

Gowel, E. (2004) *Amazing Jellies*. Bunker Hill Publishing.

Hardy, A. C. (1964) *The Open Sea: The World of Plankton*. Collins.

Haeckel, E. (1882) *The Radiolarian Atlas*. Rev. ed., *Art Forms from the Ocean*. Prestel Verlag, 2010.

Hill, R. W., G. A. Wyse, and M. Anderson (2008) *Animal Physiology*. Sinauer Associates.

Hinrichsen, D. (2011)*The Atlas of Coasts and Oceans: Ecosystems, Threatened Resources*. University of Chicago Press.

Jacques, G. (2006) *Écologie du plancton*. Tec & Doc Lavoisier.

Johnson, W. S., and D. M. Allen (2012) *Zooplankton of the Atlantic and Gulf Coasts: A Guide to Their Identification and Ecology*. John Hopkins University Press.

Karsenti E., and D. Di Meo (2012) *Tara Oceans: Chroniques d'une expédition scientifique*. Actes Sud, *Tara Expéditions*.

Keynes, R. D., ed. (2001) *Charles Darwin's Beagle Diary*. Cambridge University Press.

Kolbert, E. (2014) *The Sixth Extinction: An Unatural History*. Bloomsbury.

Kozloff, E. N. (1993) *Seashore Life of the Northern Pacific Coast*. University of Washington Press.

Kirby, R. R. (2010) *Ocean Drifters: A Secret World beneath the Waves*. Firefly Books.

Knowlton, N. (2010) *Citizens of the Sea: Wondrous Creatures from the Census of Marine Life*. National Geographic Society.

Konrad, M. W. (2011) *Life on the Dock*. Science Is Art.

Kraberg, A., M. Baumann, and C. Durselen (2010) *Coastal Phytoplankton: Photo Guide for Northern European Seas*. Verlag.

Larink, O., and W. Westheide (2012) *Coastal Plankton: Photo Guide for European Seas*. Verlag.

Lecointre, G., and H. Le Guyader (2007) *The Tree of Life: A Phylogenetic Classification*. Harvard University Press.

Loir, M. (2004) *Guide des diatomées*. Delachaux & Niestlé.

Munn, C. B. (2004) *Marine Microbiology*. Taylor and Francis Publishers.

Margulis, L., and K. V. Schwartz (1988). *Five Kingdoms*. W. H. Freeman.

Mollo, P., and A. Noury (2013) *Le manuel du plancton*. Charles Léopold Mayer Editions.

Moore, J. (2001) *An Introduction to the Invertebrates*. Cambridge University Press.

Nielsen, C. (2001) *Animal Evolution*. Oxford University Press.

Nouvian, C. (2007) *The Deep: The Extraordinary Creatures of the Abyss*. University of Chicago Press.

Pietsch, T. W. (2012) *Trees of Life*. John Hopkins University Press.

Prager, E. J. (2000) *The Oceans*. McGraw Hill.

Reynolds, C. (2006) *Ecology of Phytoplankton*. Cambridge University Press.

Ricketts, E., J. Calvin, and J. W. Hedgpeth (1968) *Between Pacific Tides*. Stanford University Press.

Segar, A. D. (2006) *Ocean Sciences*. W. W. Norton & Co.

Seguin, G., J.-C. Braconnot, and B. Elkaim (1997) *Le plancton*. Presses Universitaires de France.

Schmidt-Rhaesa, A. (2007) *The Evolution of Organ Systems*. Oxford University Press.

Smith, D. L., and K. B. Johnson (1996) *A Guide to Marine Coastal Plankton and Marine Invertebrate Larvae*. Kendall/Hunt Publishing

Snelgrove, P. V. R. (2010) *Discoveries of the Census of Marine Life*. Cambridge University Press.

Southwood, R. (2003) *The Story of Life*. Oxford University Press.

Strathmann, M. (1987) *Reproduction and Development of Marine Invertebrates of the Northern Pacific Coast.* University of Washington Press.

Thomas-Bourgneuf M., and P. Mollo (2009) *L'Enjeu Plancton: L'écologie de L'invisible.* Charles Léopold Mayer Editions.

Todd, C. D., M. S. Laverack, and G. A. Boxshall (1996) *Coastal Marine Zooplankton: a Practical Manual for Students.* Cambridge University Press.

Tomas, C. R., ed. (1997) *Identifying Marine Phytoplankton.* Academic Press.

Trégouboff, G., and M. Rose (1957) *Manuel de planctonologie méditerranéenne.* Centre National de la Recherche Scientifique.

Vogt, C. (1854) *Recherches sur les animaux inférieurs de la Méditerranée: Les siphonophores de la Mer de Nice.* H. Georg Editor.

Wilkins, A. S. (2004) *The Evolution of Developmental Pathways.* Sinauer Associates.

Willmer, P., G. Stone, and I. Johnston (2005) *Environmental Physiology of Animals.* Blackwell.

Wood, L. (2002) *Faune et flore sous-marines de la Méditerranée.* Delachaux & Niestlé.

Wrobel, D., and C. E. Mills (1998) *Pacific Coast Pelagic Invertebrates: A Guide to the Common Gelatinous Animals.* Sea Challengers and the Monterey Bay Aquarium.

Yamaji, I. (1959) *The Plankton of Japanese Coastal Waters.* Hoikusha.

GENERAL WEBSITES

Plankton Chronicles: www.planktonchronicles.org
Plankton Portal: http://blog.planktonportal.org
Plankton Net: http://planktonnet.awi.de
Tara Oceans Expeditions & Tara Oceans: http://oceans.taraexpeditions.org & www.embl.de/tara-oceans/start
Observatoire Océanologique de Villefranche-sur-Mer: www.obsvlfr.fr/gallery2/main.php
Encyclopedia of Life / Education / Tree of Life: http://eol.org & http://education.eol.org & http://tolweb.org/tree/home.pages/toleol.html
Marine species: http://species-identification.org/index.php
TED Ed plankton videos: http://ed.ted.com/lessons/how-life-begins-in-the-deep-ocean & www.ted.com/talks/the_secret_life_of_plankton.html
Census of Marine Life: www.coml.org & www.cmarz.org
Kahikai photos: www.kahikaiimages.com/home

ZOOPLANKTON SITES

David Wrobel / Jellies: http://jellieszone.com
Casey Dunn / Siphonophores: www.siphonophores.org
Steve Haddock / Bioluminescence: http://biolum.eemb.ucsb.edu
Jellyfish: www.jellywatch.org

PROTISTS SITES

Algae & phytoplankton: www.algaebase.org/search/species/
Protists: http://starcentral.mbl.edu/microscope/portal.php?pagetitle=index et www.radiolaria.org
John Dolan / Aquaparadox: http://www.obs-vlfr.fr/gallery2/v/Aquaparadox
Station Biologique de Roscoff / Phytoplankton: www.sb-roscoff.fr/Phyto/RCC/index.php

CREDITS

All photos by Christian Sardet, except:

Chantal Abergel and Jean-Michel Claverie, Centre National de la Recherche Scientifique
   (CNRS) IGS/IMMM, Marseille, 33 bottom × 2
Dennis Allen, Baruch Marine Field Laboratory, University of South Carolina, Georgetown, and Christian Sardet, 142, 197
Gary Bell, Oceanwidelmages.com, 9
Mark Boyle, 10
Jean and Monique Cachon, 84 × 4
Claude Carré, Université Pierre-et-Marie-Curie (UPMC), Paris, 96 bottom left, 97 top right
Margaux Carmichael, Station Biologique de Roscoff (SBR)-CNRS/UPMC, 49, 50, 66
Marie Joseph Chrétiennot-Dinet, CNRS Photothèque, 48 × 3
Laurent Colombet, 131 bottom
Wayne Davis, oceanaerials.com, 7
Mark Dayel, mark@dayel.com, 90
Charles Darwin, 15
Johan Decelle and Fabrice Not, SBR-CNRS/UPMC, Roscoff, 82, 84 bottom right
Johan Decelle and Sébastien Colin, Fabrice Not, Colomban de Vargas, SBR-CNRS/ UPMC, Roscoff, 83
Johan Decelle, SBR-CNRS/UPMC, Roscoff, and Christian Sardet, 79
Johan Decelle and Fabrice Not, SBR-CNRS/UPMC, Roscoff, 82
Anna Deniaud Garcia, *Tara Expeditions*, 17 top × 3, 24 × 4, 50
John Dolan, CNRS, Observatoire Océanologique de Villefranche-sur-Mer, 87, 88, 89
Guillaume Duprat and Christian Sardet, graphics, 12–13
Casey Dunn, Brown University, Providence, 96 top left, 97 top left and bottom right, 115, 129
European Molecular Biology Laboratory, Heidelberg, 15 bottom
Martina Ferraris, Observatoire Océanologique de Villefranche-sur-Mer, and Christian Sardet, 104
Per Flood, Bathybiologica A/S, 90, 113, 194 bottom, 205, 206 top
Mick Follows, Oliver Jahn, ECCO$_2$ and Darwin Project, Massachusetts Institute of Technology, Cambridge, 8, 21, 22, 24, 26, 28
Laurence Froget and Marie Chrétiennot-Dinet, CNRS Photothèque, Commissariat à l'énergie atomique et aux énergies alternatives
   (CEA), 49
Julien Girardot, *Tara Expeditions*, 16
Christoph Gerigk, 17 bottom × 2, 24 × 2, 94, 95
Kathryn Green, University of Queensland, Brisbane, 188 top
Ernst Haeckel, 15, 19, 85
Rebecca Helm, Brown University, Providence, 104
Kazuo Inaba, Tsukuba University, Shimoda, 28
Gerda Keller, Princeton University, 14
Nils Kroeger, Georgia Institute of Technology, Atlanta, and Chris Bowler, École Normale Supérieure, Paris, 63
Francis Latreille, *Tara Expeditions*, 24
Edouard Leymarie, CNRS, Observatoire Océanologique de Villefranche-sur-Mer, 196
Fabien Lombard, UPMC, Observatoire Océanologique de Villefranche-sur-Mer, 198
Imène Machouk and Charles Bachy, CNRS Photothèque, 88
Sophie Marro, Mediterranean Culture Collection of Villefranche-sur-Mer, and Christian Sardet, 58, 68
Monterey Bay Aquarium Research Institute, 115
Sharif Mirshak, Parafilms, Montreal, 22, 103, 105, 125, 175, 176, 177, 178, 185
Sharif Mirshak and Christian Sardet, 98, 100, 123, 125, 145, 149, 158, 159, 160, 161
Sharif Mirshak and Noé Sardet, Parafilms, Montreal, 175, 209
Sharif Mirshak, Noé Sardet, and Christian Sardet, 125, 163
Sharif Mirshak, Noé Sardet, Flavien Mekpoh, 210 t
Jan Michels, University of Kiel, 146
Tsuyoshi Momose and Evelyn Houliston, CNRS, BioDev Laboratory, Observatoire Océanologique de Villefranche-sur-Mer, 106, 107
NASA, 2, 48
Karen Osborn, Smithsonian National Museum of Natural History, Washington DC, 52, 177 middle left, 178 middle right, 194 top
Peter Parks, Imagequestmarine.com, 130
Frédéric Partensky, SBR-CNRS/UPMC, Roscoff, 43 bottom right
Yvan Perez, Université d'Aix-Marseille, 184
Monique Picard, 122
Marie Dominique Pizay, John Dolan, Rodolphe Lemée, Observatoire Océanologique de Villefranche-sur-Mer, 67
Christian Rouvière, and Christian Sardet, CNRS, BioDev Laboratory, Observatoire Océanologique de Villefranche-sur-Mer, 66 × 2
Noé Sardet, Parafilms, Montreal, 22 × 3, 59, 119, 125, 141, 162
Noé Sardet and Christian Sardet, 6, 71, 75, 77, 101, 171
Noé Sardet, Sharif Mirshak, Flavien Mekpoh, 210

Ulysse Sardet and Christian Sardet, 76
Alexander Semenov, 166, 167, 192
Stefan Siebert, Brown University, Providence, 114 × 2, 177 top, 179, 180–181, 202 right, 203, 208 top and bottom
Keoki Stender, MarinelifePhotography.com, 128, 131 top
Matthew Sullivan, Jennifer Brum, University of Arizona, Phoenix, 34 bottom
Tara Oceans Expedition, 120–121, 156–157, 186 middle, 200 top
Atsuko Tanaka and Chris Bowler, CNRS, École Normale Supérieure, Paris, 43 left
Eric Thompson, Philippe Ganot, and Endy Spriet, Sars Laboratory, Bergen, Norway, 206 bottom left × 2
Jeremy Young, University College, London, 49, 50, 51
Markus Weinbauer, CNRS, Observatoire Océanologique de Villefranche-sur-Mer, 33 top, 34 top × 2, 35
Wikicommons, 15 top left and top right
David Wrobel, 92, 112, 118, 201

INDEX

Christian Sardet is cofounder and research director emeritus of the Laboratory of Cell Biology at the Marine Station of Villefranche-sur-Mer, part of the Centre National de la Recherche Scientifique and Université Pierre et Marie Curie in Paris. He is also cofounder and a scientific coordinator of the *Tara Oceans Expedition*, a global voyage to study plankton, and creator of the Plankton Chronicles Project, www.planktonchronicles.org

The University of Chicago Press, Chicago 60637
The University of Chicago Press, Ltd., London
© 2015 by The University of Chicago
Prologue © 2015 by Mark D. Ohman
All rights reserved. Published 2015.
Printed in China

24 23 22 21 20 19 18 17 16 15   1 2 3 4 5

ISBN-13: 978-0-226-18871-3 (cloth)
ISBN-13: 978-0-226-26534-6 (e-book)
DOI: 10.7208/chicago/9780226265346.001.0001

Originally published in France as   .
© 2013 Les Editions Eugen Ulmer, Paris. www.editions-ulmer.fr

Library of Congress Cataloging-in-Publication Data
Sardet, Christian, author, translator.
  [Plancton. English]
  Plankton: wonders of the drifting world / Christian Sardet ; edited by Rafeal D. Rosengarten and Theodore Rosengarten ; translated from the French by Christian Sardet and Dana Sardet ; prologue by Mark Ohman.
      pages cm
  "Originally published in France as Plancton, aux origines du vivant. © 2013 Les Editions Eugen Ulmer, Paris"—Title page verso.
Includes bibliographical references and index.
    ISBN 978-0-226-18871-3 (cloth : alkaline paper) — ISBN 0-226-18871-X (cloth : alkaline paper) — ISBN 978-0-226-26534-6 (e-book) — ISBN 0-226-26534-X (e-book) 1. Plankton—Pictorial works. I. Rosengarten, Rafeal D., editor. II. Rosengarten, Theodore, editor. III. Sardet, Dana, translator. IV. Ohman, Mark, writer of preface. V. Title.
    QH90.8.P5S2713 2015
    578.77´6—dc23
                                                                      2014034445

Ouvrage publié avec le concours du Centre national du livre / This title has been published with the support of the Centre national du livre.

♾ This paper meets the requirements of ANSI/NISO Z39.48–1992 (Permanence of Paper).